Advanced Materials in Smart Building Skins for Sustainability

Julian Wang · Donglu Shi · Yehao Song
Editors

Advanced Materials in Smart Building Skins for Sustainability

From Nano to Macroscale

Springer

Editors
Julian Wang
Department of Architectural Engineering and Department of Architecture
Pennsylvania State University
University Park, PA, USA

Donglu Shi
Department of Mechanical and Materials Engineering
University of Cincinnati
Cincinnati, OH, USA

Yehao Song
School of Architecture
Tsinghua University
Beijing, China

ISBN 978-3-031-09694-5 ISBN 978-3-031-09695-2 (eBook)
https://doi.org/10.1007/978-3-031-09695-2

This Springer imprint is published by the registered company Springer Nature Switzerland AG
The registered company address is: Gewerbestrasse 11, 6330 Cham, Switzerland

Preface

Today's nanotechnologies and advanced materials have rapidly advanced into many areas, particularly in civil engineering and architectural design for the development of responsive and adaptive structures which are not only for providing occupant comfort but also for achieving building energy efficiency and sustainable environmental performance. The concept of "smart" building skins has emerged that can respond to any environmental changes, especially involving energy sources and health factors. These environmental changes can trigger a "response" manifested by a material property that can be readily measured and translated to other variables for required functionalities and applications. In this fashion, a building skin is no longer a passive physical barrier but structurally transformed into an active device system capable of multifunctional energy and environmental performances such as energy harvest and CO_2 reduction. For instance, the smart building skin can now be engineered for solar harvesting, conversion, and utilization as in the so-called Building Integrated Photovoltaic (BIPV). In BIPV, the high-rise building skins provide ideal transparent substrates for energy device architecture based on nanoscale thin films. The smart building concept has paved a new path to energy-neutral civic infrastructures that will have high societal impacts on energy conservation, public health, prosperity, and welfare.

This book is a collection of chapters that focus on key areas of smart building skins embodied in the novel advanced materials with unique structures and properties that enable multiple functions in energy efficiency, solar harvesting, and environmental greenness. These chapters provide the most up-to-date research outcomes on novel materials synthesis, structure characterization, device architecture, and their novel applications in smart building skins. The topics of these chapters cover broad interdisciplinary areas including nanotechnology, materials science, energy devices, architecture, urban planning, and civil engineering. In particular, new trends in modern architectural design are introduced with consideration of energy conservation, environmental greenness, and net-zero. All chapters are developed by experts in different fields who have been collaboratively conducting research about energy-efficient building skins.

Chapters 1–5 concentrate on the design, characteristics, development, fabrication, and performance analysis of *advanced materials*, taking the consideration of building-integrated application paradigms. The introduced materials range from nanoscale spectral selective materials to molecular scale programmable biological materials, and architectural scale wooden materials/structures. Spurred by the introduction of various new materials, Chapters 6–9 deal with *engineering solutions and technologies* to incorporate advanced materials and investigate their impacts, such as the envelope thermal engineering with smart materials, building-integrated photovoltaic technologies and their influences on microclimates, and computational technologies for energy analysis of dynamic building skins. Chapters 10–13 take the perspective of *architectural design*, illustrating and demonstrating new methodologies, processes, models, and practical principles for integrating dynamic envelopes consisting of advanced materials into the architectural scale, with considerations of material design knowledge transfer, environmental interactions, esthetics, and user testing.

We have organized this book to provide overviews of key issues on the novel applications of advanced materials in architecture that are tailored to multifunctional, energy-efficient, environmentally green building skins. All chapters are developed in a tutorial fashion for technical specialists in a variety of fields such as materials science, mechanical engineering, civil engineering, and architectural design. This book is expected to have a large readership suitable for both academic and industrial researchers, including graduate and undergraduate students, scholars, engineers, and practitioners in sustainable design and building façade engineering.

We are most grateful to all authors for contributing their most valuable works to this book. The assistance of editor Madanagopal Deenadayalan from Springer for formatting this book is very much appreciated.

University Park, USA — Julian Wang
Cincinnati, USA — Donglu Shi
Beijing, China — Yehao Song

Contents

Chapter 1
Spectral Selective Solar Harvesting and Energy Generation via Transparent Building Skin

Jou Lin, Mengyao Lyu, Yuxin Wang, Brent Webster, and Donglu Shi

Abstract Today's nanoscience has rapidly advanced into many areas of engineering, particularly in architectural design of building skins for the development of environmentally responsive structures. Not only does this improve occupant comfort, but also increases energy sustainability. Recently, advanced materials have been utilized for developing large-scale building skins with intelligent functionalities. Many current challenges focus on improving environmental safety, increasing energy efficiency, and reducing our carbon footprint. Advanced materials have played key roles in addressing these critical issues through fascinating properties that are ideal for building skin engineering. In this chapter, a series of bio-inspired green nano hybrids are introduced for solar harvesting, energy generation, and photothermally-activated building heating. These nano biohybrids can be physically retrofitted into existing residential buildings and structures, independently providing self-sustainable green energy. This chapter explores the most recent developments of nanostructure-based building skins that are smart, intelligent, adaptive, responsive and biologically inspired for energy and environmental sustainability. More specifically, we have developed smart building façades utilizing nanomaterial assemblies that have the ability to regulate and control energy consumption and generation, leading to energy neutral civic infrastructure. Novel nanostructures are designed, synthesized, and developed capable of the most efficient solar harvesting and energy generation. Fundamental studies have been carried out to identify operating mechanisms dictating the optical, thermal, and electrical properties of the thin films on building skins for required functions. Lab-scale modules can be designed to test the performance of multifunctional thin films in terms of solar harvesting, visible transmittance, photovoltaic (PV) and photothermal (PT) efficiencies. The novel concepts include Optical Thermal Insulation (OTI) without any intervention medium typically used in glazing technologies, and building skins with PV-PT dual modalities that can be seasonably altered for energy efficiency and generating solar-mediated thermal energy for building heating utilities. OTI is established by solar heating of a

J. Lin · M. Lyu · Y. Wang · B. Webster · D. Shi (✉)
The Materials Science and Engineering Program, Department of Mechanical and Materials Engineering, College of Engineering and Applied Science, University of Cincinnati, Cincinnati, OH 45221, USA
e-mail: shid@ucmail.uc.edu

J. Wang et al. (eds.), *Advanced Materials in Smart Building Skins for Sustainability*,
https://doi.org/10.1007/978-3-031-09695-2_1

photothermal coating on a window surface. by reducing the temperature difference between the window surface and room interior, the heat loss through a single pane is lowered without the air gap of the double pane. A transparent building skin can be engineered as a PV and PT device in the same surface coating with a dual modality. While the solar energy harvested can be converted to electricity via PV in the summer, the same film is photonically-activated to generate heat in the winter for reduced heat loss. The PV-PT dual-modality device can be applied as a smart building skin upon a large surface area for enhanced solar harvesting and alternative energy generation such as electrical and thermal energy. The solar light can also be wave-guided through transparent photo-thermal panels for generating high heat for building utilities. This chapter will devote much of its portions to describing the fundamental mechanisms of these new concepts as well as the engineering implementations in building skin architecture.

Keywords Advanced materials · Nano technology · Smart building skin · Nanobiohybrids · Spectral selective · Photothermal · Photovoltaic · Energy sustainability

1.1 Introduction

1.1.1 Optical Thermal Insulation via Photothermal Window Coatings

The residential and commercial building sectors accounted for about 40% (or about 40 quadrillion British thermal units) of the total U.S. energy consumption in 2018 (Arasteh et al., 2006). Therefore, there is an increasing need to develop lighter material structures (such as single pane) and energy efficient building skins. One of the key issues deals with overall building energy and material consumption including electricity, heating and cooling, and materials production due to large building sizes. Recent advancements in technology have enabled glass to be integral of both the interior and exterior architecture making it a major component of building façades (Fig. 1.1). However, great challenges remain in terms of thermal transfer, energy efficiency and lighting requirements, especially heat loss in cold climate. Although various glazing technologies have been developed including low emissivity coatings and double/triple panes, a huge consumption of energy will not be resolved until fundamentally different concepts and engineering innovations are developed.

For instance, the current technology for efficient windows relies upon the double-pane insulated glass unit (IGU) with a low-emissivity (low-e) coating. Although the double-pane IGU may meet some of the requirements in energy consumption, the single-pane windows will be highly desirable for obvious reasons including lighter weight, straightforward manufacturing, and less materials needed. The advanced single pane IGUs may ultimately replace double panes with similar performance characteristics. However, a key challenge in developing the single-pane IGUs is the

Fig. 1.1 Typical modern building with glass facades

thermal insulation without the air gap between the double panes. As air is well known for its low thermal conductivity, few materials can be used to replace air except heavier gases such as argon. There are two main aspects in the design of a single-pane IGU: (1) replace the air gap with a novel insulator so that the thermal conductivity no longer depends on the air gap spacing (i.e. double panes), and (2) replace the low emissivity coating with a new reflection system that can efficiently block the heat loss from the interior where the heating source exists.

The basic criterion for a thermally well-insulated single-pane window is the U-factor according to Energy Star certification in colder regions of the US (Environmental Protection Agency, 2015). The single-pane IGU will have to meet the following conditions: (1) a single pane without air spacing; (2) the U-factor for an entire window needs to be less than 0.30 (Environmental Protection Agency, 2015; U. E. P. Agency, 2018; Carmody et al., 2007); (3) a single-pane IGU with sufficiently high interior temperate allowing a 30% relative humidity for external temperature of −0 °C, and (4) a low emissivity coating that can selectively and efficiently reflect the thermal energy generated from the heating source. In order to meet these conditions, advanced materials will play key roles in many aspects of engineering design, optical and thermal property characterization, and architectural integration into the building structure. The current advancement of nanotechnologies will enable design, development and manufacture of new single panes that are thermally highly insulating, optically transparent, visually finesse, and energy efficient.

The goal in developing energy-efficient building skins is to achieve the same level of thermal insulation of conventional glazing technologies by using single panes without any thermal intervention medium, such as air or argon. However, this critical issue cannot be easily addressed by only replying on identifying insulating materials since most of the thermal insulators are either nontransparent or opaque, thus not

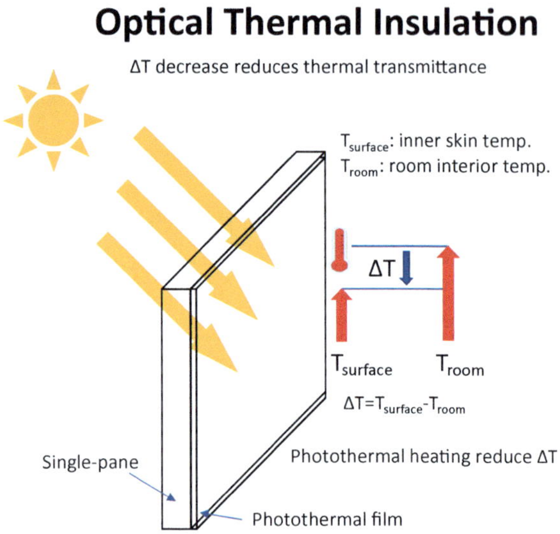

Fig. 1.2 Concept of optical thermal insulation

suitable for window applications. Recently, a novel concept of Optical Thermal Insulation (OTI) has been developed by solar heating of the window surface through a nano-scale thin film coating (Fig. 1.2) (Lin et al., 2020). Thermal losses can be significantly reduced if the window surface temperature is raised to the level approaching that of the room interior (Fig. 1.2). In this way, double panes will not be needed to reduce the heat loss, therefore saving tremendous energy and insulation materials.

The optical thermal insulation is achieved by developing a transparent photothermal film deposited on the window surface which can rapidly heat up under solar light irradiation. The thin film is designed, synthesized, and tailored to selectively absorbing solar light in the ultraviolet (UV) and near infrared (NIR) regions for converting these photon energies to thermal heat, leading to window surface temperature increase. For window applications, the film has to be rendered highly transparent. This requires manipulation of the optical properties of the thin films with high Average Visible Transmittance (AVT).

Figure 1.3 shows the absorption spectra for various materials as indicated (Tian et al., 2013; Zhao et al., 2019a). As can be seen in this figure, while gold nanoparticles exhibit strong absorption below 300 nm, there is another peak at 520 nm, contributing to considerable visible light absorption. Much stronger absorption is observed in the UV range for graphene and Fe_3O_4 nanoparticles, but they gradually decay to the minimum in the visible range up to Near-Infrared (NIR). Therefore, all conducting materials with high charge densities exhibit strong photothermal effects but lacking sufficient visible transmittance. However, chlorophyll shows a saddle-like spectrum with two peaks respectively at 400 nm (blue-violet) and 700 nm (NIR), which is consistent with its green color. It is this typical spectrum with a saddle-like shape that qualifies chlorophyll for transparent building skin application. As can also be seen in Fig. 1.3, $Fe_3O_4@Cu_{2-x}S$ is characterized with a "U" shaped spectrum showing strong

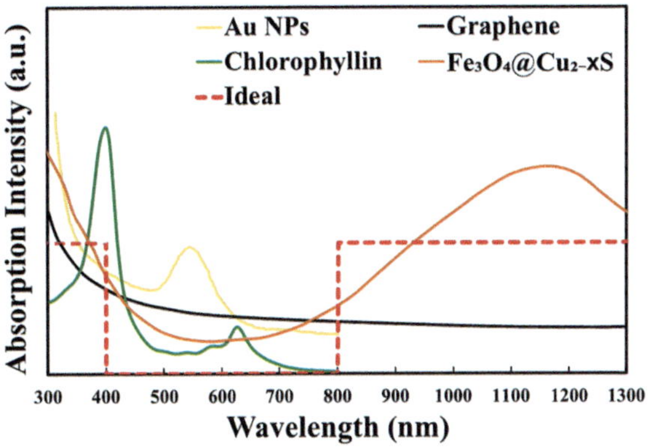

Fig. 1.3 Spectra of various photothermal materials

IR absorption extending to 1500 nm or even further (the peak is around 1200 nm). The dashed line in Fig. 1.3 represents an idealized optical characteristic with step-function like UV and IR absorptions and high visible transmittance for single-pane applications.

Based on the optical properties shown in Fig. 1.3, materials with either "U-shaped" or "saddle-like" absorption characteristics may meet the requirements in single-pane designs. Both iron oxides and porphyrin compounds are ideal materials for photothermal coatings in energy-efficient building skin applications. The porphyrin compounds such as chlorophyll and chlorophyllin share very similar structures of the porphyrin rings with the only difference in the metal atom at the ring center. Chlorophyll has a center magnesium atom; replacing it with a copper atom results in chlorophyllin. Both are highly abundant in nature and environmentally green.

Absorption behavior of Fe_3O_4 nanoparticles can be modified by coating the particle surfaces with CuS, forming a core–shell nanostructure in $Fe_3O_4@Cu_{2-x}S$ that exhibits much stronger near-infrared absorptions as shown in Fig. 1.3 (Tian et al., 2013) The thin films of both porphyrins and iron oxides can be spin-coated on glass substrates with 500 nm thickness achieving an average visible transmittance above 90%. Upon irradiation with simulated solar light, the surface temperature can rapidly increase as shown in Fig. 1.4a for $Fe_3O_4@Cu_{2-x}S$, which contributes to the optical thermal insulation by reducing the temperature difference between the window surface and room interior (Fig. 1.1). Similar heating curves have been obtained for chlorophyll (Fig. 1.4b) and chlorophyllin films (Fig. 1.4c). The current research has been focused on improving both heating efficiency and AVT in the PT coatings for OTI applications.

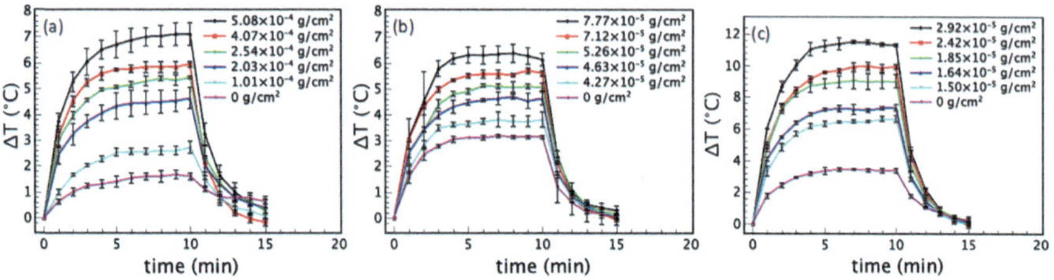

Fig. 1.4 Heating curves of **a** $Fe_3O_4@Cu_{2-x}S$, **b** chlorophyll, and **c** chlorophyllin thin films

1.1.2 Photovoltaic and Photothermal Dual-Modality Building Skins

Harvesting solar energy for various applications has been an intense research area, especially in photovoltaic solar cells. Interestingly both transparent photothermal (TPT) and transparent photovoltaic (TPV) films share a common spectral characteristic requirement: high visible transmittance (AVT) for visible transparency. However, TPT requires high absorptions in both UV and NIR bands, while only UV is preferred for TPV since IR heating results in considerable efficiency decrease (Effect of Temperature, 2021; Bjørk & Nielsen, 2015). Another important common feature of TPV and TPT lies on a large surface area requirement for efficient solar light harvest. As noted above, the current glass building skin is characterized with entire building surfaces, particularly large for high-rises. Take both advantages of free solar light and huge building surfaces, the so-called building integrated photovoltaic (BIPV) has been extensively researched and developed. However, it is possible to further increase the energy efficiency of a building skin by incorporating both TPV and TPT via the concept of PV-PT dual modality that can most efficiently harvest solar light and convert it to two forms of energy: thermal and electric energy, which can be alternated seasonably. This is highly possible if a novel transparent film on glass building skin surface can be engineering to have both PT and PV properties (Lin & Shi, 2021).

The concept of the photothermal-photovoltaic (PT-PV) dual-modality assembly is schematically illustrated in Fig. 1.5. As shown in this figure, the PT-PV film is featured with several key characteristics: (1) it is highly transparent but only absorbing UV and NIR for energy conversion; (2) upon absorbing solar light in the UV and NIR regions, energy conversion seasonably takes two paths: photon-to-heat conversion in the winter for lowering the U-factor, and photon-to-electricity in warmer seasons as a solar panel, and (3) the dual modality can be switched easily depending upon the seasonal changes and internal thermal needs. In the wintertime, the photothermal effect is utilized to heat the single-pane in order to reduce heat loss. The single-pane is switched to the PV mode in warmer seasons especially in the summer for producing electricity.

The synergetic effects of PT and PV, seasonably alternated, will significantly improve building energy efficiency. In sharp contrast to conventional windows, the

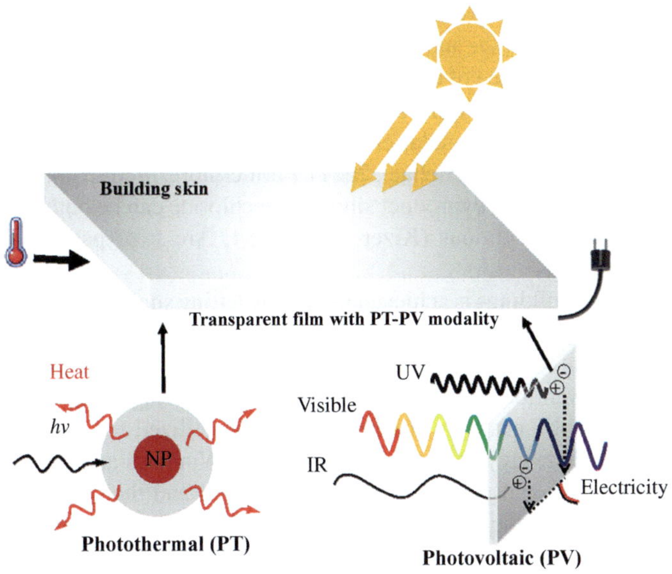

Fig. 1.5 Schematic of PV-PT dual modality building skin

building skin with the PT and PV dual modality will essentially become an energy generator and convertor rather than a source of heat loss. The advantages of this approach are several-folds: (1) no insulating medium is needed as in the traditional double-pane structure; (2) the building skin with the PT-PV thin films can be mass-produced with low cost; (3) the PT-PV thin films can alternatively generate heat and electricity with a much large surface area; (4) ΔT between the warm room interior and the inner window glass surface temperature can be spectrally tuned by adjusting the composition and additives in the PT-PV films, and (5) adhesive products built upon this approach can be applied in wide-scale energy-efficient window retrofits with low balance-of-system costs.

1.1.3 3D Solar Harvesting and Photothermal Energy Generation for Building Heating Utilities

In recent years, formation of megacities presents enormous social and environmental challenges. The sustainability of megacities depends to a large extent on how to manage the energy and material resources. In particular, the use of electricity is not only inefficient but also environmentally detrimental. The combustion of fossil fuels is responsible for the majority of emissions of carbon dioxide CO_2, sulfur dioxide SO_2 and nitrogen oxides NO_2 (Perera, 2018). To address these critical energy inefficiency issues, utilization of solar energy via PV has been the main approach to produce clean energy. However, PV requires additional utilities for solar harvesting, electricity generation, storage and distribution, which is not ideal for a net zero energy system. Although the source of energy appears "free" as it is abundant in nature, the processes in harvesting solar light can be quite costly against the net economic gains. It has been estimated that the solar energy costs approximately 8 times more than primary fuels (oil, coal, gas) (Michael, 2021). Furthermore, production of silicon PV panels is not "clean" as the byproduct silicon tetrachloride can be both occupational and environmentally hazardous (Kizer et al., 1984; Are solar panels toxic to the environment, 2021).

Solar heating in buildings is achieved mainly by letting sunlight through windows or guiding it in building structures via optical means such as solar tunnels (solar pipes) (Spacek et al., 2018; Zhang, 2002). From these solar collectors, the heat is transferred by conventional equipment to warm up the living spaces of a building. Although certain solar energy can be harvested and utilized for energy consumption, the energy efficiency has not been satisfactory. Windows and skylights can be classified as conventional passive daylighting devices, and their performances are greatly constrained by the external and internal obstructions. Many conventional equipment requires electricity-driven fans, causing an addition of energy consumption. For instance, heating, ventilation, and air conditioning (HVAC) consume 40% to 50% of the building's energy, which is mainly supplied by primary fuels (natural gas) (Irace & Brandon, 2017). The current solar technologies are categorized into two types, namely, active solar energy and passive solar energy. The former deals with actively harvesting and concentrating solar energy by various technical means and convert it to other forms of energy such as electricity (PV) (Chang et al., 2018; Chen et al., 2012; Husain et al., 2018; Traverse et al., 2017). The latter relies on daylighting devices including windows, mirrors, prismatic glazing, light shelves, atrium and light pipes for passively collecting sunlight (Irace & Brandon, 2017; Zhang, 2002). The typical active solar space heating involves collectors to absorb solar radiation in either liquid or air. Distributions of heat require either pumps or fans depending upon the heat transfer medium. For passive solar space heating, solar light is harvested mainly through architectural features such as windows, light shelves, atrium and light pipe. Thermal mass materials such as concrete or tile are typically used to absorb heat during the daytime and release it at night. However, these passive solar space heating devices are not sufficient to heat the commercial buildings and not energy-neutral.

For next generation energy-neutral civic structures, a new concept of Energy-Free Photo Thermal System (EFPHS) can be established based on transparent multilayer photothermal nano coatings (Fig. 1.6). As shown in Fig. 1.6, the solar light is harvested by the dome of a solar-tunnel (solar light collector) and directed through an array of transparent glass substrates coated with a photothermal material. Upon solar irradiation, the photothermal coating can convert photons to heat efficiently, thus raising the surface temperature of these substrates that function as an "energy-free solar radiator." The entire system uses natural sunlight, requiring no electric power, or any other forms of energy.

The current energy saving strategies heavily rely on thermal insulation that solely depend on glazing technologies and insulating materials, such as double or triple pane

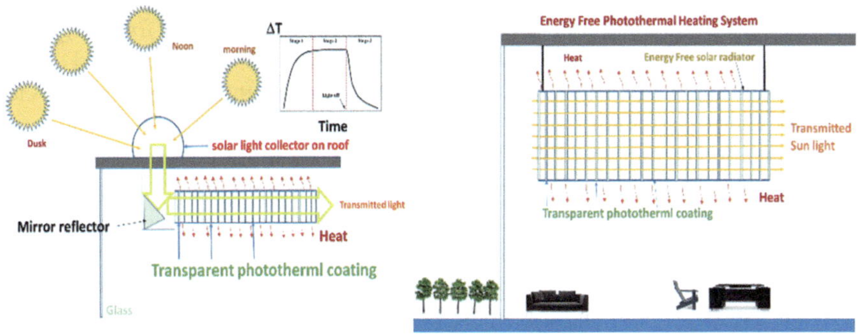

Fig. 1.6 Schematic of Energy-Free Photo Thermal System (EFPHS)

windows. Despite considerable improvements, these approaches are fundamentally incapable of reaching the state of "energy neutral" or "net zero" sustainable systems. The "Energy-Free Photothermal Heating System" (EFPHS) can be structurally and straightforwardly installed in single houses and high-rises as energy free heaters, especially for cold climates. Figure 1.7 shows one of the possible configurations of EFPHS in a building interior. As can be seen in Fig. 1.7, the sunlight is directed from the top, passing through multiple layers of transparent PT films which not only can function as a thermal radiator, but also a light source for building interior lighting. Figure 1.7 is a more realistic illustration of Fig. 1.6.

This chapter will provide the most updated research outcomes on the above topics, namely, (1) optical thermal insulation via photothermal coating of single pane

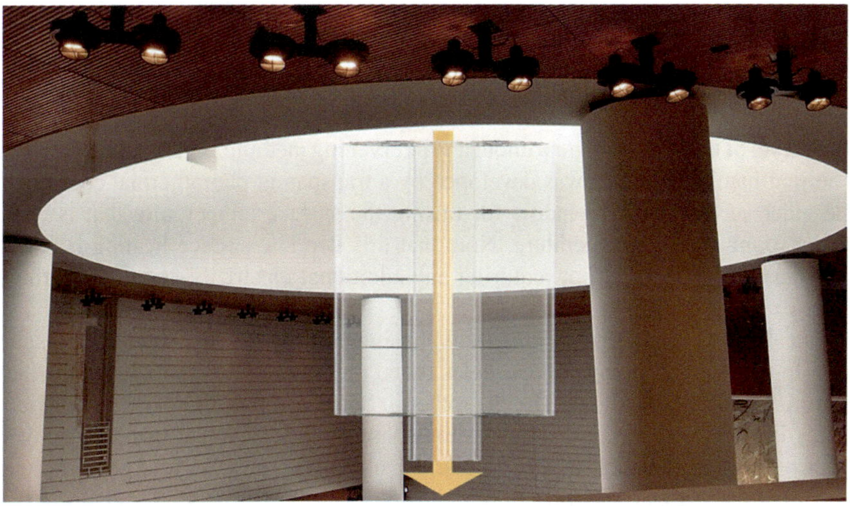

Fig. 1.7 Installation of a building solar radiator based on the concept of EFPHS

windows, (2) transparent thin film with PT-PV dual modality films for energy efficient building skin, and (3) 3D solar harvesting for generating photothermal energy used in building heating utilities. Novel synthesis routes in making nanoparticles and photothermal solutions will be described in detail with most recent publications. Film deposition methods will also be provided for coating various substrates. These multifunctional hybrid materials and films are characterized with a variety of means including UV-vis spectroscopy for absorption and transmittance, and Raman study for fundamental mechanism of structural interactions responsible for the optical and photothermal properties. Both heating and I-V curves of the films are obtained and correlated to their nano and electronic structures. Important engineering parameters including photothermal conversion efficiency and U-factor are calculated and analyzed under different conditions. Also discussed are roles of nanomaterials in next generation building skins, current challenges in multifunctional designs of large-scale surface energy devices, and future opportunities. Several new concepts are presented with their future energy applications, industrial viabilities, and commercial potentials.

1.2 Photothermal Materials for Energy Efficient Building Skin: Synthesis and Property Characterization

As noted above, the residential and commercial sectors counted for 40% of total U.S. energy consumption in 2020 (How much energy is consumed in U.S. buildings, 2021). The primary heat loss from conventional buildings is from windows since they are poor thermal insulators, which is a particular challenge of energy saving in cold climates. The conventional approach has been applying double- or triple-pane windows for reducing heat loss. However, these glazing technologies cost two to three times more than the single-pane windows (ARPA-E, 2014). To replace double-pane with single-pane, the key issue deals with reducing heat transfer of the window at the same level of the double-pane without any intervening medium. The so-called Optical Thermal Insulation (OTI) was developed by a transparent photothermal coating on the inner surface of the window for raising the surface temperature that is close to the room interior temperature. Note that this window surface heating is solely by solar irradiation, and not requiring any additional electricity, therefore entirely energy neutral.

The photothermal effect is a physical phenomenon relating to conversion of solar light to thermal energy by nanoparticles based on the so-called as localized surface plasmon resonance (LSPR). The photothermal effect of thin films has been extensively investigated for energy applications based on the heat generated by applied electromagnetic radiation in a particular wavelength range. The first photothermal thin film for energy-efficient window applications was reported by Zhao et al. (2017). The concept of the energy-efficient window involves a photothermal thin film coating on the inner surface of the window (Fig. 1.2). The window surface temperature will increase due to the photothermal effect under solar irradiation. The thermal transmittance of the window depends on the temperature difference (ΔT) between the single-pane and that of the indoor room. The reduction in ΔT effectively lowers heat transfer through the building skin via photothermal heating, characterized by a low U-factor, without double- or triple-glazing.

1.2.1 Synthesis and Characterization of the Photothermal Materials: Porphyrins and Iron Oxides

In the following, detailed synthesis routes are introduced on making a variety of photothermal materials including metals, carbon-based materials, organics and inorganics, iron oxides, and composites. Processing methods include solution synthesis and thin film deposition. Also introduced are nanoparticle techniques such as coprecipitation, hydrothermal synthesis, insert gas condensation, etc.

1.2.1.1 Synthesis of Iron Oxide and Its Composite

One pot coprecipitation synthesis is an efficient method for synthesizing nanoparticles. In our laboratory, we utilized one pot coprecipitation to synthesize photothermal materials, such as Fe_3O_4 and $Fe_3O_4@Cu_{2-x}S$. These nanoparticles were also modified by coating of poly(acrylic acid) and polystyrene to achieve hydrophilic surfaces (Lin et al., 2020; Tian et al., 2013; Zhao et al., 2017). Specifically, a given amount of oleylamine was heated to 300 °C in a nitrogen environment and stirred. A mixture of Fe $(acac)_3$ in oleylamine/N-methyl-2-pyrrolidone was injected into a preheated oleylamine. By keeping the solution at 300 °C until the solution was well mixed, the solution was cooled down to 60 °C. Fe_3O_4 nanoparticles were collected by a strong magnet and washed by methanol. The final product was dried by freeze drying. The $Fe_3O_4@Cu_{2-x}S$ nanoparticles were developed by coating $Cu_{2-x}S$ onto the Fe_3O_4 nanoparticles, forming a core–shell structure. For $Cu_{2-x}S$ coating, a specific amount of Fe_3O_4 nanoparticle solution was heated to 70 °C. A mixture of sulfur in an oleylamine/cyclohexane solution was injected into Fe_3O_4 nanoparticle solution and stirred for 10 min at 70 °C. Subsequently, a mixture of Cu $(acac)_2$ in an oleylamine/chloroform solution was injected into the reaction system and stirred for 30 min to form $Fe_3O_4@Cu_{2-x}S$ nanoparticles. $Fe_3O_4@Cu_{2-x}S$ nanoparticles were collected by a magnet and washed by methanol. The final product was dried and dispersed in toluene for thin film coating.

The mechanism of the photothermal effect has been identified as localized surface plasmon resonance (LSPR). LSPR refers to a surface plasmon confined in a nanoparticle smaller than the incident light wavelength. In contrast with surface plasmon resonance (SPR), LSPR is a particular type of SPR for nanoparticles related to nanoparticles' size and shape. The so-called dipole approximation of Mie theory describes the nanoparticles' response to the electric field vibration. The size of the scattering particles relative to the light wavelength is described by Mie scattering (Wei et al., 2013; Zaitoun et al., 2001). At plasmonic resonance, the electric fields are enhanced near the particle surface, and this enrichment drops off with distance from the surface. As a result, the optical absorption of the particle is at its maximum at the plasmon resonant frequency. This phenomenon occurs at visible wavelengths for noble metal nanoparticles. The dissipation of LSPR as thermal energy results in temperature increase.

1.2.1.2 Extraction of Chlorophyll

The extraction method of chlorophyll was reported in previous research (Zhao et al., 2019b). Fresh spinach leaves were cut into small pieces (~5 mm × 5 mm) and freeze-dried at −0 °C to dewater. The dried leaves were washed by using petroleum ether to remove the carotenoids and waxes. The chlorophyll can be extracted by immersing the washed leaves in 75% methanol and 25% petroleum ether at room temperature for 24 h. Subsequently, the solution was transferred to a separatory funnel and washed with saturated sodium chloride solution. The organic phase and aqueous phase were then separated by saturated sodium chloride solution. The organic phase was subsequently filtrated and removed by rotary evaporation to obtain the isolated film. The film was dissolved in acetone and stored at −20 °C for at least 24 h to precipitate the remaining impurities. Precipitates were collected by centrifugation, and the isolated chlorophyll was obtained by rotary evaporation. The final product was dissolved in toluene and stored at −20 °C for use.

1.2.1.3 Deposition of Photothermal Thin Films

Spin coating is a procedure used to deposit uniform thin films onto flat substrates. First, a small amount of coating material is applied to the center of the substrate, and then the substrate is rotated at a speed to spread the coating solution by centrifugal force. The liquid will be evaporated during spinning, leading to a uniform film. To form a thin film, the photothermal (PT) material solution is mixed with certain amount of polymer, such as polymethyl methacrylate (PMMA), polyethylene glycol (PEG), polyacrylic acid (PAA), or epoxy resin, etc. The viscosity of the solution can be increased by mixing PT materials with polymer, allowing the thin films to be coated on the glass substrates consistently.

PT materials are dispersed or dissolved in a solvent and mixed with a polymer, resulting in a well-mixed solution. The solution is dropped on the substrate while spinning with a specific speed, time, and amount of solution. Therefore, the spin coating parameters are essential in obtaining high-quality films, and these include spin speed, spin time, amount of materials dropped on the substrate, and the viscosity of the solutions.

In our spin coating experiments, the Fe_3O_4 and $Fe_3O_4@Cu_{2-x}S$ nanoparticles were mixed with a certain amount of PMMA in toluene. The mixture was coated evenly on a 25 mm × 25 mm glass substrate by dropping 100 μL of solution at 2000 rpm for 20 s (Tian et al., 2013). The spinning time or speed has to be optimized for different evaporation rates of the solutions. While water or some nonvolatile (such as epoxy resin) species evaporates much lower, toluene is a highly volatile chemical. Furthermore, sodium copper chlorophyllin can only be well dissolved in water and mixed with PEG (Zhao et al., 2019a). In this respect, the spinning speed will have to increase to 3000 rpm and the spinning time will be extended to 30 s in order to evaporate as much liquid as possible. Different materials require specific solvents, and it is essential to find a compatible polymer and solvent to form the films using the spin processor. Therefore, the spinning time and speed must be well controlled and optimized in the deposition of a variety of PT materials for obtaining uniform and structurally homogeneous films.

1.2.2 *Structure and Microstructure of Photothermal Materials*

The X-ray powder diffraction (XRD) patterns of both the Fe_3O_4 and $Fe_3O_4@Cu_{2-x}S$ samples are shown in Fig. 1.8. The X-ray diffraction pattern of the core–shell sample can be indexed as a mixture of the Fe_3O_4 (PDF# 01-1111) and CuS phases (PDF# 01-1281). The diffraction pattern associated with Fe_3O_4 (PDF# 01-1111) are shown in Fig. 1.8. The diffraction peaks at $2\,\theta = 30.28°$, 35.6°, 43.2°, 57.28° and 62.9° can be assigned to the (220), (311), (400), (511) and (440) facets of planes of Fe_3O_4 (PDF# 01-1111) (Tian et al., 2013). These results clearly distinguish the formation of Fe_3O_4 from the Fe_2O_3 nanocrystals through the synthesis described above. Notably, the peaks at $2\,\theta = 31.8°$ and 48.1° represent the (103) and (110) planes of CuS as shown in red line in Fig. 1.8 (Zhao et al., 2009).

Figure 1.9 shows the TEM images of the Fe_3O_4 and $Fe_3O_4@Cu_{2-x}S$ nanoparticles. As shown in this figure, while the former has an average diameter around 10 nm, the latter is slightly larger about 15 nm. The considerable size increases are

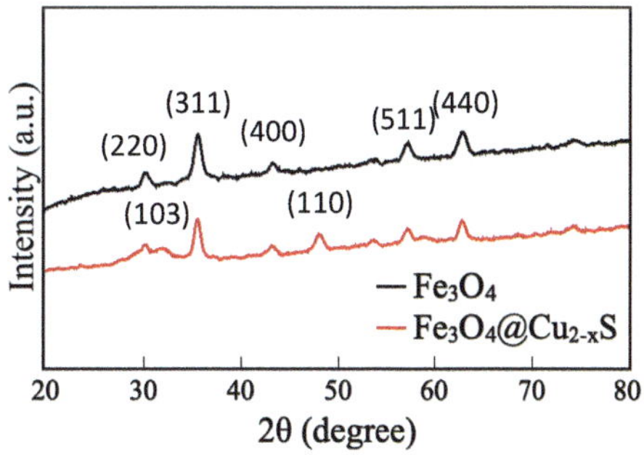

Fig. 1.8 Powder X-ray diffraction patterns of the Fe_3O_4 and $Fe_3O_4@Cu_{2-x}S$ nanoparticles, as referenced by standard Fe_3O_4 (PDF# 01-1111) and CuS (PDF# 01-1281) phases

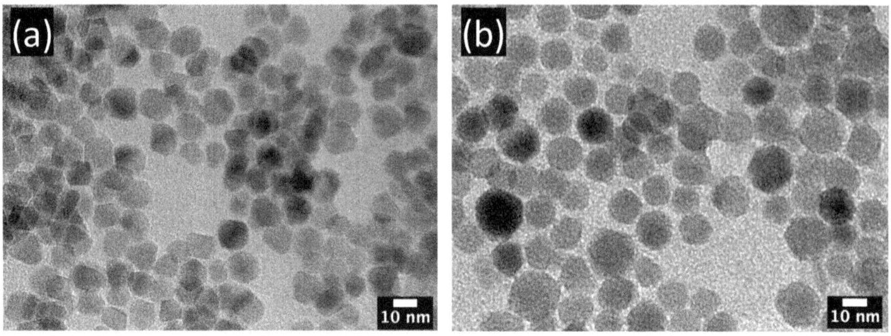

Fig. 1.9 TEM image of **a** Fe_3O_4 nanoparticles and **b** Fe_3O_4@$Cu_{2-x}S$ nanoparticles

due to the $Cu_{2-x}S$ coating on the Fe_3O_4 nanoparticles, making the microstructures quite uniquely different from a mixture of Fe_3O_4 and $Cu_{2-x}S$. Furthermore, Fe_3O_4 and Fe_3O_4@$Cu_{2-x}S$ nanoparticles appear uniform and monodispersed.

The Fe_3O_4 and Fe_3O_4@$Cu_{2-x}S$ nanoparticles were dispersed in toluene with a given amount of PMMA with good dispersity. The nanoparticle solutions containing both Fe_3O_4 and Fe_3O_4@$Cu_{2-x}S$ nanoparticles were deposited on glass substrates by using spin coating (Lin et al., 2020). Figure 1.10 shows the scanning electron microscopy (SEM) images of the cross-sectional areas of the films as indicated. As shown in this figure, all thin films exhibit smooth surfaces with varied thicknesses on an average of ~500 nm. The thickness differences were caused by various processing parameters such as spinning speed, concentration, solvent, and viscosity (Zhao et al., 2017, 2019a). As shown in Fig. 1.10a and b, the film thicknesses are respectively 464 nm and 497 nm for the Fe_3O_4 and Fe_3O_4@$Cu_{2-x}S$ thin films with similar morphology and thickness (Zhao et al., 2019a).

Figure 1.10c–f shows SEM images of the chlorophyll, chlorophyllin, hemoglobin, and phthalocyanine films; the film thicknesses are 555 nm, 515 nm, 494 nm, and 442 nm, respectively (Zhao et al., 2017). The surfaces and cross-sections of the thin films play crucial roles in both optical and photothermal properties, which can be further optimized by controlling the amount of precursor solution and spinning speed.

1.2.3 Optical Property Characterization of the Photothermal Thin Films

The spin coated thin films were characterized by using the UV–VIS-NIR spectrometer Lambda 900 (PerkinElmer Inc.) for absorption spectrum. A light transmittance meter (LS116, Linshang, Co. Ltd.) was used to measure the average visible transmittance (AVT).

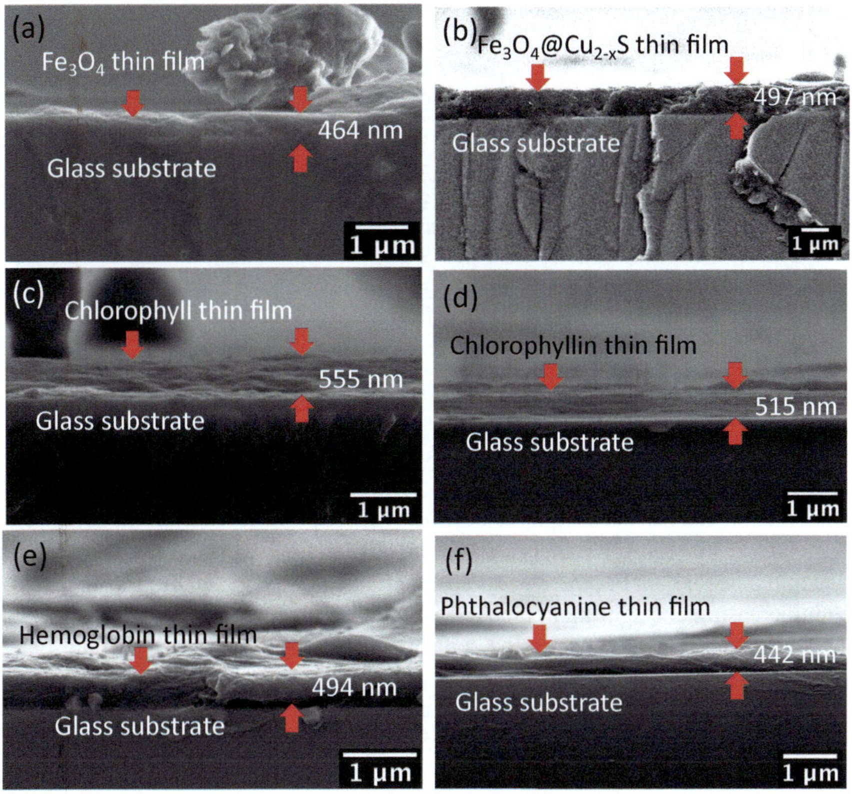

Fig. 1.10 SEM image of cross-section areas of **a** Fe_3O_4 thin film, **b** Fe_3O_4@$Cu_{2-x}S$ thin film, **c** chlorophyll thin film, **d** chlorophyllin thin film, **e** hemoglobin thin film, and **f** phthalocyanine thin film

Figure 1.11 shows optical absorptions of Fe_3O and Fe_3O_4@$Cu_{2-x}S$ nanoparticles dispersed in toluene with the identical concentration (0.1 mg/mL). The Fe_3O_4@$Cu_{2-x}S$ nanoparticle suspension exhibits a broad absorption peak at 1160 nm (green curve), while that of the Fe_3O_4 counterpart presents a similar UV absorption but none in the NIR region. The absorption of the Fe_3O_4@$Cu_{2-x}S$ solution is characterized with a U-shaped curve which is ideal for window applications. The thin film is highly transparent with an AVT above 90%, and its UV and NIR absorptions can convert photons to heat for raising the window surface temperature in the so-called optical thermal insulation. The NIR absorption of Fe_3O_4@$Cu_{2-x}S$ nanoparticles is attributed to the $Cu_{2-x}S$ core shell (Tian et al., 2013) due to the LSPR in vacancy doped semiconductors (Zhao et al., 2009).

The absorption and transmittance spectra of the Fe_3O_4 and Fe_3O_4@$Cu_{2-x}S$ thin films of various concentrations are shown in Fig. 1.12. As shown in this figure, the absorption spectra of the Fe_3O_4 (Fig. 1.12a) and Fe_3O_4@$Cu_{2-x}S$ (Fig. 1.12c) thin films are consistent with their suspension counterparts. For Fe_3O_4 suspension

Fig. 1.11 Absorption spectra of Fe_3O_4 and $Fe_3O_4@Cu_{2-x}S$ nanoparticles in solution

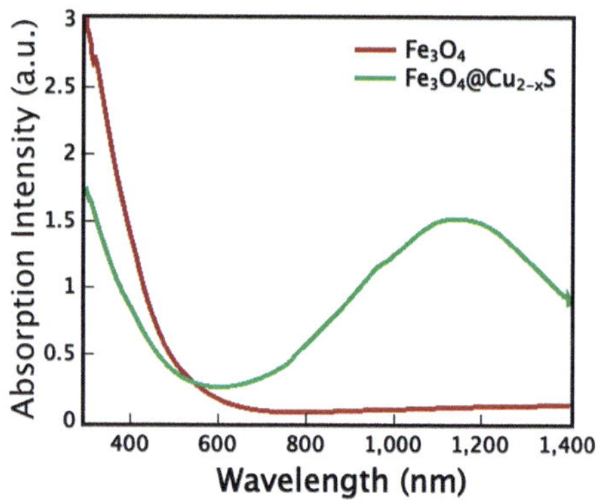

and thin films, the absorption intensities diminish beyond 700 nm. Interestingly, the absorption peaks of the $Fe_3O_4@Cu_{2-x}S$ thin films are red-shifted to 1349 nm, a phenomenon that has been explained by the SPR coupling between the nanoparticles (Kanehara et al., 2009; Lassiter et al., 2008). There are corresponding variations in transmittance for both Fe_3O_4 (Fig. 1.12b) and $Fe_3O_4@Cu_{2-x}S$ thin films (Fig. 1.12d). With increasing concentration, the AVT of the film is reduced. For comparison, the absorption and transmittance of the control film of pure PMMA are also presented in Fig. 1.12.

Porphyrin compounds are typically characterized with absorptions in the B band (380–500 nm) and Q band (500–750 nm). Figure 1.13 shows the porphyrin compounds of chlorophyll, copper chlorophyllin, hemoglobin, and phthalocyanine in solution forms of the same concentration of 0.01 mg/ml. Chlorophyll, copper chlorophyllin, hemoglobin, and phthalocyanine are dispersed in toluene, water, PBS, and DMSO, respectively. As shown in Fig. 1.13a–c, chlorophyll, chlorophyllin, and phthalocyanine all exhibit saddle like absorptions. Chlorophyll has two main peaks at 413 nm and 668 nm, the corresponding two peaks of chlorophyllin are at 402 nm and 628 nm. Phthalocyanine exhibits a strong peak at 350 nm, and broad absorption in the NIR region with two peaks at 638 nm and 730 nm. Interestingly, hemoglobin presents only one peak near UV at 405 nm, resulting in red color. Spectrally, the colors of these porphyrins vary according to the absorptions through structural replacement of the center atom in the porphyrin ring as can be seen in the porphyrin structures shown in Fig. 1.13.

Figure 1.14 shows absorption and transmittance spectra of porphyrin thin films with different concentrations. The absorption spectra of the porphyrin compound thin films are consistent with their solution counterparts. The absorption peaks are enhanced by increasing the concentration of the materials, while accordingly, the transmittances are suppressed. This figure shows that all porphyrins are quire transparent due to the saddle-like absorptions with broad valleys in the visible region. Note that there is a good correlation between the absorption intensity and the photothermal effect, as shown in Fig. 1.19. The strong absorption enhanced

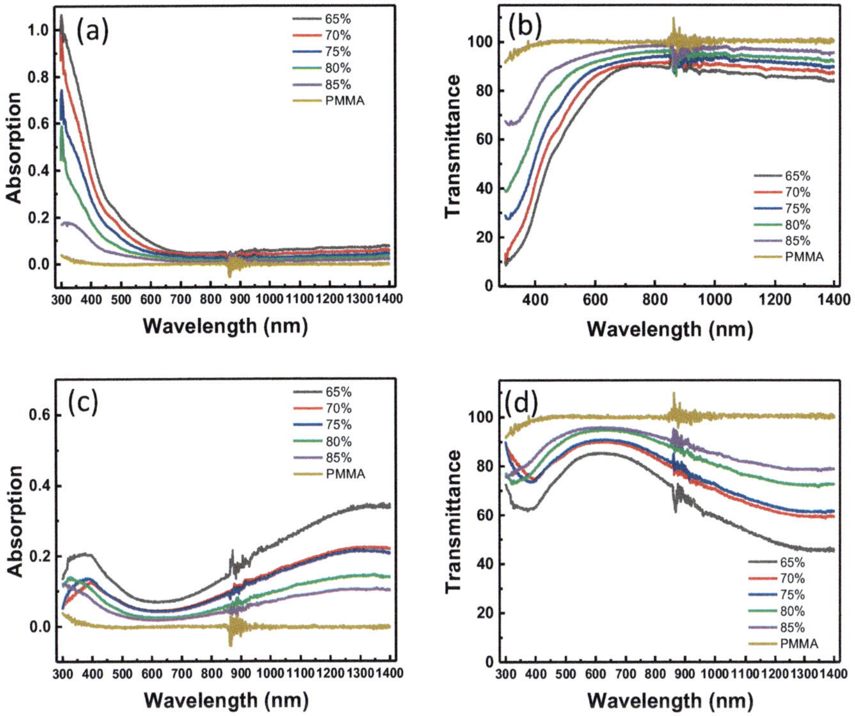

Fig. 1.12 **a** Absorption and **b** transmittance spectra the Fe_3O_4 thin films of various concentrations; **c** Absorption and **d** transmittance spectra of the $Fe_3O_4@Cu_{2-x}S$ thin films of various concentrations

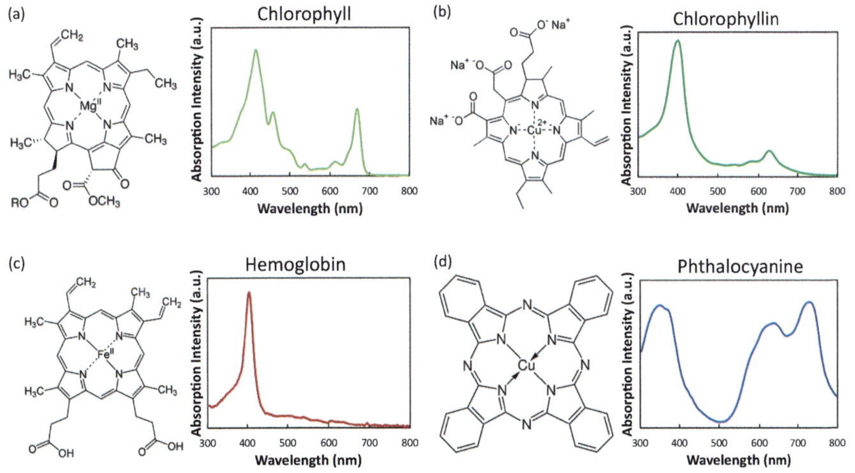

Fig. 1.13 Absorption spectra and molecular structure of **a** chlorophyll, **b** copper chlorophyllin, **c** hemoglobin, and **d** phthalocyanine (Zhao et al., 2019a)

photothermal heating. These thin-film properties are crucial in developing single-panes for window applications: strong UV and NIR absorptions for efficient solar harvesting and thermal energy conversion, but high visible transmittance for window transparency, all characteristically reflected in the saddle-shaped optical spectra.

1.3 Fundamental Studies on the Photonic and Photothermal Mechanisms

1.3.1 Raman Spectroscopy Study

LSPR mainly occurs in metals with high fractions of charge carriers. It was found that porphyrin compounds with much less charge carriers exhibit a different photothermal effect related to the molecular oscillation-induced photothermal effect (MOIPE) based on a Raman spectroscopy study (Zhao et al., 2019a). The Raman spectra of chlorophyll, chlorophyllin, hemoglobin, and phthalocyanine are shown in Fig. 1.15. The excitation laser wavelengths used were 442 nm, 514 nm and 633 nm. The molecular structures interact with the incident EM waves through vibrations of porphyrin ring's phonons and electrons, leading to an energy shift of the scattered photons. As can be seen in Fig. 1.15, the Raman peaks of chlorophyll excited by 442 nm and 514 nm laser are intensified at 1160 and 1540 cm^{-1} (Fig. 1.15a). The Raman peaks represent molecular vibrations of the porphyrin-ring bonds, corresponding to the photon absorptions. Raman scattering can be enhanced when the incident light wavelength is close to the absorption peak of the compound, which is associated with the molecular vibration of the porphyrin specifically, chlorophyll has a strong absorption peak at 413 nm, but no absorption at 514 nm. Therefore, the Raman peaks are well enhanced at 1540 cm^{-1} by the 442 nm laser but not by the 514 nm laser. Consistent results are also obtained from the Raman spectra of chlorophyllin, hemoglobin, and phthalocyanine. For chlorophyllin (Fig. 1.15b), the 422 nm laser excitation is responsible for more intensified Raman peaks at 1370 cm^{-1} and 1580 cm^{-1}. According to the absorption spectrum shown in Fig. 1.15b, chlorophyllin exhibits a stronger absorption at 400 nm, near the 442 nm laser, allowing for much higher Raman peak intensities than those of the 514 nm laser. Furthermore, chlorophyllin is characterized with the highest specific photothermal coefficient for its stronger absorptions near UV and NIR frequencies. The Raman data shown in Fig. 1.15 provides a fundamental base for the photothermal effect associated with the molecular vibrations of the porphyrins.

More evidence for the photothermal effects of porphyrins can be seen from the Raman spectra of hemoglobin and phthalocyanine. The hemoglobin exhibits a strong absorption at 405 nm, giving the Raman peaks greater intensities when excited by 442 nm laser than by 514 nm laser (Fig. 1.15c). However, the Raman scattering behaviors of phthalocyanine are quite different due to wider and stronger absorption above 600 nm (Fig. 1.15d). As expected, the Raman peaks of phthalocyanine are

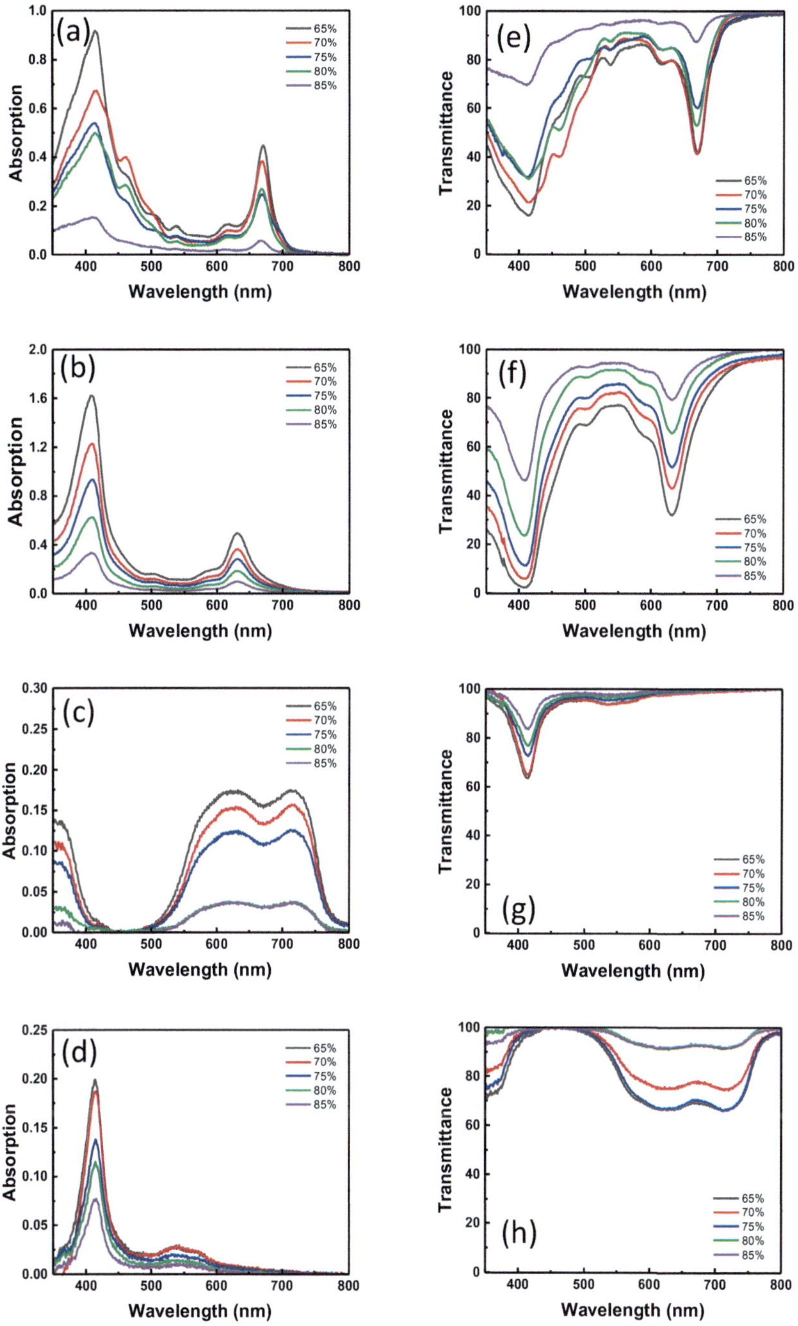

Fig. 1.14 Absorption spectra of **a** chlorophyll, **b** copper chlorophyllin, **c** hemoglobin, **d** phthalocyanine thin films of various concentrations; and transmittance spectra of **e** chlorophyll, **f** copper chlorophyllin, **g** hemoglobin, **h** phthalocyanine thin films of various concentrations

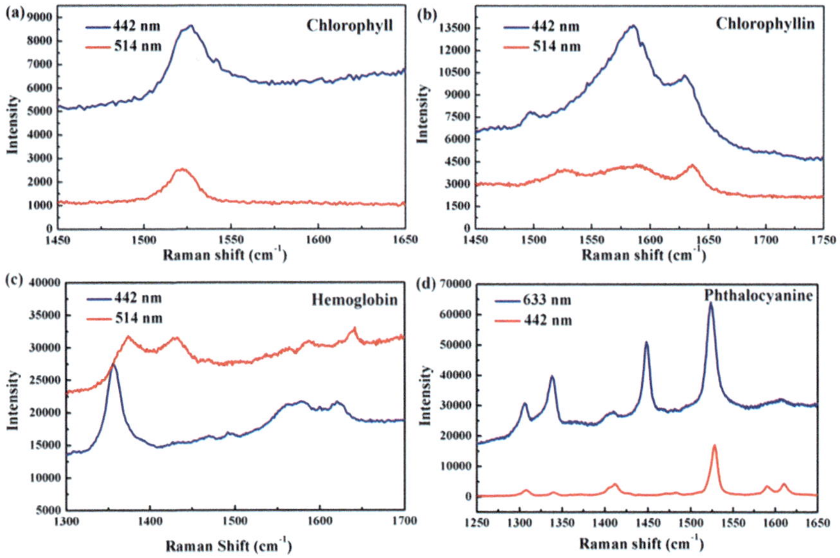

Fig. 1.15 Raman spectra of **a** chlorophyll, **b** chlorophyllin, **c** hemoglobin, and **d** phthalocyanine on aluminum foils by 442, 514, and 633 nm lasers

quite intense by excitation of the 633 nm laser than the 442 nm laser. As evidenced in the Raman spectra of porphyrins (Zhao et al., 2019a), the Raman laser light interactions with the molecular vibration of the porphyrin rings, which are associated with phonons and electrons in the porphyrin structure. These molecular vibrations due to photon excitations can dissipate heat and responsible for the photothermal effects as observed in the heating curves shown in Fig. 1.19.

Within the absorption frequency range, the photon frequency matches the vibration mode of porphyrin and oscillate to generate an electron density distribution. This creates a local electrical field with a dynamically modified charge equilibration (Fig. 1.16). With the constant irradiation by the incident light, the local electrons form a collective oscillation. The collective oscillation dissipates energy in form of heat manifested by the temperature change versus time curves. Porphyrin compounds are semiconductors with limited charge carriers, making LSPR unlikely for the photothermal effect. The molecular vibrations of porphyrins can absorb photons at various frequencies and convert photonic energy to thermal energy, responsible for the photothermal effect.

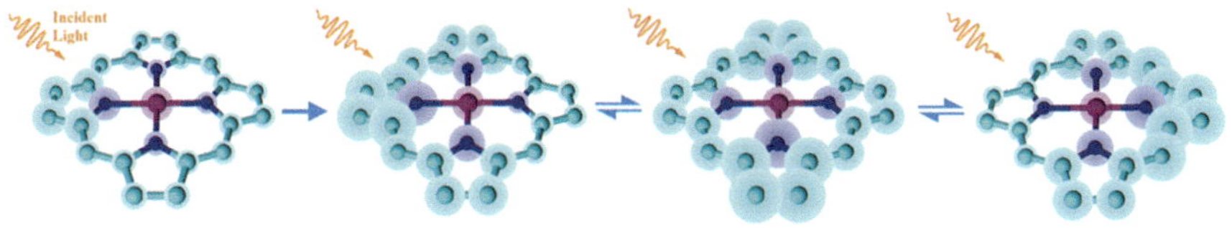

Fig. 1.16 Resonance modes of porphyrin via incident photon excitation (Zhao et al., 2019a)

1.3.2 Band Structures of Iron Oxides and Porphyrin Compounds

The fundamental absorption mechanism of iron oxide has been investigated from the defect structures of the nanoparticles. The absorption coefficient (α) can be expressed as (Singh, 2006):

$$\alpha = \alpha_o \exp[\sigma(\mathrm{E} - \mathrm{E}_o)/\mathrm{K}_B\mathrm{T}]$$

where

α_o = absorption-related constant (material-dependent constant)

E = the incident energy

E_o = the onset of absorption (material-dependent constants).

Additionally, E_u is Urbach energy can be expressed as (Rai, 2013):

$$\mathrm{E}_u = \mathrm{K}_B\mathrm{T}/\sigma$$

where

σ = the steepness parameter

K_B = the Boltzmann constant.

Urbach energy is a defect density measurement in the crystal. Based on this, the direct band gap value of the material can be estimated by plotting $(\mathrm{E}\alpha)^2$ as a function of photon energy (Fig. 1.17), where extrapolation of the linear portion of the curve to the x-axis gives the value of the direct band gap (Shi et al., 2015).

Figure 1.17 shows $(\mathrm{E}\alpha)^2$ versus photon energy for (a) Fe_3O_4, (b) $Fe_3O_4@Cu_{2-x}S$, (c) chlorophyll, (d) chlorophyllin, (e) hemoglobin, and (f) phthalocyanine samples. The direct band gap values are calculated by plotting $(\mathrm{E}\alpha)^2$ as a function of photon energy. The solid line in each figure represents the experimental absorption curve as shown in Fig. 1.14. A dashed line is the fit to the linear portion of the curve, where the intercept of the fit on the x-axis is the estimation value of direct band gap. The band gap energies of Fe_3O_4, $Fe_3O_4@Cu_{2-x}S$, chlorophyll, chlorophyllin, hemoglobin, and phthalocyanine are 3.03 eV, 2.65 eV, 1.81 eV, 1.82 eV, 2.95 eV, and 1.61 eV,

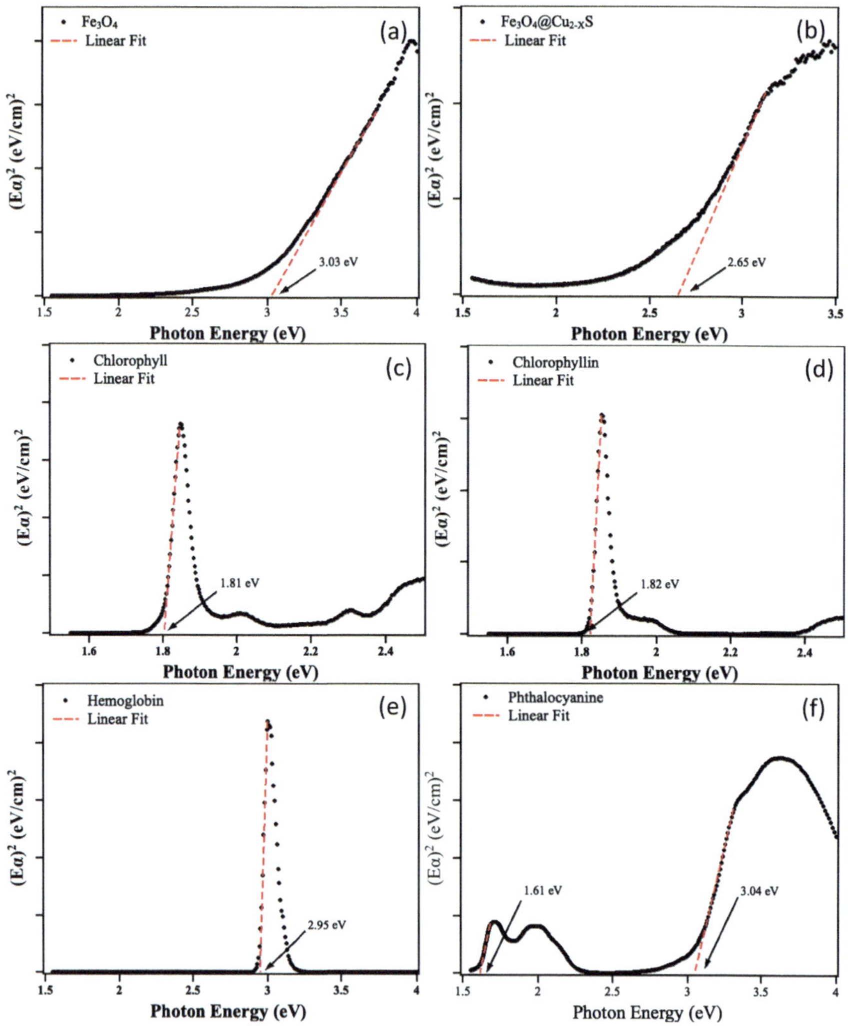

Fig. 1.17 Plot of $(E\alpha)^2$ versus photon energy, **a** Fe_3O_4, **b** $Fe_3O_4@Cu_{2-x}S$, **c** chlorophyll, **d** chlorophyllin, **e** hemoglobin, and **f** phthalocyanine, where the solid line represents the experimental absorption curve, and the dashed line—the fit to the linear portion of the data, where the intercept of that curve in the x-axis gives the estimation of direct band gap

respectively. Fe_3O_4 band energy calculated in this study (3.03 eV) corresponds with the band gap (~3.1 eV) (Zhao et al., 2017; Boxall et al., 1996; Fontijn et al., 1999) between the oxygen orbital (2P) and the octahedral site. $Fe_3O_4@Cu_{2-x}S$' band gap energy (2.65 eV) is lower than that of Fe_3O_4 due to $Cu_{2-x}S$ and Fe_3O_4 forming a core–shell structure (Lin et al., 2020; Tian et al., 2013). Note that the band gap of CuS is ~2.2 eV (Nemade & Waghuley, 2015).

1.4 Photothermal Thin Films for Energy-Efficient Windows: Optical Thermal Insulation

Zhao et al. first reported the photothermal effect of thin films via irradiation of simulated solar light (Zhao et al., 2017). The heating curves were obtained from both Fe_3O_4 nanoparticle solutions and thin films under simulated solar light. The Fe_3O_4 nanoparticles were surface modified by polymer for water solubility. They found that the photothermal effects of both Fe_3O_4 thin films and suspension quite comparable. Subsequently, Zhao et al. (2019b) investigated the photothermal characteristics of chlorophyll and found that its promising photothermal behaviors in both solutions and thin films under irradiation of white light and monochromic lasers (660 nm and 785 nm). They studied the correlations between absorption spectra and photothermal effects in both solutions and thin films and found that the chlorophyll solution exhibited a stronger photothermal response when irradiated by a 660 nm monochromic laser than by a 785 nm monochromic laser. The enhanced photothermal effect is attributed to chlorophyll having an absorption peak close to 660 nm, resulting in more photon conversion to heat. The AVTs of the chlorophyll thin films can be varied by adding different layers, respectively, 59%, 65%, 70%, 76%, 79%, and 84% for number of layers of 6, 5, 4, 3, 2, and 1 (Fig. 1.18). Each layer is approximately 500 nm. Therefore, AVT can be controlled by the number of coating layers on the same substrate.

Accordingly, there is a thickness dependence on the temperature increase in the multilayer films. For instance, the chlorophyll thin film can increase 6.7 °C from room temperature for 6 layers of coating with an AVT of 59%, while only 3.13 °C increase for the single layer coating with an AVT of 84%. These data indicate the thermal energy generated is an extensive property that depends on the volume of the materials.

Figure 1.19 shows the photothermal behaviors of chlorophyll, chlorophyllin, hemoglobin, and phthalocyanine thin films with varied AVTs under white light irradiation (0.1 W/cm^2). As shown in this figure, the temperature increase is rapid in the first 3 min for all samples with various concentrations due to strong photothermal heating. The heating curves are then leveled off at approximately 4 min, and thereafter forming plateaus, as a result of heat loss through the surfaces of the films. As the photothermal heating and heat loss are being balanced, the heating curves exhibit plateaus extending for prolong periods until the light source is turned off. As shown in Fig. 1.19, the light is turned off at 10 min, resulting in a temperature decrease following Newtonian cooling. As shown in this figure, the cooling curves indicate the amount of heat transferring to the atmosphere.

The photothermal effects have been observed to be dependent on both AVT and concentration. As shown in Fig. 1.19, ΔT of the chlorophyll thin film is 6.3 °C, corresponding to 65% AVT and concentration of 7.77×10^{-5} g/cm^2, while only 3.8 °C for the thin film of 85% AVT and concentration of 4.27×10^{-5} g/cm^2. Similar behaviors can be seen in chlorophyllin, hemoglobin, and phthalocyanine. The plots of

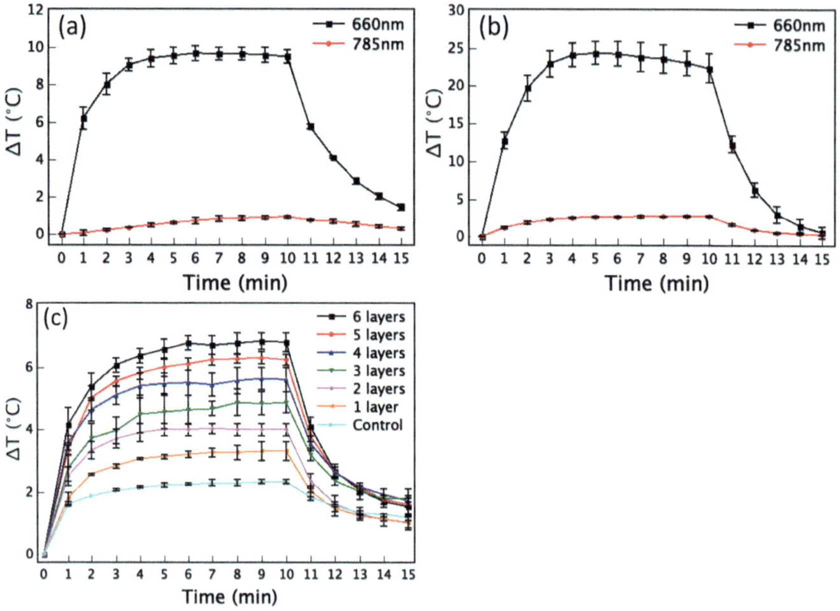

Fig. 1.18 **a** ΔT versus time for chlorophyll in 1-butanol irradiated by 661 and 785 nm lasers. **b** ΔT versus time for chlorophyll coated glass irradiated by 661 and 785 nm lasers. **c** ΔT versus time for chlorophyll coated glass with different layers of chlorophyll coating irradiated by simulated solar light

ΔT versus AVT in Fig. 1.19 for different concentrations provide valuable references for specific energy applications regarding transparency.

The similar photothermal behaviors were observed by Lin et al. (2020) in the thin films of iron oxides as shown in Fig. 1.20e–h. Thin films of Fe_3O_4 and $Fe_3O_4@Cu_{2-x}S$ were prepared with various AVTs (65%, 70%, 75%, 80%, and 85%, respectively) and their photothermal performances were investigated under irradiations of white light (0.1 W/cm^2) and monochromic laser (785 nm, 0.1 W/cm^2). Figure 1.20a shows the heating and cooling curves of the Fe_3O_4 with various concentrations under white light. As can be seen in these figures, ΔT_{max} of Fe_3O_4 thin film with 65% AVT (concentration of 4.57×10^{-4} g/cm^2) reaches 6.53 °C under white light irradiation and 5.10 °C under the 785 nm laser irradiation. These data indicate that white light irradiation results in greater photothermal effects of the films, therefore heating to higher temperatures as compared to those irradiated by the 785 nm laser. This behavior is attributable to the films being able to absorb photons in a wide frequency range under white light irradiation, while the laser is confined in a much narrow frequency. The maximum temperatures of these heating curves are denoted as ΔT_{max} and plotted against AVT for Fe_3O_4 thin films irradiated by white light

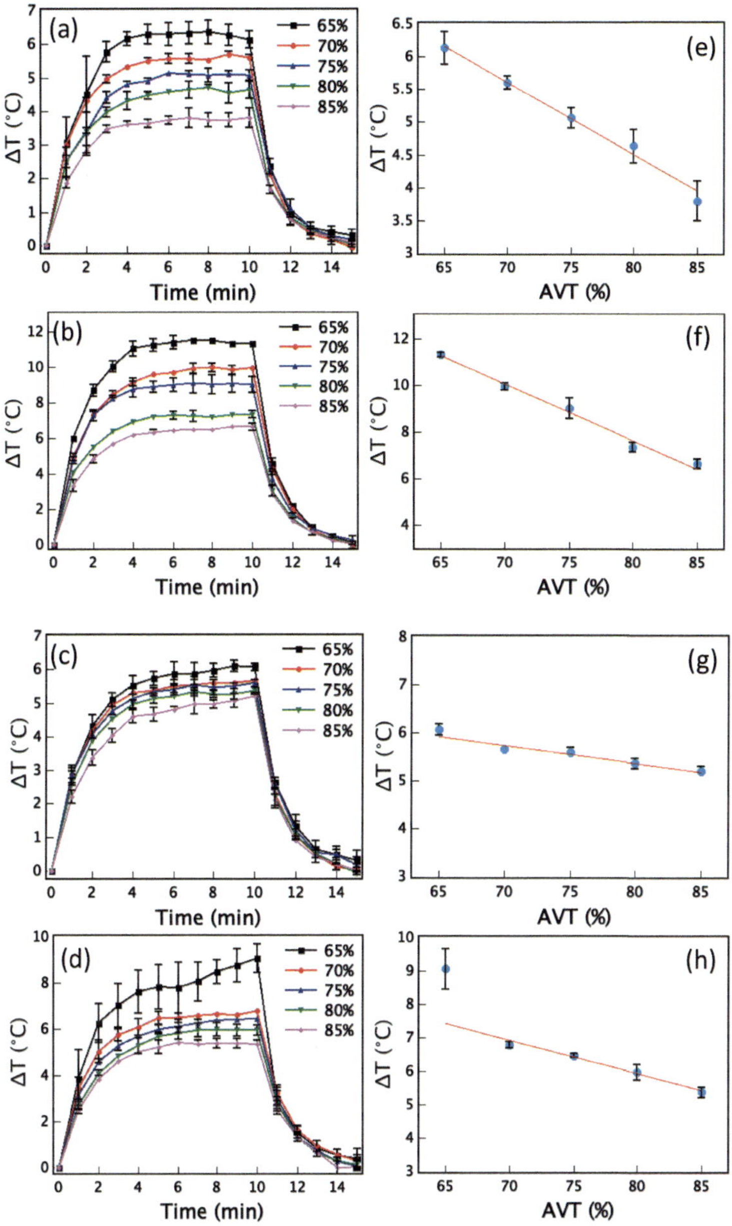

Fig. 1.19 ΔT versus time for the **a** chlorophyll samples, **b** chlorophyllin, **c** hemoglobin, and **d** phthalocyanine, with concentrations indicated; and linear fitting of the ΔT_{max} versus AVT data for the **e** chlorophyll samples, **f** chlorophyllin, **g** hemoglobin, and **h** phthalocyanine. All samples are irradiated by simulated solar light (0.1 W/cm^2) (Zhao et al., 2019a)

(Fig. 1.20b) and a 785 nm laser (Fig. 1.20d). Consistently, there are linear relationships between ΔT_{max} and AVT for both white light and laser irradiations as shown in Fig. 1.20b and d.

Similar behaviors were absorbed in the $Fe_3O_4@Cu_{2-x}S$ films as shown in Fig. 1.20e–h. As shown in Fig. 1.20e, ΔT_{max} of $Fe_3O_4@Cu_{2-x}S$ thin films is 7.77 °C by 785 nm laser irradiation higher than that of white light irradiation (7.10 °C) as shown in Fig. 1.20g. The highest temperature of the $Fe_3O_4@Cu_{2-x}S$ thin film (65% AVT) can reach 7.1 °C while that of the Fe_3O_4 counterpart is 6.53 °C) under 0.1 W/cm^2 white light irradiation (Fig. 1.20a). The enhanced photothermal effects observed from the $Fe_3O_4@Cu_{2-x}S$ thin films are associated with their unique optical characteristics, spectrally, strong absorptions in the IR regions as shown in Fig. 1.12. As a result of significant absorptions in a wide range of IR regions, large photon energies can be converted to thermal heat as reflected in Fig. 1.20e–h. This unique optical property originates from the core–shell structure of the $Fe_3O_4@Cu_{2-x}S$ nanoparticles. It has been reported that the CuS shell on the Fe_3O_4 core is responsible for significant IR absorptions (Tian et al., 2013). In Fig. 1.20f and h, we again observe the linear relationships between ΔT_{max} and AVT for white light and laser irradiations.

1.5 Photothermal Properties and Engineering Parameters

1.5.1 Photothermal (PT) Conversion Efficiency, η

The photothermal (PT) conversion efficiency (η) refers to the heating abilities of the materials in a solution when irradiated by either with a monochromic or white light. The values of the PT conversion efficiency can be calculated by following expression (Liu et al., 2014):

$$\eta = \frac{hS(\Delta T_{\text{max}} - \Delta T_{\text{amb}}) - Q_{\text{Dis}}}{I\left(1 - 10^{A_\lambda}\right)} \tag{1.1}$$

where

h = heat transfer coefficient from the solution to ambient (W/m^{2}K)

S = surface area of the well plate (m)

ΔT_{max} = maximum temperature (°C)

ΔT_{amb} = ambient temperature (°C)

Q_{Dis} = heat dissipated (W/m)

I = incident light intensity (W/m^2)

A_λ = absorbance of the incident light wavelength.

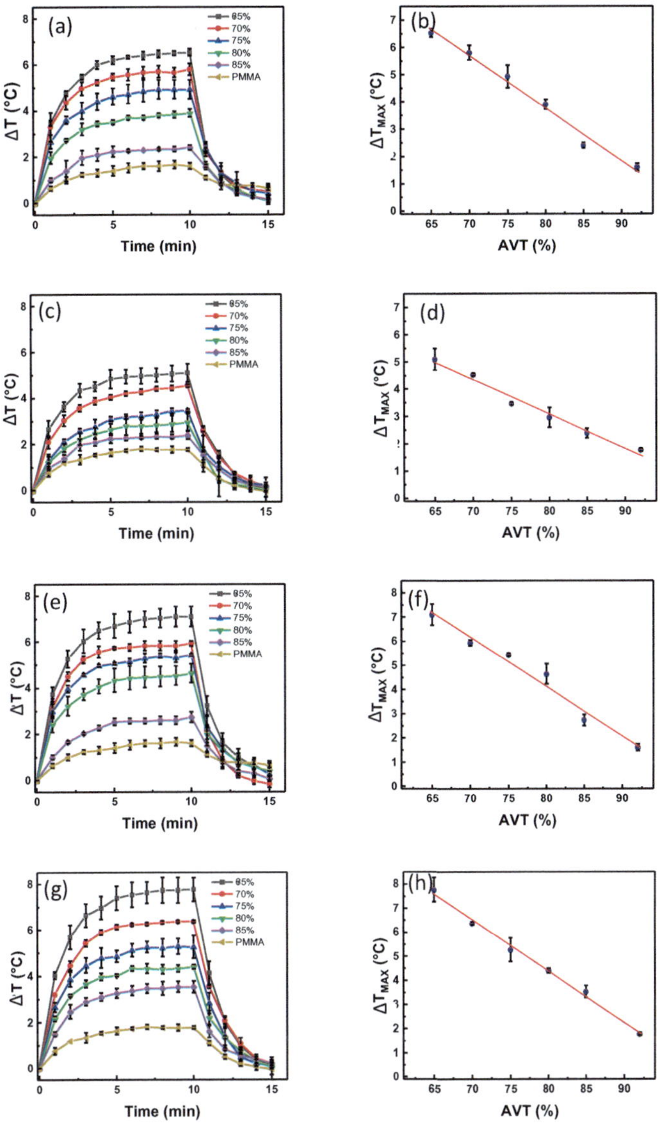

Fig. 1.20 **a** The change in temperature increase, ΔT, versus time curves under white light irradiation for the Fe_3O_4 thin films of various concentrations. **b** The maximum change in temperature increase, ΔT_{max}, versus average visible transmittance (AVT) for the Fe_3O_4 thin films of various concentrations under white light irradiation. **c** ΔT versus time curves and **d** ΔT_{max} versus AVT for the Fe_3O_4 thin films of various concentrations under the 785 nm laser irradiation. **e** ΔT versus time curves and **f** ΔT_{max} versus AVT for the $Fe_3O_4@Cu_{2-x}S$ thin films of various concentrations under white light irradiation. **g** ΔT versus time curves and **h** ΔT_{max} versus AVT for the $Fe_3O_4@Cu_{2-x}S$ thin films of various concentrations under the 785 nm laser irradiation. Intensities of the white light and 785 nm laser are both 0.1 W/cm^2. The environment temperature is 25 °C

The value of (hS) can be calculated from the relation:

$$\mathrm{hS} = \mathrm{m}_D * \mathrm{C}_D/\tau_s$$

where (m_D) is the combined mass of nanoparticles and solvent, (C_D) is the combined heat capacity that is regarded as the heat capacity of nanoparticles and solvent, and (τ_S) is a time constant.

The time constant (τ_S) can be obtained by following (Jin et al., 2016):

$$\tau_s = -\frac{t}{\ln\left(\frac{T - T_{\mathrm{amb}}}{T_{\mathrm{Max}} - T_{\mathrm{amb}}}\right)} \tag{1.2}$$

where

T = temperature (°C)

t = time at the beginning of the cooling (s).

However, Eq. 1.1 was developed only for measurement under the monochromic light in solution. Jin et al. (Zhao et al., 2019c) simplified the formula of the PT conversion efficiency for solar light with the simulated data. This formula was applied to thin films using the following equation:

$$\eta = \frac{C_P \cdot m \cdot \Delta T_{\mathrm{max}}}{I A \Delta t} \tag{1.3}$$

where

$\mathrm{C_p}$ = specific heat capacity (J/(g·K))

m = mass of the sample (g)

$\Delta \mathrm{T}_{\mathrm{max}}$ = maximum temperature change of the sample (°C)

I = incident light power density ($\mathrm{W/cm^2}$)

A = sample's surface area ($\mathrm{cm^2}$)

Δt = time to reach the maximum temperature ($\mathrm{T}_{\mathrm{max}}$) for the sample (s).

1.5.2 *Specific Photothermal Coefficient, μ*

The specific photothermal coefficient (SPC, μ) is defined as the heat generated by the PT effect per the amount of the materials. The SPC for thin films was developed by Zhao et al. and given by the following expression (Zhao et al., 2019a):

$$\mu = \frac{\dot{Q}}{m} \tag{1.4}$$

where

$\dot{Q}$ = heat generated by PT effect in a period of time (J/s)

m = mass of the materials (g).

SPC was developed from the cooling part of the temperature versus time curve. The SPC value reveals the ability of the photon-to-heat conversion of a material. Specifically, high SPC implies a material having a greater ability to convert photons to thermal energy.

1.5.3 The Solar Photothermal Efficiency

The solar PT efficiency (η) is defined as the ratio of the energy increase inside the sample to the incoming photon energy of simulated solar light. Figure 1.21 shows a schematic diagram of the temperature difference ΔT versus time curve for analysis (Zhao et al., 2019a; Liu et al., 2014; Roper et al., 2007). When the sample is heated in stage 1, $\dot{Q} > UA(T - T_0)$, indicating that heat gained is greater than the heat lost. In stage 2, the sample temperature reaches a steady state as the heat loss to the environment is balanced by the heat gain. In this stage, we can write: $mC_p\frac{dT}{dt} = \dot{Q} - UA(T - T_0) = 0$ and $\dot{Q} = UA(T - T_0)$. In stage 3, $\dot{Q} = 0$.

$$\frac{dT_i}{dt} = \frac{\dot{Q}}{mC_p} - \frac{UA}{mC_p}(T - T_0) = -\frac{UA}{mC_p}\left[(T - T_0) - \frac{\dot{Q}}{UA}\right] \tag{1.5}$$

$$\frac{1}{(T - T_0) - \frac{\dot{Q}}{UA}}dT = -\frac{UA}{mC_p}dt \tag{1.6}$$

where

T = sample temperature (°C)

T_0 = initial temperature of the sample (°C)

m = sample mass (2.145 g)

C_p = sample specific heat capacity of quartz (0.74 J/(g·K))

t = heating time (s)

U = heat transfer coefficient of the sample (W/m^2·K)

A = sample surface area (2.54 × 2.54 cm^2)

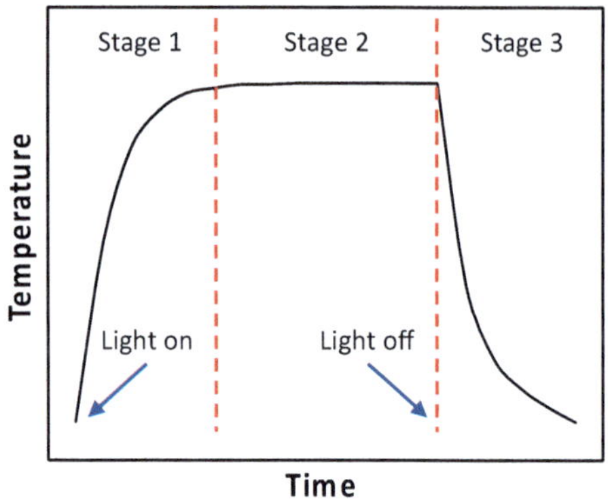

Fig. 1.21 Schematic diagram showing the temperature versus time curve

$\dot{Q}$ = heat produced by the PT effect in unit of (J/s).

Integrating the differential equation:

$$\int \frac{1}{(T - T_0) - \frac{\dot{Q}}{UA}} dT = -\int \frac{UA}{mC_p} dt \tag{1.7}$$

yields:

$$\ln\left|(T - T_0) - \frac{\dot{Q}}{UA}\right| = -\frac{UA}{mC_P} \cdot t + C \tag{1.8}$$

$$(T - T_0) - \frac{\dot{Q}}{UA} = e^{-\frac{UA}{mC_p}t} \cdot e^c = Ce^{-\frac{UA}{mC_p}t} \tag{1.9}$$

where C is constant.

$$T - T_0 = Ce^{-\frac{U}{mC_p} \cdot t} + \frac{\dot{Q}}{UA} \tag{1.10}$$

Since $\dot{Q} = 0$ in Stage 3, the equation can be written as:

$$T - T_0 = Ce^{-\frac{UA}{mC_p}t} \tag{1.11}$$

Rewriting Eq. (1.6), $T - T_0 = \Delta T$, gives:

$$\Delta T = Ce^{-\frac{UA}{mC_p}t} \tag{1.12}$$

By fitting the exponential function, the equation can be written as:

$$\Delta T = C_1 e^{C_2 t} \tag{1.13}$$

Since C_1 is constant and $C_2 = -\frac{UA}{mC_p}$, C_1 and C_2 can be obtained by fitting the exponential curve, and U can be calculated through C_2. Then,

$$\dot{Q} = UA(T - T_0) \tag{1.14}$$

Based on above equations, the PT efficiency can be written as:

$$\eta = \frac{\dot{Q}}{Q_{\text{in}}} \tag{1.15}$$

where

$\dot{Q}_{\text{in}}$ = the input energy from light source (J/s), and thus (Tables 1.1, 1.2, 1.3 and 1.4),

$$\eta = \frac{\sum \dot{Q}_i}{\sum Q_{\text{in}}} \tag{1.16}$$

Table 1.1 Parameters of the chlorophyll films under white light irradiation (0.1 W/cm^2)

	AVT (%)	ρ_{pm} (g/cm^2)	ΔT_{Max} (°C)	U (W/m^{2}K)	$\dot{Q}$ (J/s)	Efficiency, η (%)
Chlorophyll-1	64.7	7.77×10^{-5}	6.13	15.68	0.135	20.90
Chlorophyll-2	69.8	7.13×10^{-5}	5.60	16.86	0.132	20.52
Chlorophyll-3	75.3	5.21×10^{-5}	5.07	13.54	0.096	14.91
Chlorophyll-4	79.8	4.64×10^{-5}	4.63	15.63	0.102	15.73
Chlorophyll-5	84.3	4.28×10^{-5}	3.80	13.84	0.074	11.43

*The photothermal material density on the glass, ρ_{pm}, represents the mass of thin film per coating area (g/cm^2). ΔT_{Max} is the maximum steady temperature difference; U is the heat transfer coefficient (W/m^{2}K); $\dot{Q}$ is the heat produced via photothermal effect in unit of time (J/s), and η is the photothermal efficiency

Table 1.2 Parameters of the chlorophyllin films under white light irradiation (0.1 W/cm^2)

	AVT (%)	ρ_{pm} (g/cm^2)	ΔT_{Max} (°C)	U (W/m^{2}K)	$\dot{Q}$ (J/s)	Efficiency, η (%)
Chlorophyllin-1	64.5	2.92×10^{-5}	11.30	15.02	0.238	36.89
Chlorophyllin-2	70.8	2.42×10^{-5}	9.93	14.54	0.203	31.40
Chlorophyllin-3	74.7	1.85×10^{-5}	9.00	14.54	0.183	28.44
Chlorophyllin-4	80.1	1.64×10^{-5}	7.33	14.65	0.151	23.35
Chlorophyllin-5	85.2	1.49×10^{-5}	6.63	14.43	0.134	20.81

Table 1.3 Parameters of the hemoglobin films under white light irradiation (0.1 W/cm^2)

	AVT (%)	ρ_{pm}(g/cm^2)	ΔT_{Max} (°C)	U (W/m^{2}K)	$\dot{Q}$ (J/s)	Efficiency, η (%)
Hemoglobin-1	64.8	4.42×10^{-5}	6.07	12.23	0.104	16.13
Hemoglobin-2	69.7	4.13×10^{-5}	5.67	14.03	0.111	17.27
Hemoglobin-3	75.4	3.64×10^{-5}	5.60	12.07	0.095	14.69
Hemoglobin-4	80.4	1.57×10^{-5}	5.37	13.51	0.102	15.76
Hemoglobin-5	84.5	1.36×10^{-5}	5.20	14.16	0.103	16.00

Table 1.4 Parameters of the phthalocyanine films under white light irradiation (0.1 W/cm^2)

	AVT (%)	ρ_{pm} (g/cm^2)	ΔT_{Max} (°C)	U (W/m^{2}K)	$\dot{Q}$(J/s)	Efficiency, η (%)
Phthalocyanine-1	64.5	2.28×10^{-5}	6.53	11.67	0.107	16.56
Phthalocyanine-2	69.7	2.00×10^{-5}	6.03	11.37	0.096	14.90
Phthalocyanine-3	75.1	1.42×10^{-5}	5.67	12.00	0.095	14.79
Phthalocyanine-4	80.5	1.35×10^{-5}	4.87	10.91	0.075	11.55
Phthalocyanine-5	85.5	5.69×10^{-5}	4.53	11.02	0.070	10.85

1.5.4 U-factor, U

U-factor is related to thermal transmittance, therefore indicating the insulation ability standard for window manufactures proposed by the National Fenestration Rating Council (NFRC) (Rating & Council: Procedure for Determining Fenestration Product U-factors, 2013). It refers to the energy efficiency of the window assembly. A lower U-factor generally indicates a good thermal insulation. As the price of energy increases, energy efficiency and sustainable materials for thermal insulators are becoming increasingly important. According to the US Advanced Research Projects Agency-Energy (ARPA-E) (2014), buildings' energy and elements consumption can be improved by installing single-pane windows instead of a double-pane windows in cold climate areas of the US. The general specification of the U-factor for windows has been reported by NFRC (National Fenestration Rating Council Incorporated, 2020). Despite the U-factor requirement varying for diverse climate zones("Energy Star performance criteria for windows, doors, & skylights", Environmental Protection Agency, 2015), the minimum requirement of U-factor is < 1.7 W/m^{2}K. Therefore, it is crucial to reduce the U-factor for windows. In accordance to Standard Test Method for Measuring the Steady-State Thermal Transmittance (ASTM, 2014), U-factor is:

$$U = \frac{1}{\frac{1}{h_h} + \frac{1}{h_c} + \frac{1}{U_L}} \tag{1.17}$$

where

h_h = interior heat coefficient

h_c = exterior heat coefficient

U_L = heat transfer coefficient.

Equation 1.17 has been modified as (Zhao et al., 2019a).

$$U = \frac{1}{\left(\frac{T_{in}-T_{out}}{T_{in}-T_{ori}}\right) \times \frac{1}{1.46\times\left[\frac{(T_{in}-T_{ori})}{L}\right]^{0.25} + \sigma e\left[\frac{\left(T_{in}^4-T_{ori}^4\right)}{(T_{in}-T_{ori})}\right]} + \frac{1}{\sqrt{\left(0.84\times(T_{ori}-T_{out})^{1/3}\right)^2 + \left(2.38\times v^{0.89}\right)^2}}} \tag{1.18}$$

where

T_{in} = interior room temperature (°C)

T_{out} = exterior window surface temperature (°C)

T_{ori} = original interior window surface temperature (°C)

L = window's height (m)

σ = Stefan-Boltzmann constant ($5.670374419 \times 10^{-8}$ $W \cdot m^{-2} \cdot K^{-4}$).

v = outdoor wind speed (m/s).

Equation 1.18 can be applied to analyze the temperature changes on the original interior window surface (T_{ori}) upon coating with a photothermal film. Based on Eq. 1.18, Lin et al. (2020) calculated the U-factors of Fe_3O_4 and $Fe_3O_4@Cu_{2-x}S$ films, based on the hypothetical estimation parameters collected from NFRC 100–2017 (National Fenestration Rating Council Incorporated, 2020); these are: T_{in} is 21.11 °C, T_{out} is −17.78 °C, v is 5.5 m/s, L is 1.50 m, and e is 0.84. This is assuming that the original interior window temperature is 5 °C, and the Tg is calculated to be 278.15 K + ΔT_{max}. Using the above parameters, the U-factors of Fe_3O_4 and $Fe_3O_4@Cu_{2-x}S$ films are calculated, respectively, 1.54 and 1.46 (W/m^2K) (Lin et al., 2020). Furthermore, Zhao et al. (2019b) published the data from the multilayer chlorophyll films for energy-efficient window application. In this work, the calculated U-factor was found to be significantly lowered, from 1.99 to 1.42 (W/m^2K), by using chlorophyll film with 59% AVT (Fig. 1.18).

The U-factor is reduced by 30.32% when the Fe_3O_4 nanoparticles are added in the film (Table 1.5); and further reduced by 33.94% by adding $Fe_3O_4@Cu_{2-x}S$ nanoparticles (Table 1.6). These results indicate that Fe_3O_4 and $Fe_3O_4@Cu_{2-x}S$ can both significantly reduce U-factor. The U-factor data shown in Tables 1.5 and 1.6 indicate great potential of single panes via photothermal coating under white light based on the concept of OTI (Lin et al., 2020).

Table 1.5 Parameters used for calculation of U-factors for the Fe_3O_4 thin films under simulated solar irradiation (0.1 W/cm^2) (Lin et al., 2020)

Sample	ρ_{pm} ($\times 10^{-4}$ g/cm^2)	ΔT_{max} (°C)	AVT (%)	η(%)	SAR (W/g)	U (W/m^2·K)
Fe_3O_4-1	4.57	6.53	64.8	2.56	1.88	1.54
Fe_3O_4-2	4.07	5.80	70.4	2.27	1.80	1.65
Fe_3O_4-3	3.55	4.93	75.2	1.93	1.64	1.77
Fe_3O_4-4	2.54	3.90	80.4	1.53	1.58	1.91
Fe_3O_4-5	1.52	2.40	85.1	0.94	0.92	2.11
Control	0	1.60	91.0	0.63	0.00	2.21

*The photothermal material area density on the glass, ρ_{pm}, represents the mass of thin film per coating area (g/cm^2), ΔT_{max} is the maximum steady temperature difference of the sample, U is the heat transfer coefficient (W/m^2·K), and η is the photothermal conversion efficiency

Table 1.6 Parameters used for calculation of U-factors for the $Fe_3O_4@Cu_{2-x}S$ thin films under simulated solar irradiation (0.1 W/cm^2) (Lin et al., 2020)

Sample	ρ_{pm} ($\times 10^{-4}$ g/cm^2)	ΔT_{max}(°C)	AVT (%)	η (%)	SAR (W/g)	U (W/m^2·K)
$Fe_3O_4@Cu_2$-xS -1	5.08	7.10	65.2	2.78	1.89	1.46
$Fe_3O_4@Cu_2$-xS -2	4.07	5.93	69.7	2.33	1.86	1.63
$Fe_3O_4@Cu_2$-xS -3	2.54	5.43	74.9	2.13	2.64	1.70
$Fe_3O_4@Cu_2$-xS -4	2.03	4.63	80.5	1.82	2.62	1.81
$Fe_3O_4@Cu_2$-xS -5	1.01	2.73	85.0	1.07	1.95	2.07
Control	0	1.60	91.0	0.63	0.00	2.21

1.5.5 Angle Dependence of Solar Harvesting and Thermal Energy Generation

In window applications of the photothermal coatings, one important factor deals with the angle dependence of the incident solar light. Despite significant thermal loss reduction by OTI, another critical issue is associated with seasonal changes that can significantly alter the solar harvesting efficiency as solar light incident angle varies. Seasonal changes on earth dictate the incident angle of the sunlight; Chicago, for instance has a north latitude of 41.8° at winter solstice, giving an incident angle of $\alpha = 24.70°$. This low angle of α provides more sun-exposed area of the window. However, at the summer solstice, $\alpha = 71.70°$, resulting in less sun rays through the windows. Therefore, we expect more pronounced photothermal effects from the surface coating during the winter season, but much less in the summertime. Therefore, it is crucial to investigate the angle dependence of solar harvesting and thermal energy generation for single-pane applications (Lyu et al., 2020).

The thin films of different PT materials (including Fe_3O_4, $Fe_3O_4@Cu_{2-x}S$, chlorophyll, and chlorophyllin) were deposited on glass substrates by spin coating method and characterized by their microstructural and optical properties. These thin films exhibit uniform microstructures (Fig. 1.10) with thicknesses around 500 nm to 800 nm, which allow for high visible transmittances above 90% for all photothermal materials studied in wide spectra. Both absorption peaks in the UV and NIR regions were observed in the UV–Vis optical characterization of chlorophyll and chlorophyllin thin films. These absorption behaviors are ideal in both efficient solar light harvesting and enhanced transparency. It was also found that $Fe_3O_4@Cu_{2-x}S$ exhibited pronounced NIR absorption in a broad range which largely contributes to the increased photothermal effect (Fig. 1.12) compared to Fe_3O_4.

1.5.5.1 Incident Light Angle Dependence Measurement

Photothermal experiments were conducted with 0.4 W/cm^2 of white light using a Newport 150 W solar simulator (Lamp model 67,005) to irradiate the samples. An infrared camera (FLIR E6) was used to record the temperature. The power density of the solar simulator was calibrated by an optical power meter (Coherent Inc.). The heating curves were measured for 20 min, and then the light source is turned off to cool for 15 min.

The incident light angle (θ) dependence experimental set up is shown in Fig. 1.22. As shown in Fig. 1.22a, the simulated light is directed from top down. The sample, with the coated photothermal film surface directly facing the solar-simulated light source, can be rotated from $\theta = 0°$ to 90° and is placed below the light while the photothermal curves are recorded at a given angle. The infrared camera shown in Fig. 1.22a measures the temperature variations as a function of time (the heating curves). The light reflectance and transmittance (Fig. 1.22b) were measured by using a light power density meter.

The reflected light should increase with an increasing angle of incidence, α, that will reach its maximum when the light is near parallel to the film surface (El-Sebaii & Khallaf, 2020; Khan & Mustafa 2019). The transmitted light, just opposite to the reflected light, is maximized at $\alpha = 0$, which indicates that more photons pass through the thin film sample in the vertical direction, therefore generating the highest photothermal effect. Both reflectance and transmittance can be directly correlated to the heating curves in this way (Lin et al., 2020; Zhao et al., 2019a).

1.5.5.2 Experimental Results

The photothermal heating curves were measured by using the set up shown in Fig. 1.22a. For measurements at all angles of incidence, the distance between the center of the film sample and the light source was kept the same to maintain the

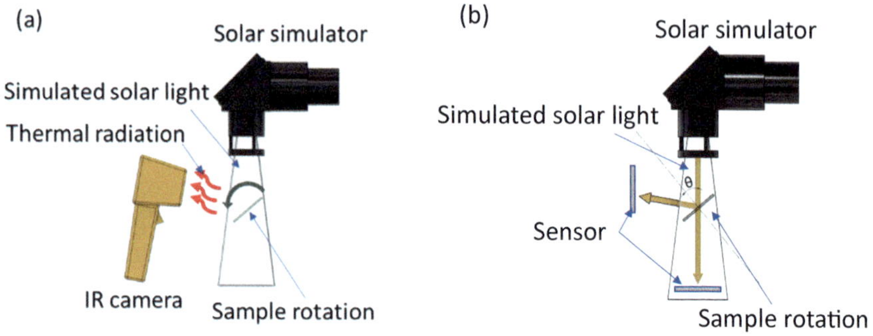

Fig. 1.22 **a** Angle dependence measurement set up and **b** reflectance and transmittance measurements

same level of incident light intensity. Figure 1.23 shows the heating curves of (a) $Fe_3O_4@Cu_{2-x}S$, (b) Fe_3O_4, (c) chlorophyll, and (d) chlorophyllin samples with 85% AVT (concentration of each sample: $Fe_3O_4@Cu_{2-x}S$: 1.1 mg/cm^2, Fe_3O_4: 1.6 mg/cm^2, chlorophyll: 0.9 mg/cm^2, chlorophyllin: 0.7 mg/cm^2). All samples show incident light angle dependence with different effects, as can be seen in Fig. 1.23. When the angle of incidence is $\theta = 0$, as shown in Fig. 1.23, all samples reach the highest temperatures: 52.3 °C for $Fe_3O_4@Cu_{2-x}S$, 47.6 °C for Fe_3O_4, 32.1 °C for chlorophyll, and 34.1 °C for chlorophyllin. For $\theta = 90°$, the highest temperatures of the four samples dropps to: 30.5 °C for $Fe_3O_4@Cu_{2-x}S$, to 29.5 °C for Fe_3O_4, 24.2 °C for chlorophyll, and to 24.3 °C for chlorophyllin. Figure 1.23e shows the maximum temperature versus angle of incidence for all samples.

The angle dependence of the maximum temperature was found to be related to the incident light being reflected at different θ (Fig. 1.23f). The reflection and transmission for different samples is shown in Fig. 1.23f. All samples follow general trends, in which the light reflectance increases as θ increases; transmittance decreases as θ increases. The reflectance and transmittance can be measured between $\theta = 10°$ and $\theta = 80°$ (Fig. 1.22b).

The reflection of light at an interface between different optical media has been well described by the Fresnel equations (Miyazaki, 2014). When considering the light traveling through air and striking on an interface between a PT thin film and a glass substrate (PMMA film containing various PT particles has a greater refractive index), we can treat it as an interface between air ($n = 1$) and glass ($n \sim 1.5$) since the film is extremely thin (~500–800 nm) and has a high AVT (85%). The amount of light reflected at random incidence angles follows the conservation law, $R = 1-T$, where reflectance R = Reflected Power/Incident Power and T is transmittance. The absorbance is negligible due to the very thin thickness of the film and high AVT of the sample. For cases involving an ideal air/glass interface, Fresnel equations determine that R of the unpolarized light is not very angle dependent between from 0° to 50°. It then increases significantly above 70° and approaches 1 at $\theta = 90°$ (Hecht, 1998). However, the light reflected by the PT film in this study steadily increases with the

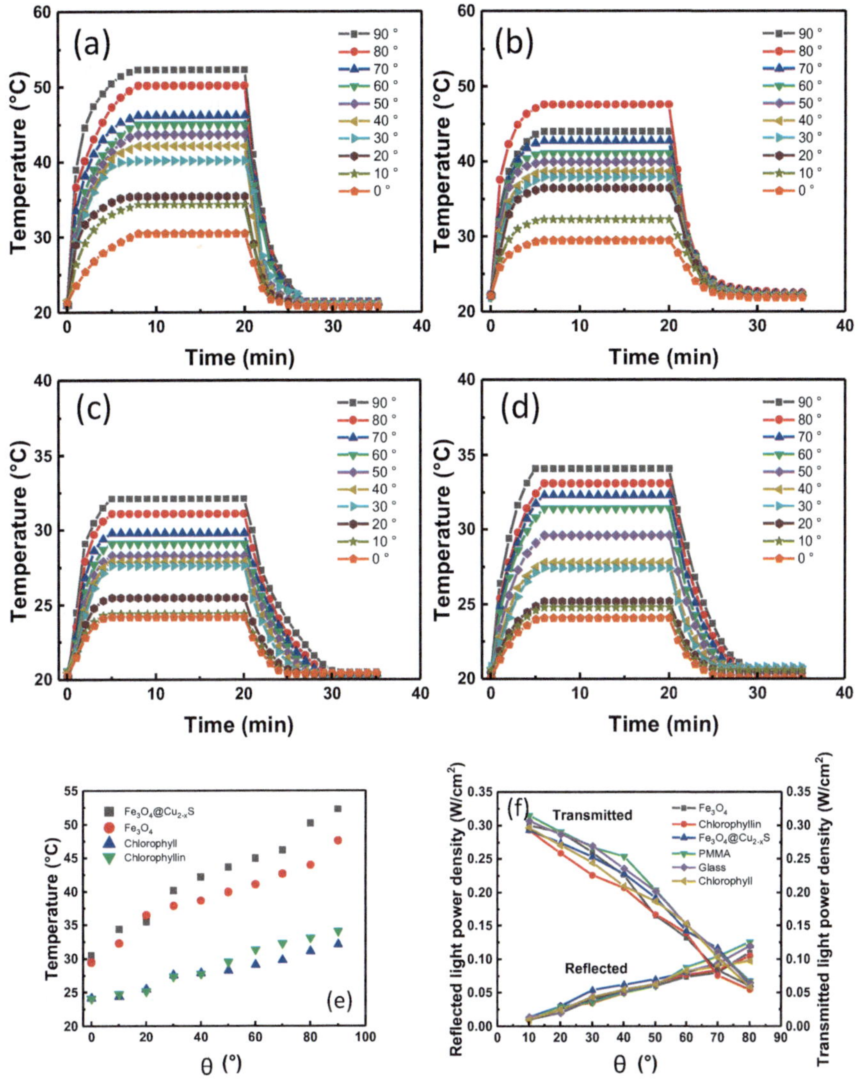

Fig. 1.23 Heating curves of **a** the 85% AVT $Fe_3O_4@Cu_{2-x}S$ film (room temperature 20.9 ± 0.4 °C, humidity 25.6 ± 0.4%), **b** the 85% AVT Fe_3O_4 film (room temperature 22.1 ± 0.4 °C, humidity 27.8 ± 0.3%), **c** the 85% AVT chlorophyll film (room temperature 22.1 ± 0.4 °C, humidity 25.5 ± 0.2%), **d** the 85% AVT chlorophyllin film (room temperature 22.1 ± 0.4 °C, humidity 29.2 ± 0.3%), **e** maximum temperature versus angle of incidence for the samples as indicated. All film samples are tested with a 0.4 W/cm^2 power density solar simulator at different incident angles from 0° to 90°, **f** reflected and transmitted light power density at different light incident angles

angle of incidence, between 10° and 80°, and the transmitted light decreases with increasing angle of incidence.

Compared to an ideal interface, as described by the Fresnel equations, this angle dependence was found to be directly associated with the light being reflected by the thin film surfaces in a more gradual fashion. The reduced light reflection can be explained by light scattering on non-ideal surfaces of the deposited thin films. The transmitted light contributes to the temperature rise caused by the photon interactions with the electronic structure of Fe_3O_4, $Fe_3O_4@Cu_{2-x}S$ (Zhao et al., 2017), and molecular structure of chlorophyll (Wang et al., 2009) and chlorophyllin (Attia et al., 2012). The photon-to-heat conversion attributes to the molecular vibrations in the porphyrin structures of chlorophyll and chlorophyllin (Zhao et al., 2019a) while the photothermal mechanism of iron-oxide based materials has been identified as the localized plasmonic surface resonance (LPSR) (Lin et al., 2020; Zhao et al., 2017).

The R & T measurements are well-correlated to the heating curves (Fig. 1.23a–d), especially with the transmitted light. The photothermal effect indicates that as an amount of light enters the film, interacting with the electrons of the PT molecules, and converting photons to thermal energy, leading to the temperature increases. The reflected light makes no contribution to photothermal effect. The difference between data shown in Fig. 1.23f and Fresnel curves (Hecht, 1998) is due to light scattering taking place on the rough film surfaces (diffuse reflection), which causes a significant reduction in light reflected (especially at high angles of incidence) (Atkinson & Hancock, 2007; Maradudin & Méndez, 2007; Nolan et al., 2015; Stavenga, 2014).

In this study, light is reflected at two surfaces, top and bottom surfaces of the thin film. The light reflected by the bottom surface can interfere with light reflected by the top surface. The interference may be in or out of phase depending on the thickness of the film, which has a greater refractive index than that of air (Stavenga, 2014). The difference in path lengths of the light reflected by both surfaces is negligible for thin films with a thickness of 500–800 nm. However, there is a phase change when light travels through the film (PMMA has a refractive index of 1.4905 at 589.3 nm and a visible transmittance of 92%). The photothermal effect can be enhanced as light is internally reflected multiple times at both surfaces.

The sunlight incident angle is $\alpha = \theta$ (the photothermal experimental angle of incidence), as depicted in Fig. 1.22. The highest temperatures are reached when light is normal to the plane of interface ($\theta = 0°$), according to the results on the angle (α) dependence of the photothermal effect (Fig. 1.23). At the winter solstice in Chicago $\alpha = \theta = 24.70°$. Figure 1.23 shows the maximum temperatures achieved, by the PT films at an angle of 20° are: 46.2°C, 42.7 °C, 29.8 °C, and 32.3 °C, respectively, for $Fe_3O_4@Cu_{2-x}S$, Fe_3O_4, chlorophyll, and chlorophyllin. The ΔT gained via photothermal thin film coatings will significantly reduce the thermal loss through single-pane windows. At the summer solstice $\alpha = \theta = 71.70°$, we find temperatures reached by the PT films at 70° in Fig. 1.23 are: 35.5 °C, 36.5 °C, and 25.5 °C, respectively, for $Fe_3O_4@Cu_{2-x}S$, Fe_3O_4, chlorophyll, and chlorophyllin. These temperatures are considerably lower compared to those at 20°. Therefore, the incoming sunlight in the summer season at solar noon does not directly strike the window glass, and the photothermal effect is reduced. The window surfaces receive most of the sunlight at winter solstice for enhanced photothermal effect and significant reduction of heat loss via raising the window surface temperature. The angle dependence data (Fig. 1.24) indicates important factors in the design of PT thin film coated windows. These experimental results are proof of concept of OTI for a viable approach in implementing single pane windows without any intervening gas media.

In conclusion, the photothermal effects of various thin films are found to be dependent on incident light angle. The angle dependence is mainly associated to the light reflected and transmitted at the film interfaces, and also affected by other factors such

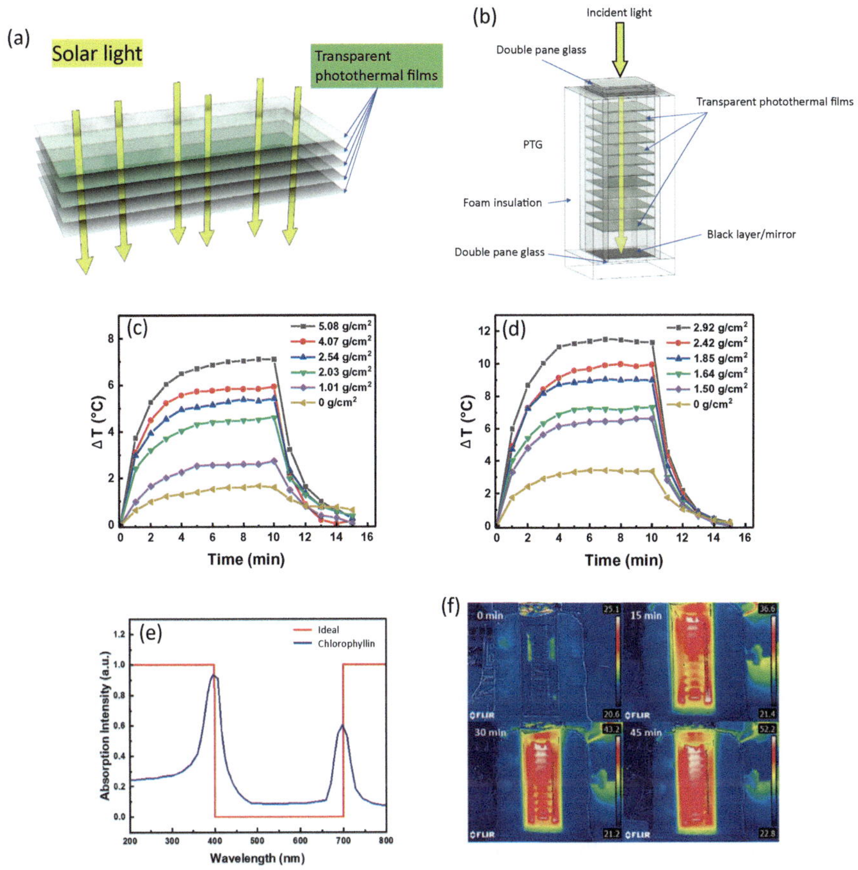

Fig. 1.24 Schematic diagrams of **a** solar light harvesting via multilayer transparent photothermal thin films, **b** the photo-thermal-generator (PTG), **c** heating curves of the $Fe_3O_4@Cu_{2-x}S$, **d** chlorophyllin thin films of various concentrations irradiated by white light, **e** schematic diagrams of absorption spectra of different compounds, and **f** infrared thermal photographs of the photo-thermal generator (one side of insulation is removed) with 10 layers of $Fe_3O_4@Cu_{2-x}S$ films (85% AVT) illuminated by solar simulator (0.4 W/cm^2 power density) for different times

as refractive indices, constructive/destructive interferences, film thickness, and the wavelength of the incident light. These experimental data provide a fundamental base for engineering design of single pane windows based on the concept of optical thermal insulation which is a novel approach in reducing heat loss via the photothermal thin film coating without intervening media.

1.6 Multilayer Solar Harvesting and Energy Generation

1.6.1 Photothermal Generator (PTG)

Photothermal (PT) materials have strong photon absorptions in a wide frequency range which dictate their abilities to convert photons to thermal energy (Zhao et al., 2019a; Attia et al., 2012; Lyu et al., 2021; Arras et al., 2012; Pérez-Hernández et al., 2015; Gao et al., 2019; Ding et al., 2018). Solar harvesting has been, however, limited to 2D surfaces such as PV. There will be great advantages if solar harvesting can be extended to 3D through multiple transparent thin films with desirable optical characteristics. As is well-known, most of the photothermal materials (such as gold and graphene) are not transparent, which are not suitable for multilayer 3D solar harvesting. A novel concept has been proposed with multilayer 3D solar harvesting, analogous to multilayer capacitors for increasing the total surface area. As shown in Fig. 1.24a, the solar light can pass through several transparent substrates coated with PT materials, therefore significantly increasing the light harvesting surface area. Upon light irradiation, heat is generated on each layer, generating a large amount of thermal energy accumulatively in a photothermal generator (PTG). In this fashion, the PTG is a 3D solar harvester, photon to thermal energy converter, and generator (Fig. 1.24b), which is much more efficient compared to the 2D films.

As shown in Fig. 1.24b, a right rectangular prism is constructed with 10 layers of transparent PT layers with dimensions of 25.4 $\times$ 25.4 $\times$ 1.4 mm^3. These PT-coated transparent substrates are placed in parallel in a cuboid (5 $\times$ 5 $\times$ 15 cm^3) with thermally insulated walls. The inner bottom surface of the cuboid is covered with a mirror (reflectivity 90%) to reflect the incident light. Similar to an *optical cavity* (Ding et al., 2018) all cuboid inner surfaces (except the top inner surface) can be made of highly reflective mirrors to reflect light multiple times, maximizing the photothermal energy generated on all layers.

The cuboid is thermally insulated with polystyrene foam (thickness: 35 mm) surrounding the four side faces, as shown in Fig. 1.24b. The top is insulated with a double-pane glass with a 20 mm air spacing to allow the incident light into the cuboid. When the simulated solar light source (0.4 W/cm^2 power density) is turned on, the light enters from the top of cuboid, penetrates the top double-pane, then goes through all PT substrates, and reaches the bottom layer (Fig. 1.24b). The bottom of the cuboid can be designed with a mirror which is placed at the inner bottom to reflect the incident light to the opposite direction for passing through PT films multiple times in order to generate more thermal energy. When light passes through each PT layer, the photons are activated to generate thermal energy which raises the local thin film temperature characterized by the heating curves.

Solutions of $Fe_3O_4@Cu_{2-x}S$ and chlorophyllin were deposited on quartz substrates, as described in Sect. 1.2.1.3. Deposition of photothermal thin films. The PT effects of the $Fe_3O_4@Cu_{2-x}S$ and chlorophyllin thin films of different concentrations were characterized by obtaining the heating curves, as shown in Fig. 1.24c and d. The heating curves shown in Fig. 1.24c and d were obtained on one film sample in air without the cuboid (see Methods). Figure 1.24c and d show the changes in temperature (ΔT) as a function of time for the $Fe_3O_4@Cu_{2-x}S$ and chlorophyllin thin films at different concentrations. The PT films were irradiated by simulated solar light at 0.1 $W \cdot cm^{-2}$ power density. The temperature increases rapidly at the beginning but reaches a plateau after 5 min. This plateau is due to heat loss through the environment being balanced by the heat generated by the PT films. The light is turned off at 10 min resulting in rapid temperature decrease for samples of all concentrations. As shown in Fig. 1.24c and d, ΔT reaches 7 °C for $Fe_3O_4@Cu_{2-x}S$ at a concentration of 5.08×10^{-5} g/cm^2 and 11.5 °C for chlorophyllin at a concentration of 2.92×10^{-5} g/cm^2. Note that these are extremely low concentrations, and ΔT can be further increased at higher concentrations. However, transparency will be compromised at higher concentrations, therefore hindering light penetration.

As the thermal energy generated in PTG is cumulative, the maximum temperature of a PTG can reach an appreciable level depending on the light power intensity, thermal insulation, and other design parameters. A PTG is also defined as a photothermal amplifier (PTA), a device that amplifies thermal energy by adding additional layers. As shown in Fig. 1.24f, the infrared images show the temperature distributions of the PTG (with 10 layers of the 85% AVT $Fe_3O_4@Cu_{2-x}S$ thin films) before/after the light is turned on for different time intervals. The PTG is well insulated with minimum heat loss, as shown in Fig. 1.24f. When light is turned on for 15 min, the top of the PTG is heated up more because light enters the PTG from the top. The lower part becomes more heated at 30 min, and the temperature distribution in the PTG becomes even at 45 min exhibiting a cumulative PT effect. The PTG can generate thermal energy with an appreciable conversion efficiency. The maximum temperature is expected to increase with more incident light power density, heating time, concentration of PT film, AVT, and thermal insulation.

1.6.2 Characterization of the Transparent Photothermal Thin Films

1.6.2.1 Optical Properties of Photothermal Thin Films

Figure 1.25a shows the absorption and transmittance spectra of the $Fe_3O_4@Cu_{2-x}S$ and chlorophyllin thin films, both with high AVTs. Figure 1.25b shows the

(a)
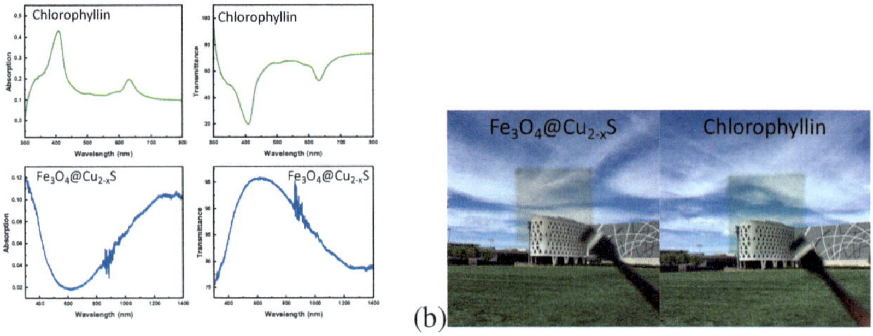

Fig. 1.25 **a** Absorption and transmittance spectra of the $Fe_3O_4@Cu_{2-x}S$ and chlorophyllin thin films, and **b** photographs of the $Fe_3O_4@Cu_{2-x}S$ and chlorophyllin thin films in front of a campus building

photographs of the $Fe_3O_4@Cu_{2-x}S$ and chlorophyllin films in front of a building on the main campus of the University of Cincinnati. As can be seen in these photographs, both films are highly transparent due to their high AVTs, displaying clear images of the building behind the films. High light transmittance is one of the fundamental features of PTG, enabling solar light harvesting through multiple layers in a confined 3D space.

The heating curve measurements were conducted by using the photo-thermal-generator shown in Fig. 1.24, following the procedures described in the Methods section. The heating efficiency at each layer is important especially at the lower portion of the PTG. This is due to the fact that the incident light will be weakened after penetrating several layers. Each layer has a $Fe_3O_4@Cu_{2-x}S$ or a chlorophyllin thin film deposited on a quartz substrate. The simulated solar light is directed from the top of the system, penetrating the top double-pane and the PT layers. As all layers are coated with the PT thin films, each layer generates heat raising the surface temperature as reflected in the heating/cooling curve.

1.6.2.2 Photothermal Experimental Details

The photothermal experiments on the single layer thin films were carried out in air under ambient conditions. A Newport 150 W solar simulator (Lamp model 67,005) was used to obtain the heating/cooling curves. The experimental design is displayed in Fig. 1.26. A light power density of 0.1 W/cm^2 was used for the heating curves shown in Fig. 1.24c and d. For PTG experiments (Fig. 1.24b), a light power density of 0.4 W/cm^2 was used to obtain the heating curves. The temperatures of PTG layers were monitored using thermal couples. The simulated solar light power density was measured by using an optical power meter (Newport Inc.) on the sample surfaces (Fig. 1.26) (Yun et al., 2019; Liu et al., 2019).

The design of PTG is shown in Fig. 1.24b. Each layer consists of a $Fe_3O_4@Cu_{2-x}S$, or a chlorophyllin film coated on the quartz substrate. AVT of the layer is adjusted to

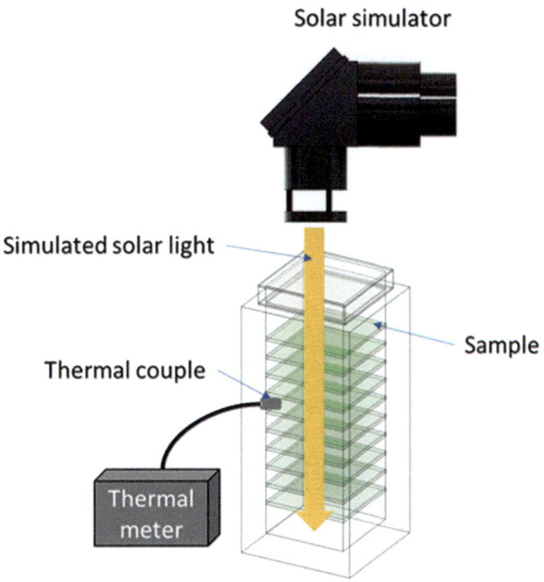

Fig. 1.26 Schematic diagram showing the apparatus of the photothermal experiment

85% by optimizing the PT film concentrations: 1.6 mg/cm^2 for $Fe_3O_4@Cu_{2-x}S$, and 0.7 mg/cm^2 for chlorophyllin. The PTG designed in this study can accommodate a maximum of 10 PT layers; the distance between each layer is 5 mm, and the size of each quartz substrate is 25.4 × 25.4 × 1.4 mm^3. Simulated solar light is directed from the top of the system, penetrating all PT layers, as shown in Fig. 1.24b.

1.6.2.3 Thermal Insulation of PTG

For PT characterization, the heating and cooling curves were obtained from the multilayer thin films with and without thermal insulation. The heating curves shown in Figs. 1.27 and 1.28 were obtained with PTG thermally insulated with polystyrene foam (thickness: 35 mm, shown in Fig. 1.24b). The heating curves shown in Fig. 1.29 were obtained without thermal insulation.

1.6.2.4 The Solar Photothermal Efficiency

The solar PT efficiencies are calculated using Eq. 1.16 for the multilayer thin films in the PTG and their values are listed in Table 1.7.

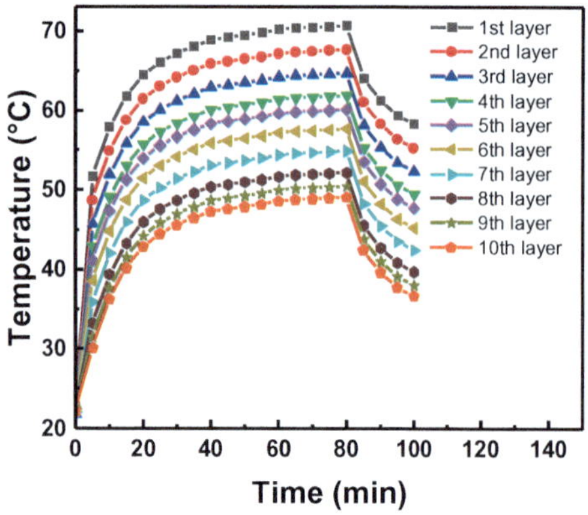

Fig. 1.27 Heating curves of chlorophyllin thin films in thermally insulated PTG with 10 layers

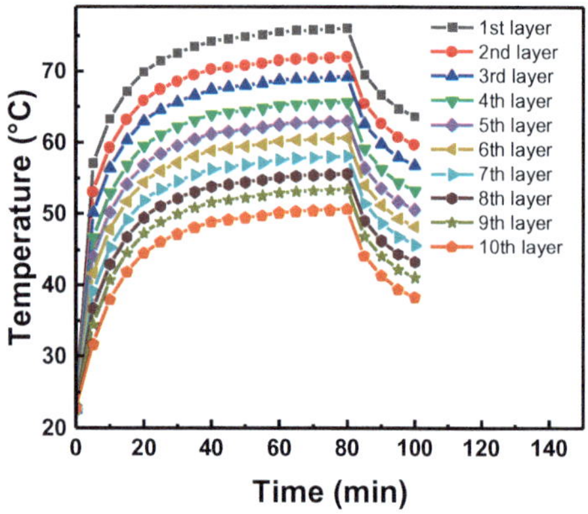

Fig. 1.28 Heating curves of $Fe_3O_4@Cu_{2-x}S$ thin films in thermally insulated PTG with 10 layers

1.6.2.5 Thermal Energy Generation

Thermal energy generated is proportional to the number of layers, as the incoming light passes through a number of PT thin films. In this study, thermal energy Q is defined by: $Q = \dot{Q} \times \Delta t$, where $\dot{Q}$ is the heat produced per unit time in units of (J/s), and Δt is the heating time. Using this equation, the thermal energy Q can be calculated based on the heating curves shown in Figs. 1.29 and 1.30, from which $\dot{Q}$ can be determined.

PTG was thermally insulated with the polystyrene foam surrounding the four lateral faces (shown Fig. 1.24b) for thermal energy generation. Two transparent double-pane glasses were used to seal the top for the incident light and the bottom. Heating/cooling experiments were conducted in the same way as those without thermal insulation. To determine the thermal energy generated by multiple layers, each heating/cooling curve was obtained for a given number of layers, as shown in Fig. 1.27 for chlorophyllin (concentration: 0.7 mg/cm^2) and Fig. 1.28 for $Fe_3O_4@Cu_{2-x}S$ (1.6 mg/cm^2). The maximum temperature difference (ΔT_{max}) of

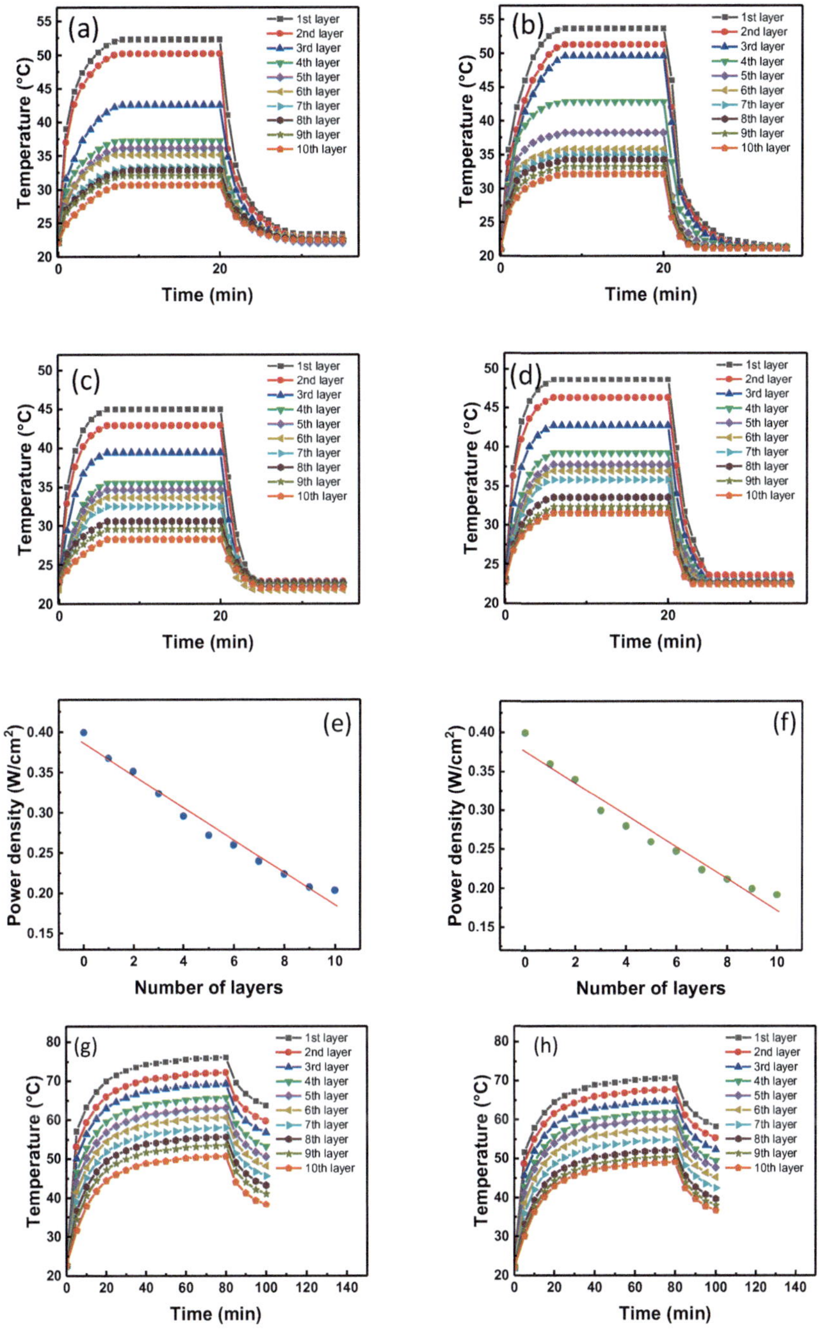
(a)
(b)
(c)
(d)
(e)
(f)
(g)
(h)
1st layer
2nd layer
3rd layer
4th layer
5th layer
6th layer
7th layer
8th layer
9th layer
10th layer
Temperature (°C)
Time (min)
Power density (W/cm²)
Number of layers

◀**Fig. 1.29** **a** Heating curves of the $Fe_3O_4@Cu_{2-x}S$ thin films in PTG with neither thermal insulation, nor a base mirror at the bottom; **b** $Fe_3O_4@Cu_{2-x}S$ with thin films in PTG without thermal insulation, but with a base mirror at the bottom (see Fig. 1.22); **c** Heating curves of the chlorophyllin thin films in PTG with neither thermal insulation, nor a base mirror at the bottom; **d** the chlorophyllin thin films in PTG without thermal insulation, but with a base mirror at the bottom (see Fig. 1.22); **e** light power density attenuation for the $Fe_3O_4@Cu_{2-x}S$ thin films, **f** light power density attenuation for the chlorophyllin thin films; **g** heating curves of the $Fe_3O_4@Cu_{2-x}S$ thin films in PTG with the external thermal insulation, and **h** heating curves of the chlorophyllin thin films in PTG with the external thermal insulation

Table 1.7 The photothermal efficiencies of the photothermal generator (Fig. 1.25b) with different numbers of layers, calculated by using Eq. 1.16 for both $Fe_3O_4@Cu_{2-x}S$ and chlorophyllin thin films

Number of layer(s)	1	2	3	4	5	6	7	8	9	10
Chlorophyllin η (%)	30.9	43.7	46.4	52.4	53.5	56.9	57.7	58.9	60.7	61.6
$Fe_3O_4@Cu_{2-x}S$ η (%)	40.9	42.9	44.2	46.3	48.0	49.6	57.2	58.2	60.6	63.4

the heating curve is used in the equation $\dot{Q} = UA(T - T_0)$ to calculate the thermal energy.

1.6.3 Heating Curves of Multilayer Photothermal Thin Films

The heating curves of the $Fe_3O_4@Cu_{2-x}S$ (85% AVT, concentration: 1.6 mg/cm^2) and chlorophyllin (85% AVT, concentration: 0.7 mg/cm^2) thin films deposited on quartz in PTG without any thermal insulation are shown in Fig. 1.29a–d. As shown in Fig. 1.29a, the $Fe_3O_4@Cu_{2-x}S$ films reach a maximum temperature of 52.5 °C (from 21.9 °C) on the top layer. The temperature raised at each layer gradually decreases due to less light power densities at those at the low portion of PTG (Fig. 1.29e and f). The light power density attenuation is associated with several factors, including the AVT of the film, light reflection/scattering, and light absorption at each layer (Rating & Council: Procedure for Determining Fenestration Product U-factors, 2013; Inoue et al., 1994).

To improve the PT effect, a mirror with 90% reflectivity was placed on the bottom inner surface of the cuboid (shown in Fig. 1.24) to reflect the incident light. The temperatures of all heating curves were considerably increased, especially at the 10th layer (as shown in Fig. 1.29b). The temperatures at the 3rd and 4th layers increase to the higher levels, indicating the reflected light reaches these layers. These experimental results show the possibility of solar harvesting in the 3D fashion through multiple transparent layers.

The chlorophyllin film samples show similar heating characteristics compared to those of the $Fe_3O_4@Cu_{2-x}S$ counterpart, as shown in Fig. 1.29c and d. However, the heating curves appear to be more evenly distributed, indicating uniform visible transmittance of the films on all layers. The cooling curves of the $Fe_3O_4@Cu_{2-x}S$ coated layers are more gradual. As shown in Fig. 1.29a–d, the temperatures drop rapidly for all cooling curves when the light is turned off at 20 min. Cooling of the $Fe_3O_4@Cu_{2-x}S$ layer is slowed down at the temperatures below 35 °C until 30 min. In contrast, the temperatures of the chlorophyllin layers continue to decrease rapidly all the way down to room temperature after the light is turned off. This difference in cooling curves is due to specific heat differences between these compounds. Iron oxides generally have greater heat capacity than chlorophyllin, making them contain more heat per film sample.

To prove the concept of PTG, the cuboid is thermally insulated with the polystyrene foam, as shown in Fig. 1.29. A transparent double-pane glass was placed on the top of PTG for the incident light and at bottom for the transmitted light. The heating and cooling curves were obtained in the same way as described in **Methods**. The heating curves are significantly improved with thermal insulation, and the highest temperatures reached 76.1 °C for $Fe_3O_4@Cu_{2-x}S$ and 71.6 °C for chlorophyllin. As shown in Fig. 1.29g and h, the heating plateau is changed to a tilted curve, indicating continual increasing temperature. Without thermal insulation, all heating curves reach balances between the PT heating and the heat loss through the environment. The heat loss of PTG is reduced when it is thermally insulated (though not ideally), resulting in an upward tilting plateau (as shown in Fig. 1.29g and h). The heating time is extended to 80 min for reaching even higher temperatures in PTG. Since the thermal energy generated by PTG is cumulative, it can be stored for different applications.

The top layer of PTG with $Fe_3O_4@Cu_{2-x}S$ reaches 76.1 °C after 80 min while the bottom layer temperature reaches 50.7 °C, as shown in Fig. 1.29g. For the chlorophyllin films shown in Fig. 1.29h, the heating curves are within similar temperature ranges (70.7 °C for the top layer and layer 49.1 °C for the bottom layer). The cooling curves are much more gradual after the light is switched off at 80 min due to thermal insulation.

These large differences between the heating curves of Fig. 1.29a–d and g–h show the possibility of generating thermal energy via a constant light source based on the concept of PTG. Photons can be collected via multiple transparent layers and converted to heat and stored as thermal energy in a 3D space, provided that a constant light source is available. More layers can be added in a significantly larger volume to achieve even greater solar light harvesting.

1.6.4 Photothermal Energy Generation and Amplification via Multilayers

As described above, the PTG is a photothermal amplifier (PTA) for increasing thermal energy through adding more multiple transparent PT films. The energy amplification can be determined by comparing the thermal energy generated between one single layer and multiple layers. In this study, we define the thermal energy $Q = \dot{Q} \cdot \Delta t$, where $\dot{Q}$ is the output power of the PTG, Δt is heating time period (Zhao et al., 2019a). The thermal energy is calculated by using the procedures shown in 5.3 the solar photothermal efficiency (Zhao et al., 2019a), plotted against the number of layers (as shown in Fig. 1.30a).

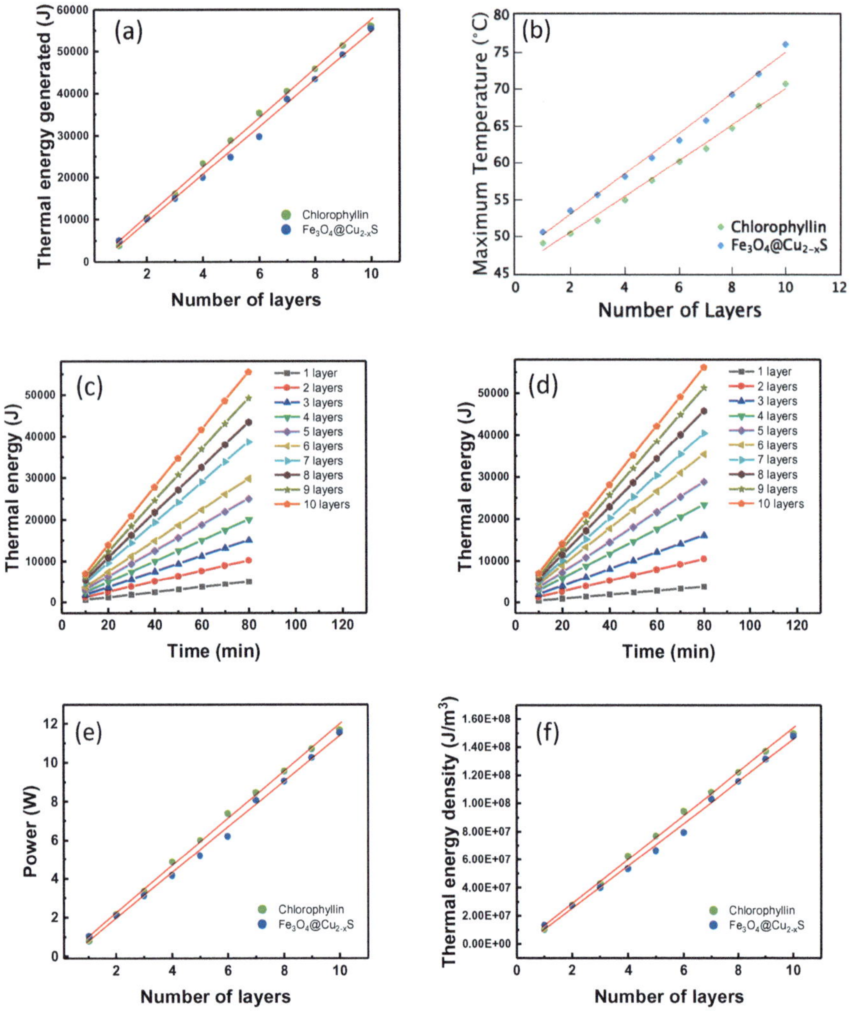

Fig. 1.30 **a** Thermal energy versus number of layers, **b** maximum temperature versus number of layers, **c** thermal energy versus heating time for different layers of the $Fe_3O_4@Cu_{2-x}S$ films, **d** thermal energy versus heating time for different layers of the chlorophyllin films, **e** power ($\Delta Q/\Delta t$) versus number of layers for the $Fe_3O_4@Cu_{2-x}S$ and chlorophyllin thin films during the 10 to 20 min heating time period, and **f** thermal energy density versus number of layers for both $Fe_3O_4@Cu_{2-x}S$ and chlorophyllin thin films

A linear relationship between the thermal energy and number of layers is shown in Fig. 1.30a. As shown in this figure, the thermal energy generated increases from 5,074.3 J (one layer) to 55,465.8 J (ten layers) for the $Fe_3O_4@Cu_{2-x}S$ PTG; and increases from 3,826.3 J (one layer) to 56,143.9 J (ten layers) for chlorophyllin PTG; both exhibiting a more than tenfold increase. The thermal energy generated can be increased by enhanced film AVT, thermal insulation, and the PT coefficient for a given geometry of the PTG (which achieves higher PTG energy density). Figure 1.30b shows the temperature versus number of layers for $Fe_3O_4@Cu_{2-x}S$ and chlorophyllin thin films. The maximum temperatures can reach over 100 °C if the PTG parameters are optimized. Figure 1.30c and d show the thermal energy generated at different layers, respectively for the $Fe_3O_4@Cu_{2-x}S$ and chlorophyllin films. As shown in these figures, thermal energy increases linearly between 10 and 80 min. These results correspond to the heating curves shown in Fig. 1.29g and h, with more gradual increase of temperature due to heat loss through the cuboid surfaces.

If ideally insulated, both the heating (Fig. 1.29g and h) and thermal energy curves should continue to increase. The slopes of the thermal energy versus heating time indicate the heating power of the system (Power = $\Delta Q/\Delta t$). The slopes between 10 and 20 min of the Q versus t curves (Fig. 1.30c and d) are plotted against the number of layers, as shown in Fig. 1.30e. The power of the PTG increases, from 1.1 W to 11.6 W for $Fe_3O_4@Cu_{2-x}S$ and from 0.8 W to 11.7 W for chlorophyllin, as the number of layers increases from 1 to 10. As shown in these figures, thermal energy generation can be significantly improved by adding multiple PT layers.

Energy densities of PTG are plotted against the number of layers in Fig. 1.30f. Based on the data shown in Fig. 1.30c and d, energy density inside the PTG increases, from 1.35×10^7 J/m^3 to 1.48×10^8 J/m^3 for $Fe_3O_4@Cu_{2-x}S$ and from 1.02×10^7 J/m^3 to 1.49×10^8 J/m^3 for chlorophyllin, as the number of layers increase from 1 to 10.

Compared to the single layer system, the PT energy generation of the multi-layer system depends on several factors. As light passes through multiple layers, it attenuates due to several factors (see Fig. 1.29e and f), including: AVT of the film, light reflection by films and substrates, refraction at the interfaces between film and substrate, and the scattering/absorption at each layer. The PTG energy generation depends on PT efficiency (η) of the films, which is the ratio of the energy increase of the PT film to the simulated solar light. η is calculated by using the previously published procedures (Zhao et al., 2019a). Table 1.1 shows that η increases from 40.9% (one layer) to 63.4% (10 layers) for the $Fe_3O_4@Cu_{2-x}S$ PTG, and from 30.9% (one layer) to 61.6% (ten layers) for the chlorophyll PTG. The efficiency of the multilayer PT system is much higher compared to that of single layer PVs. For example, an average silicon PV efficiency is on the order of 25% (Inoue et al., 1994). The solar PT efficiencies of the multilayer system have approached around 60% and are expected to further increase as more layers are added.

In this study, multilayer solar harvesting has been achieved via transparent PT thin films. These spectral-selective thin films spatially expand solar energy harvesing to a 3D space, which is fundamentally different from 2D solar panels. 3D multilayer PV systems can potentially be designed with significantly increased surface area per volume in a similar fashion. Conversion of solar light to thermal energy in 3D systems is continuous and cumulative, capable of reaching high temperatures and large thermal masses. Two PT materials, $Fe_3O_4@Cu_{2-x}S$ and chlorophyllin, have been synthesized and deposited on quartz substrates for 3D solar light harvesting. These thin films exhibit magnificent PT effects as well as high AVT, making them ideal candidates for 3D multilayer systems. Light is able to pass through multiple tranparent films, genrate heat at each layer, and reach a maximum tempeature of 76.1 °C. A 12- fold increase in thermal energy is obtained through the multilayer system, leading to significantly increased enery density compared to a single layer. The solar PT efficiency (η) of the PT films also show significant increases when the number of layers is increased from 1 to 10, which indicates the strong dependence on the surface area exposed to light. This can be achieved by the spectral selective and transparent PT thin films. The increase in thermal energy generated can be further improved with the enhanced PT efficiency of the thin films. The concept of the "Photo-Thermal Generator" will shed light upon the research of energy-neutral systems.

As described in Sect. 1.1.3 3D solar harvesting and photothermal energy generation for building heating utilities, and dipiced in Fig. 1.7, the PTG concept can be utlized for generating thermal energy for building heating utilities by 3D solar harvest. Units of PTGs can be retrofited into building structures for both heating and lighting purposes Combining the concepts of both optical thermal insulation and photothemal generator, next-generation architecture designs will be implimented for a net-zero civic infrustructure.

1.7 PT-PV Dual-Modality Building Skins

The glass-based high-rise building skins provide ideal transparent substrates for device architecture of energy harvesting nanoscale thin films. A simple building skin coating can be engineered to offer two major functions: photovoltaic or photothermal, switched alternatively depending on the seasonal needs. In summer, the photovoltaic effect of the coating consumes most of the solar infrared therefore less cooling is required. In winter, the slight increase in skin temperature by the photothermal coating can lead to lowered heat loss from room interior. Compared with multi-pane glazing, single-panes are practically not viable due to rapid heat transfer in winter. If a spectral-selective thin film is applied on a window surface, the skin surface temperature can be increased from 25 °C to > 50 °C via the photothermal effect. This will in turn effectively reduce the thermal energy loss from the interior. In this way, thermal insulation can be achieved optically without intervention medium. On the other hand, the undesirable solar infrared in summer can be compensated by the same thin film but in a different modality: photovoltaic. Absorption of large infrared irradiation not only reduces cooling energy but generates electricity for other appliances. The outcomes of the research activities will address the national needs in energy sustainability by entirely transforming the landscape of architectural engineering, civic system design, and energy saving strategy.

As mentioned previously, OTI and PTG are developed based on 3D solar light for heat conversion while maintaining visible transparency. Thus, a high AVT would be essential for window applications. Regarding some semiconducting materials that exhibit both PT and PV effects, the PT-PV dual modality is possible if the dual-modality is well structurally designed. Transparent PV and PT films share some of the spectral requirements. The key requirement of 2D PV and PT panels lies on a large surface area to optimize the solar light harvesting. A modern building skin with glass façades can be engineered into PV and PT panels (Fig. 1.5) for energy harvest and generation while maintaining an aesthetical visual effect (Lin & Shi, 2021). This approach utilizes large building transparent skins as substrates for developing PV and PT panels which can function in a dual modality, and seasonably altered for energy efficiency.

As described above, the PT coated single-pane window can raise surface temperature so that heat transfer through the building skin is significantly reduced in cold weather, based on the concept of OTI. With OTI, transparent building skins can be well optically insulated without any intervention medium. In the summer season, the building skin will be switched to the PV mode via dual modality. The dual-modality building skin constantly absorbing most of the infrared radiation which can reduce cooling energy, and at the same time generate electricity for additional appliances.

Various materials have been identified, developed, and engineered for making PV deices. These devices include Si-based solar cells, perovskite solar cells (PSCs), and dye-sensitized solar cells (DSSCs). Among others, the porphyrin compounds have shown promise in the development of photosensitizers in DSSCs. Furthermore, due to functionalization of macrocyclic dyes in the molecular structure, the porphyrin compounds provide ideal absorptions in the UV–vis region (O'regan & Grätzel, 1991; Zeng et al., 2020; Pistner et al., 2012).

A DSSC consists of photoanode, dye, electrolyte, and counter electrode (Fig. 1.31a). The working principle of DSSCs (Fig. 1.31b) can be described as: (i) when light hits the device, the electrons in the dye molecules will be excited from the ground state (HOMO) into an excited state (LUMO), (ii) the excited dye molecules are oxidized (loss of electrons) and electrons are injected into the conduction band of the semiconductor, (iii) the oxidized dye molecule is restored by electrons contributed by a redox mediator (I^-), and two I^- ions are oxidized as iodine (I_3^-), (iv) the I_3^- diffuses toward the cathode and then it is reduced to I^- by gaining electrons from the cathode. Although DSSC has low photon-to-electricity performance, it has many advantages such as environmentally friendly, cost-efficient, a simple manufacturing processes, and visible transparency.

The first application of Cu-2-α-oxymesoisochlorin as a photosensitizer for DSSC with an overall Power Conversion Efficiency (PCE) of 2.6% was published in 1993 by Kay and Gratzel (Kay & Graetzel, 1993). Subsequently, the PCE of DSSC was increased up to 12.3% in 2011 (Yella et al., 2011) and further improved to 13% in 2014 (Mathew et al., 2014) by using Zn–porphyrin derivatives in the same group. Despite the constant improvement in PCE from 2009 to 2014 (Lu et al., 2018), challenges

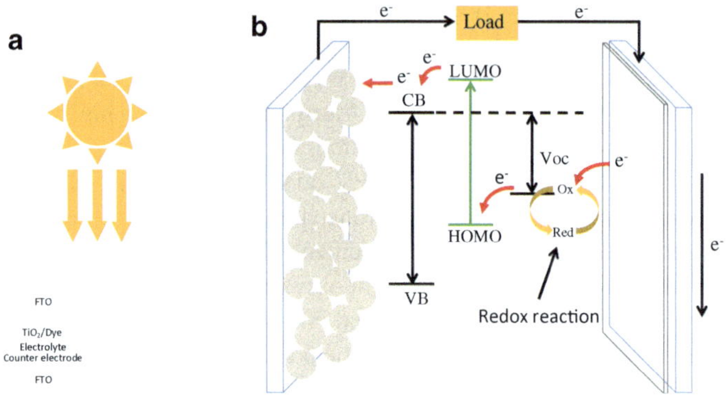

Fig. 1.31 **a** DSSC structure and **b** Schematic working principle of DSSC

are still to be addressed in the molecular design, mainly on the aging resistance and inefficient photon collection in the visible light region. The porphyrin-based DSSCs have been developed with multiple dyes or molecular-modified porphyrin compounds to broaden the light-absorbing area for higher PCE (Li et al., 2019; Otaka et al., 2004; Ogura et al., 2009).

Transparent PV and PT films have recently gained significant attention for their potential in efficient solar harvesting and energy applications. The transparent organic photovoltaic (TOPV) thin films have been extensively studied (Husain et al., 2018; Traverse et al., 2017; Li et al., 2020; Colsmann et al., 2011). The reported efficiency of TOPV is about 10% (Li et al., 2020), while the average PCE is 15 to 22% for the commercial silicon based solar panels (Barbose et al., 2021). Recently, the so-called building integrated photovoltaic (BIPV) has been developed for building skins, taking advantage of large surface areas without conflicting with appearance requirements (Colsmann et al., 2011). The amorphous silicon BIPV windowpane was developed in 2013 (Yoon et al., 2013). The studies on BIPV show that the internal surface temperatures of inclined and horizontal BIPV double-panes are about 2 °C higher than the average windowpanes in winter. Further, enhancement of windows' thermal performance via CdTe BIPV was proposed in 2020 (Alrashidi et al., 2020). The research analyzed the surface temperature with the single-pane and reported that the U-factor of CdTe BIPV is lower than the single-pane.

Based on the aforementioned research, the PT-PV dual-modality was recently proposed by Lin and Shi (2021). The building skins with large surface area provide device substrate for depositing transparent PT-PV thin films with dual modality. Based on the spectral selective characteristics of porphyrins, they can be ideal candidates for developing PV-PT dual-modality building skins. However, while the transparent organic photothermal (TOPT) thin films require intense UV and IR absorptions for photon-to-thermal conversion, the IR absorption should be eliminated for TOPV since heating of the PV panel can considerably reduce efficiency. Therefore, fundamental research in materials science will have to be carried out in order to engineer the PT-PV dual-modality building skin with tailored optical properties. Multilayer architecture of the building skin will be the key in the development of the PT-PV dual modality building skin that can spectral selectively utilize the full solar spectral for efficiently generating thermal energy in the cold climate and electricity in the warm season. The concept of PT-PV dual modality building skin paves a new path to next-generation device-based building skins that will have profound impact in many areas including energy sustainability, environmental control, living comfort, utility efficiencies, and economic concerns (Lin & Shi, 2021).

1.8 Conclusion

A major global energy challenge is the formation of megacities with a rapid sprawling of high-rise buildings. The huge areas and high populations of megacities present enormous social and environmental demands. One of the key issues deals with overall building energy and material consumptions including electricity, heating and cooling, and materials production simply due to their large sheer sizes. A new trend in architecture has been transforming traditional windows to glazed building façade. However, great challenges remain in terms of thermal transfer, energy efficiency and lighting requirements, especially heat loss in cold climate.

To address critical energy issues in civic structures, we have developed a novel concept of Optical Thermal Insulation (OTI) without relying on conventional thermal intervention medium, such as air or argon, as often used in conventional window systems. We have identified and synthesized new spectral-selective material systems, such as porphyrin compounds and iron oxide nanoparticles that exhibit strong UV and NIR absorptions, but high visible transmittance. Pronounced photothermal (PT) effect is optically-activated to increase the window surface temperature, thereby effectively lowering the heat transfer through the building skin, and achieving energy saving without double- or triple- glazing. The identified PT materials are abundant in nature and environmentally green.

We have also formulated a pioneering concept of PT-PV modality building skins. In this new approach, a building skin is considered a multifunctional active device for solar harvesting, conversion, and utilization. The glass-based high-rise building skins provide the ideal transparent substrates for device architecture of energy harvesting nanoscale thin films. A simple building skin coating can be engineered to offer two major functions: PV and PT, switched alternatively depending on the seasonal energy needs. In summer, the window coating functions as PV and consumes most of the solar infrared therefore less cooling is required. In cold winter, the slight increases skin temperature by the photothermal coating which can lower heat loss from room interior. Thus, the same glass-based building skin can be engineered to serve multiple functions including solar harvesting, conversion, and generation. The synergetic effects of PT and PV, seasonably alternated, will significantly improve building energy efficiency. In sharp contrast to conventional windows, the building skin with the PT and PV dual modality will essentially become an energy generator (or convertor) rather than a source of heat loss. The advantages of this approach are several-folds: (1) no insulating medium is needed as in the traditional double-pane structure; (2) the building skin with the PT-PV thin films can be mass-produced with low cost; (3) the PT-PV thin films can alternatively generate heat and electricity with a much large surface area; (4) the optical properties of the PT-PV building skin can be spectrally tuned by adjusting the composition and additives in the films, and (5) adhesive products built upon this approach can be applied in wide-scale energy-efficient window retrofits with low balance-of-system costs.

The novel solar harvesting system is capable of converting light to thermal energy directly for heating unities without any additional energy source and complicated electric circuitry for energy storage and distribution. The solar harvesting system is able to waveguide solar light through transparent multilayer photothermal thin films, generating sufficient heat at each layer collectively for efficient thermal energy generation and conversion. This is fundamentally different from the current 2D PV panels with limited surface areas in solar harvesting. The multilayer structure profoundly revolutionizes the current 2D solar harvesting by multilayer transparent PT thin films, therefore significantly increasing the solar harvesting area and energy density in a confined 3D space. Based on this novel concept, a photothermal generator is theoretically and experimentally possible to generate thermal energy via constant incoming solar light for various energy applications including utility heating. Fundamental studies on the relationships between the material structures and optical properties will introduce a new dimension of nanotechnology to civil engineering not only for fundamental science but also novel designs of the next generation energy-neutral civil infrastructure.

Acknowledgements We acknowledge the financial support from National Science Foundation CMMI-1635089 and CMMI-1953009.

References

Alrashidi, H., Ghosh, A., Issa, W., Sellami, N., Mallick, T. K., & Sundaram, S. (2020). Thermal performance of semitransparent CdTe BIPV window at temperate climate. *Solar Energy, 195*, 536–543.

Arasteh, D., Selkowitz, S., Apte, J., & LaFrance, M. (2006). *Zero energy windows* (No. LBNL-60049). Ernest Orlando Lawrence Berkeley NationalLaboratory, Berkeley, CA (US).

ARPA-E. (2014). Single-Pan Highly Insulating Efficient Lucid Designs (SHIELD) program overview.

Are solar panels toxic to the environment? Retrieved December 30, 2021, from https://news.energysage.com/solar-panels-toxic-environment/

Arras, R., Calmels, L., & Warot-Fonrose, B. (2012). Half-metallicity, magnetic moments, and gap states in oxygen-deficient magnetite for spintronic applications. *Applied Physics Letters, 100*(3), 032403.

ASTM C1199-14. (2014). Standard test method for measuring the steady-state thermal transmittance of fenestration systems using hot box methods. ASTM International: West Conshohocken, PA. www.astm.org

Atkinson, G. A., & Hancock, E. R. (2007). Shape estimation using polarization and shading from two views. *IEEE Transactions on Pattern Analysis and Machine Intelligence, 29*(11), 2001–2017.

Attia, S., Gratia, E., De Herde, A., & Hensen, J. L. (2012). Simulation-based decision support tool for early stages of zero-energy building design. *Energy and Buildings, 49*, 2–15.

Barbose, G., Darghouth, N., O'Shaughnessy, E., & Forrester, S. (2021). Tracking the sun pricing and design trends for distributed photovoltaic systems in the United States. https://emp.lbl.gov/sites/default/files/2_tracking_the_sun_2021_report.pdf.

Bjørk, R., & Nielsen, K. K. (2015). The performance of a combined solar photovoltaic (PV) and thermoelectric generator (TEG) system. *Solar Energy, 120*, 187–194.

Boxall, C., Kelsall, G., & Zhang, Z. (1996). Photoelectrophoresis of colloidal iron oxides: Part 2—Magnetite (Fe3O4). *Journal of the Chemical Society, Faraday Transactions, 92*(5), 791–802.

Carmody, J., Arasteh, D., & Heschong, L. (2007). *Residential windows: A guide to new technologies and energy performance.* WW Norton & Company.

Chang, S. Y., Cheng, P., Li, G., & Yang, Y. (2018). Transparent polymer photovoltaics for solar energy harvesting and beyond. *Joule, 2*(6), 1039–1054.

Chen, C. C., Dou, L., Zhu, R., Chung, C. H., Song, T. B., Zheng, Y. B., Hawks, S., Li, G., Weiss, P. S., & Yang, Y. (2012). Visibly transparent polymer solar cells produced by solution processing. *ACS Nano, 6*(8), 7185–7190.

Colsmann, A., Puetz, A., Bauer, A., Hanisch, J., Ahlswede, E., & Lemmer, U. (2011). Efficient semi-transparent organic solar cells with good transparency color perception and rendering properties. *Advanced Energy Materials, 1*(4), 599–603.

Ding, T., Zhu, L., Wang, X. Q., Chan, K. H., Lu, X., Cheng, Y., & Ho, G. W. (2018). Hybrid photothermal pyroelectric and thermogalvanic generator for multisituation low grade heat harvesting. *Advanced Energy Materials, 8*(33), 1802397.

Effect of Temperature. (2021). Retrieved December 30, 2021 from https://www.pveducation.org/pvcdrom/solar-cell-operation/effect-of-temperature

Environmental Protection Agency. (2015). Energy Star performance criteria for windows, doors, and skylights. Retrieved May 26, 2015.

El-Sebaii, A., & Khallaf, A. E. M. (2020). Mathematical modeling and experimental validation for square pyramid solar still. *Environmental Science and Pollution Research, 27*(26), 32283–32295.

Fontijn, W. F. J., Van der Zaag, P. J., Feiner, L. F., Metselaar, R., & Devillers, M. A. C. (1999). A consistent interpretation of the magneto-optical spectra of spinel type ferrites. *Journal of Applied Physics, 85*(8), 5100–5105.

Gao, M., Zhu, L., Peh, C. K., & Ho, G. W. (2019). Solar absorber material and system designs for photothermal water vaporization towards clean water and energy production. *Energy & Environmental Science, 12*(3), 841–864.

Hecht, E. (1998). *Optics*, 3rd ed.

How much energy is consumed in U.S. buildings? Retreieved 3 May 2021. https://www.eia.gov/tools/faqs/faq.php?id=86&t=1.

Husain, A. A., Hasan, W. Z. W., Shafie, S., Hamidon, M. N., & Pandey, S. S. (2018). A review of transparent solar photovoltaic technologies. *Renewable and Sustainable Energy Reviews, 94*, 779–791.

Inoue, H., Yamashita, H., Furuya, K., Nonomura, Y., Yoshioka, N., & Lib, S. (1994). Determination of copper (II) chlorophyllin by reversed-phase high-performance liquid chromatography. *Journal of Chromatography A, 679*(1), 99–104.

Irace, P., & Brandon, H. (2017). Solar heating in commercial buildings.

Jin, H., Lin, G., Bai, L., Amjad, M., Bandarra Filho, E. P., & Wen, D. (2016). Photothermal conversion efficiency of nanofluids: An experimental and numerical study. *Solar Energy, 139*, 278–289.

Kanehara, M., Koike, H., Yoshinaga, T., & Teranishi, T. (2009). Indium tin oxide nanoparticles with compositionally tunable surface plasmon resonance frequencies in the near-IR region. *Journal of the American Chemical Society, 131*(49), 17736–17737.

Kay, A., & Graetzel, M. (1993). Artificial photosynthesis 1: Photosensitization of titania solar cells with chlorophyll derivatives and related natural porphyrins. *The Journal of Physical Chemistry, 97*(23), 6272–6277.

Khan, M., & Mustafa, M. (2019). Solar still distillate productivity enhancement by using reflector and design optimization. *Innovative Energy & Research, 8*(1), 2576–1463.

Kizer, K.W., Garb, L.G. and Hine, C.H., 1984. Health effects of silicon tetrachloride. Report of an urban accident. *Journal of occupational medicine.: official publication of the Industrial Medical Association, 26*(1), 33–36.

Lassiter, J. B., Aizpurua, J., Hernandez, L. I., Brandl, D. W., Romero, I., Lal, S., Hafner, J. H., Nordlander, P., & Halas, N. J. (2008). Close encounters between two nanoshells. *Nano Letters, 8*(4), 1212–1218.

Li, Y., Guo, X., Peng, Z., Qu, B., Yan, H., Ade, H., Zhang, M., & Forrest, S. R. (2020). Color-neutral, semitransparent organic photovoltaics for power window applications. *Proceedings of the National Academy of Sciences, 117*(35), 21147–21154.

Li, W., Elkhaklifa, M., & He, H. (2019). Design, engineering, and evaluation of porphyrins for dye-sensitized solar cells. In *Nanostructured materials for next-generation energy storage and conversion* (pp. 351–381). Springer, Berlin, Heidelberg.

Lin, J., & Shi, D. (2021). Photothermal and photovoltaic properties of transparent thin films of porphyrin compounds for energy applications. *Applied Physics Reviews, 8*(1), 011302.

Lin, J., Zhao, Y., & Shi, D. (2020). Optical thermal insulation via the photothermal effects of Fe_3O_4 and Fe_3O_4@ Cu2−xS thin films for energy-efficient single-pane windows. *MRS Communications, 10*(1), 155–163.

Liu, X., Li, B., Fu, F., Xu, K., Zou, R., Wang, Q., Zhang, B., Chen, Z., & Hu, J. (2014). Facile synthesis of biocompatible cysteine-coated CuS nanoparticles with high photothermal conversion efficiency for cancer therapy. *Dalton Transactions, 43*(30), 11709–11715.

Liu, Y. Q., Wei, D., Cui, H. L., & Wang, D. Q. (2019). Photovoltaic effect related to methylammonium cation orientation and carrier transport properties in high-performance perovskite solar cells. *ACS Applied Materials & Interfaces, 12*(3), 3563–3571.

Lu, J., Liu, S., & Wang, M. (2018). Push-pull zinc porphyrins as light-harvesters for efficient dye-sensitized solar cells. *Frontiers in Chemistry, 6*, 541.

Lyu, M., Lin, J., Krupczak, J., & Shi, D. (2020). Light angle dependence of photothermal properties in oxide and porphyrin thin films for energy-efficient window applications. *MRS Communications, 10*(3), 439–448.

Lyu, M., Lin, J., Krupczak, J., & Shi, D. (2021). Solar harvesting through multilayer spectral selective iron oxide and porphyrin transparent thin films for photothermal energy generation. *Advanced Sustainable Systems*, 2100006.

Maradudin, A. A., & Méndez, E. R. (2007). Light scattering from randomly rough surfaces. *Science Progress, 90*(4), 161–221.

Mathew, S., Yella, A., Gao, P., Baker, R.H., Curchod, B.F.E., Astani, N.A., Tavernelli, I., Rothlisberger, U., Nazeeruddin, M., & Gratzel, M. (2014). *Nature Chemistry 6*, 242.

Michael S. D. (2021). Understanding the cost of solar energy. Retrieved December 30, 2021 from https://greenecon.net/understanding-the-cost-of-solar-energy/energy_economics.html.

Miyazaki, D. (2014). Fresnel equations.

National Fenestration Rating Council Incorporated. (2020). ANSI/NFRC 100-2017 [E0A2]: Procedure for Determining Fenestration Product U-factors. National Fenestration Rating Council Incorporated: National Fenestration Rating Council Incorporated. Accessed 18 July 2020.

National Fenestration Rating Council. (2013). Procedure for Determining Fenestration Product U-factors Retrieved March 10, 2020 from https://www.nfrccommunity.org/store/ViewProduct.aspx?id=1380591.

Nemade, K. R., & Waghuley, D. S. (2015). Band gap engineering of CuS nanoparticles for artificial photosynthesis. *Materials Science in Semiconductor Processing, 39*, 781–785.

Nolan, D., Senaratne, W., Baker, D., & Liu, L. (2015). Optical Scattering from Nanostructured Glass Surfaces. *International Journal of Applied Glass Science, 6*(4), 345–355.

Ogura, R. Y., Nakane, S., Morooka, M., Orihashi, M., Suzuki, Y., & Noda, K. (2009). High-performance dye-sensitized solar cell with a multiple dye system. *Applied Physics Letters, 94*(7), 54.
O'regan, B., & Grätzel, M. (1991). A low-cost, high-efficiency solar cell based on dye-sensitized colloidal TiO 2 films. *Nature, 353*(6346), 737–740.
Otaka, H., Kira, M., Yano, K., Ito, S., Mitekura, H., Kawata, T., & Matsui, F. (2004). Multi-colored dye-sensitized solar cells. *Journal of Photochemistry and Photobiology a: Chemistry, 164*(1–3), 67–73.
Perera, F. (2018). Pollution from fossil-fuel combustion is the leading environmental threat to global pediatric health and equity: Solutions exist. *International Journal of Environmental Research and Public Health, 15*(1), 16.
Pérez-Hernández, M., Del Pino, P., Mitchell, S. G., Moros, M., Stepien, G., Pelaz, B., Parak, W. J., Gálvez, E. M., Pardo, J., & de la Fuente, J. M. (2015). Dissecting the molecular mechanism of apoptosis during photothermal therapy using gold nanoprisms. *ACS Nano, 9*(1), 52–61.
Pistner, A. J., Yap, G. P., & Rosenthal, J. (2012). A tetrapyrrole macrocycle displaying a multi-electron redox chemistry and tunable absorbance profile. *The Journal of Physical Chemistry C, 116*(32), 16918–16924.
Rai, R. C. (2013). Analysis of the Urbach tails in absorption spectra of undoped ZnO thin films. *Journal of Applied Physics, 113*(15), 153508.
Roper, D. K., Ahn, W., & Hoepfner, M. (2007). Microscale heat transfer transduced by surface plasmon resonant gold nanoparticles. *The Journal of Physical Chemistry C, 111*(9), 3636–3641.
Shi, D., Sadat, M. E., Dunn, A. W., & Mast, D. B. (2015). Photo-fluorescent and magnetic properties of iron oxide nanoparticles for biomedical applications. *Nanoscale, 7*(18), 8209–8232.
Singh, J. (ed.) (2006). *Optical properties of condensed matter and applications* (Vol. 6). John Wiley & Sons.
Spacek, A. D., Neto, J. M., Biléssimo, L. D., Ando Junior, O. H., Santana, M. V. F. D., & Malfatti, C. D. F. (2018). Proposal of the tubular daylight system using acrylonitrile butadiene styrene (abs) metalized with aluminum for reflective tube structure. *Energies, 11*(1), 199.
Stavenga, D. G. (2014). Thin film and multilayer optics cause structural colors of many insects and birds. *Materials Today: Proceedings, 1*, 109–121.
Tian, Q., Hu, J., Zhu, Y., Zou, R., Chen, Z., Yang, S., Li, R., Su, Q., Han, Y., & Liu, X. (2013). Sub-10 nm Fe3O4@ Cu2–x S core–shell nanoparticles for dual-modal imaging and photothermal therapy. *Journal of the American Chemical Society, 135*(23), 8571–8577.
Traverse, C. J., Pandey, R., Barr, M. C., & Lunt, R. R. (2017). Emergence of highly transparent photovoltaics for distributed applications. *Nature Energy, 2*(11), 849–860.
U. E. P. Agency. (2018). Windows, Doors & Skylights Key Product Criteria. Retrieved March 19, 2018, from https://www.energystar.gov/products/building_products/residential_windows_doors_and_skylights/key_product_criteria.
Wang, L., Gwilliam, J., & Jones, P. (2009). Case study of zero energy house design in UK. *Energy and Buildings, 41*(11), 1215–1222.
Wei, T., Liu, Y., Dong, W., Zhang, Y., Huang, C., Sun, Y., Chen, X., & Dai, N. (2013). Surface-dependent localized surface plasmon resonances in CuS nanodisks. *ACS Applied Materials & Interfaces, 5*(21), 10473–10477.
Yella, A., Lee, H. W., Tsao, H. N., Yi, C., Chandiran, A. K., Nazeeruddin, M. K., Diau, E. W. G., Yeh, C. Y., Zakeeruddin, S. M., & Grätzel, M. (2011). Porphyrin-sensitized solar cells with cobalt (II/III)–based redox electrolyte exceed 12 percent efficiency. *Science, 334*(6056), 629–634.
Yoon, J. H., Shim, S. R., An, Y. S., & Lee, K. H. (2013). An experimental study on the annual surface temperature characteristics of amorphous silicon BIPV window. *Energy and Buildings, 62*, 166–175.
Yun, M. J., Sim, Y. H., Cha, S. I., Seo, S. H., & Lee, D. Y. (2019). 3-Dimensional dye sensitized solar cell sub-module with oblique angled cell array for enhanced power and energy density output utilizing non-linear relation in cosine law of light incident angle. *Solar Energy, 177*, 355–363.

Zaitoun, M. A., Mason, W. R., & Lin, C.T. (2001). Magnetic circular dichroism spectra for colloidal gold nanoparticles in xerogels at 5.5 K. *The Journal of Physical Chemistry B, 105*(29), 6780–6784.
Zeng, K., Tong, Z., Ma, L., Zhu, W. H., Wu, W., & Xie, Y. (2020). Molecular engineering strategies for fabricating efficient porphyrin-based dye-sensitized solar cells. *Energy & Environmental Science, 13*(6), 1617–1657.
Zhang, X. (2002). *Daylighting performance of tubular solar light pipes: measurement, modelling and validation.* Doctoral dissertation, Edinburgh Napier University.
Zhao, Y., Pan, H., Lou, Y., Qiu, X., Zhu, J., & Burda, C. (2009). Plasmonic Cu2− x S nanocrystals: Optical and structural properties of copper-deficient copper (I) sulfides. *Journal of the American Chemical Society, 131*(12), 4253–4261.
Zhao, Y., Sadat, M. E., Dunn, A., Xu, H., Chen, C. H., Nakasuga, W., Ewing, R. C., & Shi, D. (2017). Photothermal effect on Fe_3O_4 nanoparticles irradiated by white-light for energy-efficient window applications. *Solar Energy Materials and Solar Cells, 161*, 247–254.
Zhao, Y., Lin, J., Kundrat, D. M., Bonmarin, M., Krupczak, J., Jr., Thomas, S. V., Lyu, M., & Shi, D. (2019a). Photonically-activated molecular excitations for thermal energy conversion in porphyrinic compounds. *The Journal of Physical Chemistry C, 124*(2), 1575–1584.
Zhao, Y., Dunn, A. W., & Shi, D. (2019b). Effective reduction of building heat loss without insulation materials via the photothermal effect of a chlorophyll thin film coated "Green Window." *MRS Communications, 9*(2), 675–681.
Zhao, Y., Dunn, A., Lin, J., & Shi, D. (2019c). Photothermal effect of nanomaterials for efficient energy applications. In *Novel nanomaterials for biomedical, environmental and energy applications* (pp. 415–434). Elsevier.

Chapter 2
Low Energy Adaptive Biological Material Skins from Nature to Buildings

Laia Mogas-Soldevila

Abstract This chapter reviews emergent work in large-scale interactive building skins that use biological materials derived from abundant, renewable, biodegradable sources like silk, algae, wood, cellulose, chitin, fungi, or bacteria. They are surveyed as new interactive systems for material-driven environmental sensing and response within the outer layer of architectural applications. Programmed at the molecular scale, they respond to their surroundings at the building scale by; self-healing cracks, performing programmed decay, tuning flexibility and opacity depending on sunlight and rain, changing color to diagnose health markers, shapeshifting with humidity changes, digesting waste into structure, cooling and cleaning air, or transforming city pollutants into fuel and aliments. Demonstrators are often in testing phases, but critical in signaling a future for sustainable material systems offering adaptive solutions at the intersection of building construction and biotechnology that are elegant in both their efficiency and new aesthetics.

Keywords Bio-composites · Adaptive materials · Programmed matter · Responsive skins

2.1 Introduction: Nature to Buildings

Biomimicry in architecture—in other words, design inspired by how functional challenges have been solved in biology (Pawlyn, 2019)—has broadened its scope in recent year's research. Solutions in buildings emerge that expand beyond traditional advances in technology inspired by biological strategies, and towards including living and bio-based matter itself. This is because bio-based, biological, or even living material solutions present a dual advantage to common construction materials such as glass, concrete, ceramics, steel, or aluminum. On one hand they involve much

L. Mogas-Soldevila (✉)
DumoLab Research Director, Stuart Weitzman School of Design, Department of Architecture, University of Pennsylvania, 210 South 34th Street, Suite 207B, Philadelphia, PA 19104, USA
e-mail: laiams@design.upenn.edu
URL: https://www.design.upenn.edu/dumolab

J. Wang et al. (eds.), *Advanced Materials in Smart Building Skins for Sustainability*,
https://doi.org/10.1007/978-3-031-09695-2_2

lower construction waste and energy consumption in extraction, synthesis, and end-of-life (Ashby, 2021). On the other hand, they are programmable with new function enhancing sustainability as well as multiplicity of unprecedented function as reviewed later in the text (Shtein & Shoseyov, 2017). New man-made composites from synthetic materials also present large ecological footprints in their production and are difficult to disassemble for recycling (Bechthold & Weaver, 2017). Many naturally grown materials like cellulose, silk, or chitin are lightweight but strong, stiff, and tough, while made at ambient conditions, processed with low energy and low waste, and composed of simple molecules that biodegrade without toxicity (Vincent, 2012; Wegst & Ashby, 2007). Not only these materials present desirable environmental friendliness, but they also have inherent capacity to interact and adapt to their surroundings with passive strategies able to be programmed for specific functions. Today, in the field of building construction, and following similar avenues in materials science and biomedical engineering, mimicking Nature's intelligence will ensure superior function and efficient reuse of resources.

Biological materials are indeed gaining interest in the architectural community with large-scale solutions being developed as demonstrators of a new paradigm. A paradigm observing matter as a design element to be described to the molecule (Mogas-Soldevila, 2021), a task enabled by the fact that natural matter is in constant re-design during growth by responding to the environment and to the forces acting upon it, making material properties vary across species, within the same species, and throughout the same organism. For instance, organic composite materials give living tissues the ability to adapt by rearranging their material configurations towards optimized ones. The leaf closing of *Mimosa pudica* in response to touch, heliotropism in sunflowers following sunlight, skin color changes in *Loliginidae squid*, catapult seed actions of some fruits, variable-stiffness collagen in marine animals, opening and closing of pine cones, and the hinged operation of ice plant seed capsules, are just a few examples of this extraordinary ability (Ball, 2012; Bechthold & Weaver, 2017; Jeronimidis, 2009).

Such adaptation that helps trees, insects, and humans survive, can be re-programmed to solve specific building problems, or deliver desired signals as described next. Reviewed below, recent solutions show efforts to invent systems that respond to changes in the environment performing adaptation of mechanical, optical, and chemical material properties. Examples chosen are larger than meter-scale, geared towards façade implementation, and made of raw bio-based materials derived from wood, silk, or chitin, or made of living organism assemblies like fungi networks, bacterial colonies, mosses, or microalgae.

2.2 Methodologies in Practice: The Active Skin

The function of a building skin is to shelter and enclose human activity by filtering light, radiation, dirt, moisture, temperature, and pathogens. This emphasis in protection, renders traditional buildings static for decades hardly interacting with their occupants and the environment. Mechanically responsive facade systems are emerging in the field and typically poses some of the following features embedded within man-made actuated technology; energy storing, natural ventilation, radiation control, or automated management of plants on the building skin (Romano et al., 2018). In nature, skin, shell, or cuticle do protect organisms by filtering light, radiation, dirt, moisture, temperature, and pathogens, and do store energy and manage systems automatically like these new facades do, but they perform outstanding added functions of self-repair, shape-shift, sensing, and color change to adapt to their changing surroundings.

It is the goal of many of the solutions presented in this section to achieve these superior levels of interaction by mimicking naturally grown skins. Biological material systems and biology-driven strategies in the demonstrator examples reviewed below aim to create buildings that respond to their surroundings while being environmentally friendly (Sandak et al., 2019). Proposals can perform programmed decay, change flexibility and opacity responding to sunlight and rain, self-heal cracks using bacteria, change color in response to health markers, curl and shapeshif with humidity changes, capture carbon dioxide, digest waste into structural members, react to hot weather with evaporative and radiative cooling, clean city air, or transform air pollutants and water contaminants into fuel and food (Figs. 2.1 and 2.2).

2.2.1 *Wood*

Most advanced large scale biological material solutions for building skins are based on timber and wood composites and are a promising resource because of their renewability, sustainability, and versatility. Producing wood for buildings uses about 10% of the energy required to produce equivalent amount of steel. It is transformed with much simpler tooling while enabling prefabrication, fast installation, favorable weight-to-load-bearing capacity ratio, and low thermal conductivity increasing its applicability in façades (Sandak et al., 2019). However, many efforts aim at keeping wood systems static and avoiding their natural tendency to deform and interact with environmental humidity. Examples below harness that ability, instead of suppressing it, to achieve higher order of performance.

In nature, structural anisotropy in the organization of cellulose wood fibers can induce movement through water absorption and differential swelling, as observed in pinecones when they passively open their seed pods. Larger scale hydromorphic effects can be programmed in plywood selecting and arranging wood in specific grain directions using its expanding behavior as an actuator. The ICD at the University of

Fig. 2.1 Column one; (top) Hygroskin by the ICD (Menges & Reichert, 2015) and (bottom) Hydroculus by Thermal Architecture Lab (Aviv et al., 2020). Column two; (top) HiFy by The Living (The Living NY, 2014) and (bottom) Growing Pavilion by Company New Heroes (Biobased Creations, 2019). Column three; (top) Sportplaza Mercator by VenhoevenCS (VenhoevenCS architecture + urbanism, 2006) and (bottom) Bioreceptive concrete by BiotA Lab (Cruz & Beckett, 2016). Column four; (top) Aguahoja1 by The Mediated Matter Group (Duro-Royo et al., 2018) and (bottom) Hidaka Ohmu by Julia Lohmann (Lohmann, 2017; Toivola, 2020). Column five; (top) HORTUS XL by ecoLogicStudio (Valenti & Pasquero, 2021) and (bottom) Bioconcrete by Jonkers Lab (Jonkers, 2011)

Stuttgart has developed a series of studies for thin wood veneer façade systems that shape-shift with humidity changes during building use. For instance, Hygroskin uses flat petal-like units arranged in pentagons of about half a meter wide that stay stiff and closed in dry weather but become flexible and open in high humidity situations to ensure ventilation. Their deformation is reversible in several thousand cycles. (Krieg & Menges, 2013; Menges & Reichert, 2015) (Fig. 2.2, column one). New prototypes by the same group envision; additively-manufacturing wood-like systems using cellulose and other biopolymers to match similar motion rates with larger design freedom, or obtaining doubly-curved 9 m-tall assemblies using the same principle during panel formation (Correa et al., 2015; Grönquist & Bechert, 2020).

Also made of wood and in partnership with water, a recent project called Hydroculus by the Thermal Architecture Lab at the University of Pennsylvania can

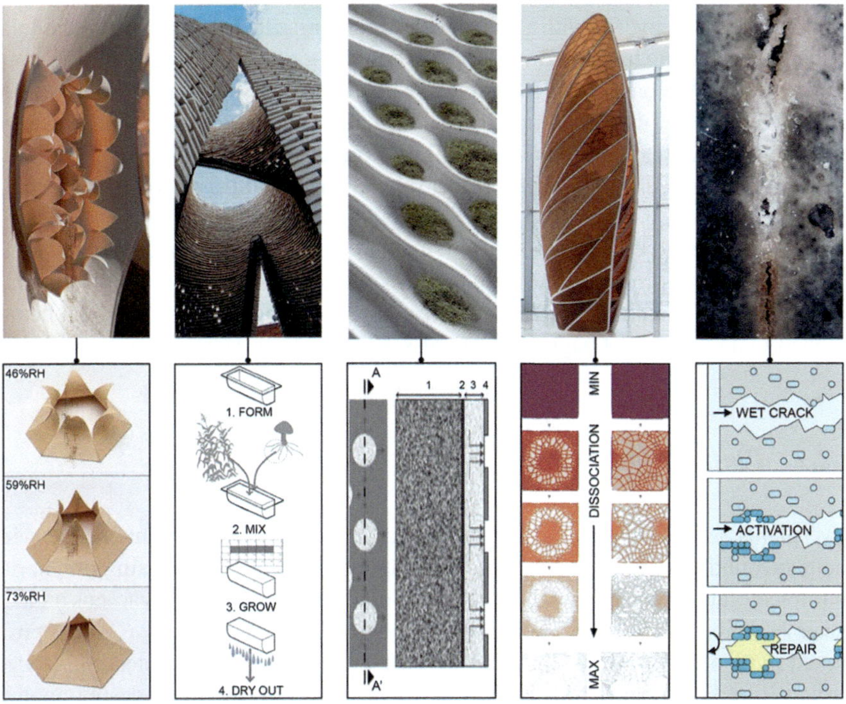

Fig. 2.2 Column one; (top) Hygroskin structure by ICD, Institute for Computational Design at the University of Stuttgart (Menges & Reichert, 2015) and (bottom) Hygroskin veneer-composite unit shape changing by hygroscopic response to relative humidity change (adapted from (Correa et al., 2015)). Column two; (top) HiFy by The Living (The Living NY, 2014) and (bottom) mycelium brick formation steps including 5-day agricultural waste digestion and heat treatment before use (adapted from (Holcim Foundation Awards, 2015)). Column three; (top) Bioreceptive concrete by BiotA Lab and (bottom) section of bioreceptive panel depicting (1) structural Portland Cement, (2) anchoring and sealing interface, (3) bioreceptive mortar with water retention capabilities, (4) water absorbing coating (adapted from (Cruz & Beckett, 2016)). Column four; (top) Aguahoja1 by The Mediated Matter Group (Duro-Royo et al., 2018) and (bottom) simulation of swelling and decay by effects of weathering based on measurements of printed patches following Aguahoja's materialization technique (adapted from (Tai et al., 2018)). Column five; (top) Bioconcrete by Jonkers Lab (Jonkers, 2011) and (bottom) steps of crack-healing by concrete-immobilized bacteria (in teal) activated due to water penetration through cracks and precipitating repairing minerals (in yellow) to protect the steel reinforcement from further external chemical attack (adapted from (Jonkers, 2007))

cool air in hot-dry climates by tapping into new technology and vernacular knowledge. Hydroculus is a prototype for a combined evaporative and radiative cooling chimney integrated into a building's envelope in hot-dry climates. It uses hygroscopic materials to generate cooled airflow: a hydrogel membrane is embedded in the wooden funnel-shaped top of the chimney, which acts as a wind catcher. The hydrogel stores water, which is diffused into the incoming wind, inducing evaporative cooling and downdraft flow (Aviv et al., 2020). The chimney structure constitutes waffle timber ribs covered with a thinly coated membrane that reflects shortwave solar radiation, protecting high-thermal-capacity liquids stored in the envelope from overheating during the day. During the night, when the night sky temperature drops to below freezing, the photonic properties of the membrane allow for radiation exchange between the liquids and the sky in the longwave range, thus providing additional free cooling to be stored by the chimney's envelope across a diurnal cycle (Aviv et al., 2020).

Additionally, exposure to wood in living and working environments is linked to reduction of stress-related illness and improved moods. Using wood in skins with human interaction certainly creates positive psychophysiological effects and health impacts that we must not neglect (Burnard & Kutnar, 2020; Mcsweeney et al., 2015).

2.2.2 *Plants and Mosses*

Live plants on building skins -or vertical gardens- are inherently interactive with their surroundings and introduce numerous applications providing a single-material multiple-solution paradigm, as Nature does. Some benefits of vertical gardens include water retention, air filtering, wind gust dampening, heat gain reduction, and if positioned in front of openings, they can provide shading and noise protection as well as light-filtering. Wonderwall by Patrick Blanc is a vertical garden design applied to the Sportplaza Mercator building in Amsterdam (VenhoevenCS architecture + urbanism, 2006). The wall consists of a steel frame attached and separated from the roof construction of the building, then a 'growing wall' system made of metal, plastic and a felt fleece with notches and small buckets for each plant to grow. A significant number of different plants is maintained by a rain and feeding system with hoses and sensors. The authors explain that every wall has its own climate and demands therefore different kinds of plants (VenhoevenCS architecture + urbanism, 2006). This technology has vibrant aesthetic effects, indoor comfort benefits, and contributes to the thermoregulation and carbon sink capacities of the city.

There are other organisms like mosses or lichens that form what is called the cryptogamic crust and reproduce with spores, without flowers or seeds providing minimal root systems thus accounting for low weights. They cover large surfaces in forest substrates, and over bark or rock, and fix carbon dioxide and nitrogen from the atmosphere. Compared to vertical gardens, moss façade systems present low tech and low maintenance advantages while providing similar benefits. Mosses in building skins have been proven to increase outdoor air quality, provide indoor insulation, and help cool the city. They provide efficient solutions for large-scale applications due to their low requirements in substrates, nutrients, and water, and to their high desiccation tolerance. They do not need maintenance and irrigation is provided by rainfall (Cruz & Beckett, 2016; Perini et al., 2020). Bio receptive surfaces for moss and microalgae have been explored at the BiotA lab Bartlett School of Architecture, UCL. A series of projects aim at growing microorganisms directly on the surface of façade panels to overcome many of the limitations of existing green walls, particularly the need for mechanical irrigation systems and expensive maintenance as mentioned above (Cruz & Beckett, 2016). Low pH magnesium phosphate concrete is used instead of traditional Portland cement which would be too acidic to allow mosses, lichen, or algae to grow. Via digitally designed molds, surfaces made of layered concrete casting acquire fissures and depressions emulating tree bark and produce shaded areas as well as channels to guide rainwater to specific growth areas. Then panels are seeded with a mix of algae cells and moss spores, photosynthetic organisms that collect water from weather events, absorb radiation, CO_2, and other pollutants and produce oxygen (Fig. 2.2, column three). Researchers identify future work on moss façade systems towards water distribution and retention during dry periods and adhesion of moss mixture onto facade materials (Birch, 2016).

2.2.3 Fungi

Combining living biological systems with methods in materials science and nanotechnology is enabling superior tuning of material properties by guiding growth instead of de-novo engineering matter from atoms and molecules (Niemeyer, 2001). An example is engineered fungi materials. Fungi can provide function beyond the repertoire of plant-derived materials, and they have been studied to make pigments, construction materials, packaging, or paints. Mycelia, the vegetative tubular filament networks of fungi, contribute to circular economies by transforming local residual flows into fibrous, natural composite materials with controllable physical properties that can be produced in large quantities for carbon-negative buildings and applicable to architecture facades (Almpani-Lekka et al., 2021; Haneef et al., 2017).

The following two examples use mycelium to make compression-based building skins with bricks and panels manufactured under Ecovative's license and method (Fig. 2.2, column two) (Bayer & McIntyre, 2012; Holt et al., 2012). The method entails mixing local agricultural waste, such as stumps and branches, with *Gonoderma Mushroom* spores kept in a dark room in a covered mold ensuring minimal oxygen supply. After one week, spores have broken down the agricultural mass forming a lightweight solid using a natural digestion process that is then halted with heat (Leboucq et al., 2019). The heat treatment of mycelial composites stops growth of mycelium in the mold rendering an inert material (Almpani-Lekka et al., 2021), however, controlling living fungi materials on skins could offer new active properties such as self-healing, self-repairing, and partial self-organization.

The Hi-Fy tower by The Living and Arup for MoMA PS1 uses grown-to-shape fungi roots that digest waste into structure. It is the largest construction project with mycelium composite materials to date forming a 13 m-tall structure made of 10,000 lightweight bricks combining corn stalk waste and living mushrooms, and a timber substructure framing to ensure stability. The structure provides shade and cooling through openings and reflective brick coatings at the coronation. After use, it was biodegraded by shredding and soil composting during two months (The Living NY, 2014; Attias et al., 2019).

The Growing Pavilion a collaboration of Company New Heroes, the Dutch Design Foundation, and Eric Klarenbeek (Biobased Creations, 2019), is an entirely bio-based cylindrical structure made of wooden frame and mycelial cladding panels measuring 2 × 0.7 m each. Panels are coated to increase their weather resistance and demonstrate their use as façade elements. During the life of the structure, mycelium mediated the sound qualities of the pavilion's interior environment by insulating indoor musical performances from outdoor noise (Almpani-Lekka et al., 2021).

2.2.4 Biopolymers

During their life cycles, green plants, animals, bacteria, and fungi produce biopolymers. They are easily biodegradable and include; animal protein-based biopolymers - such as wool, silk, casein, gelatin, and collagen-, as well as polysaccharides - such as chitin, cellulose, pectin, and starch-, or carbohydrate polymers produced by bacteria and fungi -such as xanthan, dextran or cellulose (Yadav et al., 2015). Biopolymers systems called biomaterials are used emergently in advanced research within life sciences and biomedical disciplines to improve human health in drug delivery and tissue scaffold applications for regenerative medicine (Ratner, 2013). In this section, biopolymers are defined as based on naturally occurring polymeric materials (Plank, 2005) and pioneer projects made entirely of them are reviewed.

As the architecture field has become more aware of the impacts of plastic products in human safety (Faircloth, 2015), promising research has emerged in the last decade identifying strategies and material opportunities to replace man-made fuel-based polymers with bioplastics synthesized by organisms. For instance, Julia Lohmann inspects the capacity of brown kelp to form large-scale skins (Lohmann, 2017). She borrows a Japanese cuisine ingredient—a natural soup glutamate—as a material for "making instead of eating". It is not unusual that designers look at food industry and biomedical fields for plastic-like materials that our body can naturally digest with the hope that our planet will too (Mogas, 2018; Mogas et al., 2021). This seaweed can make fertilizer and turn into bioplastic, biofuel, dyes, veneer, and textiles. It grows yearly up to 6 m long and 30 cm wide while cleaning the ocean by filtering toxic farm run offs and fish feces from sea water. Hidaka Ohmu is a six-meter-wide structure with a seaweed skin in tension within a birch plywood and rattan frame. Seaweed has been treated with an environmental method to remain flexible (Toivola, 2020) by trapping water within its molecule chains. This is key, as most biopolymers, like kelp, lignin, or collagen, are hydrated in nature within the living bodies of organisms, plants, or animals, and will inevitably dehydrate when becoming non-living materials for buildings.

Tunning of intramolecular water absorption can be made interactive by programming the behavior of biopolymers. In Aguahoja1, a six-meter tall chitin and cellulose composites tower by The Mediated Matter Group at MIT (Duro-Royo et al., 2018), we observe control of mechanical, optical, and chemical properties within the system's skin. This is because biopolymer blend compositions are graded differentially throughout it. Opacity, color, and strength can be varied in correspondence to chitin-cellulose and pectin-calcium mixtures that are 2.5-dimensionally printed in large panels assembled onto a biodegradable polymer skeleton. Interestingly, the interaction of this system with its environment allows for programmed decay based on two design parameters: the hydrophilic property of biopolymers which swell in the presence of high humidity and ultimately disintegrate, and their density in the structure enabled by geometric distribution via computational tool pathing accounting for open and closed cell configurations (Fig. 2.2, column four).

Water and geometry were also crucial in the development of Lachesis by The Silklab at Tufts University (Mogas, Matzeu & Omenetto, 2018; Matzeu et al., 2020). Lachesis are a series of silk fabric tapestries imprinted with silk protein-based inks that embed a chemical reporter able to sense and diagnose its environment. Inks are water-based and formulated for screen-printing applications by combining an algae biopolymer as thickener, a plasticizer, and regenerated silk fibroin which is the fibrous protein found in silk cocoons. This formulation is made responsive through the addition of pH sensing molecules that are encapsulated within the fibroin matrix, making them interactive with their surroundings and able to change color in rich accurate palettes in the presence of weather events like saturation or rain. By changing color, the tapestries display the acidity of rain, for instance. This system is derived from biomedical research where such sensors work over a reduced pH range, with low sensitivity, and with local sensing restricted to small patches. However, Lachesis 3 m-tall tapestries can withstand repeated wetting, dry cleaning, and reversibility to diagnose their surroundings. Screen-printing of robust bioactive inks like these on a large scale opens a promising direction toward mass-production of responsivee interfaces for distributed environmental sensing in buildings with applications in façade or roof canopies (Matzeu et al., 2020).

2.2.5 Microorganisms

Engineered Living Materials (ELMs) merge the fields of materials science and synthetic biology and study generation of biologically active materials with tailorable properties providing new function across fields in, for instance, medical, electrical, and construction applications (Srubar, 2021). There are a few examples of successful engineered building materials where living cells both give structure and modulate performance. BioMASON makes bacterial derived mortars and bricks (Dosier, 2011), Henk Jonkers group at TU Delft and Basilisk produce concrete able to self-heal its cracks by selectively activating its bacterial composition (Roy et al., 2020), the Srubar Research Group at UC Boulder derives cementitious materials that regenerate themselves by using photosynthetic microorganisms to biomineralize inert sand-gelatin scaffolds (Heveran et al., 2020), or ecoLogicStudio investigates photosynthetic living microalgae building skins and products (Valenti & Pasquero, 2021).

Self-healing bio-concrete is made of concrete infused with dormant bacteria. These microorganisms allow it to self-heal its cracks and prevent steel rebars from corroding. This is enabled by certain strains of bacteria that metabolically mediate the precipitation of minerals such as limestone when exposed to the right environmental conditions. In this case, when dormant spores of alkali-resistant bacteria are in contact with outside water entering concrete through a crack, they multiply and precipitate minerals such as calcite thus sealing small cracks and autonomously remediating concrete before steel reinforcement is damage (Fig. 2.2, column five). In order to bring living concrete to industry, Basilisk is continuing research on robust bacteria able to withstand construction times for longer than a few months, and evaluating if bio-precipitation can be sustained over decades (Jonkers, 2007, 2011).

BioMASON's bio-cements are assembled and cured at ambient conditions by bacteria reducing energy costs and resulting in zero carbon emissions. In particular, the technology uses sand aggregate, infused with calcium ions and water, and then seeded in cycles with a broth of robust bacteria strands that produce urease enzymes. Molded bricks or parts are allowed to harden to ASTM specification. Hardening is mediated with the help of bacteria in the creation of calcium carbonate from calcium ions which fills the bonds between loose pieces of aggregate forming a solid construction material (Dosier, 2011). Interestingly, matter can be tailored to different requirements of porosity, lightness, or insulation, and still maintain properties comparable to traditional masonry materials.

In the IBA Hamburg in 2013 Arup and SSC built the BIQ five-story housing building, the first algae-powered building in the world. Bioreactor panels fill the south façade and autonomously cultivate photosynthetic microalgae -by feeding them CO_2 and nutrients- to generate energy and biofuel as renewable energy resources (Elrayies, 2018). Several ecoLogicstudio implementations look at distribution and maintenance of similar photosynthetic microalgae systems in soft building skins. Photo.Synth.Etica, installed in Dublin in 2018 is perhaps the largest system they have investigated. It features a bioplastic envelope façade with aqueous solution pockets filled with living microalgae and nutrients. The architects explain that city air enters the system and CO_2 molecules and atmospheric pollutants are captured and stored by algae before being transformed into biomass, which can be then collected and used in the production of new bioplastics. In the meantime, oxygen expelled by photosynthesis is free to leave and return into the city (Valenti & Pasquero, 2021). These living microorganism systems can produce outputs in the form of biomass for human nutrition, biofuel for green energy, or inputs for other organisms to perform new functions in symbiosis. In all these examples, replenishment and survival of living systems is critical, and comprises most of the groups' future research.

2.3 Outlook: Challenges in Disguise

The aim of this text is to inspire consideration of a future for biological and living material building skins at the intersection of construction and biotechnology. A future where facades interact and adapt to changes in their environments providing passive cooling and autonomous ventilation, health diagnostics and programmed decay, and even embedding and supporting other organisms able to perform sophisticated functions of self-healing, biomass production, and air cleaning. All of this by inherently increasing resource efficiency when using abundant local materials instead of global supply chains, shifting from fossil-fuels to solar power, and from linear materials and energy use to a circular economy producing more growth and no waste.

It is true that, compared with traditional building materials, biological materials possess properties that have traditionally been difficult to control, however, mastering strategies applied to these properties can spark solutions to old problems.

Water absorption in wood and natural fibers (5 to 40% of their dry weight) can lead to deformation via shrinkage and swelling and to fungal growth and decay (Sandak et al., 2019). Paradoxically, hygroscopicity is one of the most interactive properties of bio-based materials as it enables autonomous hydration-dehydration cycles in tune to daily and seasonal changes. For instance, laminar wood surfaces can curl and relax linked to atmospheric humidity, which can be harnessed to tune light and ventilation through building skins (Krieg & Menges, 2013).

Deformation in response to changing loads are characteristic of living bone or wood, and generally penalized in building code. However, this property could help existing building structures, for instance, "grow" more matter where highest wind or traffic forces are applied. Such beneficial dynamics are yet to be regulated in architectural handbooks, but they are part of ancient techniques making living root bridges in Meghalaya (Ravishankar & Ji, 2021), or grown-to-shape wood chairs by Full Grown (Munro and Munro, 2006), and are becoming within reach as designers and architects push for their implementation (Joachim & Silver, 2017).

Time scales can be a limiting factor to produce engineered living materials for construction applications as organisms output matter at different time scales, from hours to make bacterial cellulose, to decades to make wood (Srubar, 2021). However, it is also true that time cycles can help bio-based material systems achieve their optimal outputs, such as programmed decay in biopolymers (Duro-Royo et al., 2018), self-repair in bacterial cements (Dosier, 2011; Heveran et al., 2020; Jonkers, 2011), or self-regulation in cooling hydrogels (Aviv et al., 2020).

Replenishment is another challenge of biological and living materials systems in building construction applications. Research is advancing so that mosses adhere to bio receptive concrete for decades without intervention (Cruz & Beckett, 2016), bacteria powering self-healing concrete wait dormant for years before being activated by water penetrating the cracks (Roy et al., 2020), or photosynthetic microalgae in façade bioreactors receive enough nutrients to perform their function. In nature, when something stops working, death and new life is preferred. Ensuring that our buildings, codes, and aesthetic expectations account for autonomous cyclical renewal is key to the adoption of active and adaptive biological material construction.

Current industrial certification processes sustain biological material solutions to traditional testing methods and compliance rates in deformation, fire, or water resistance. However, biological function such as self-regulation, adaptation, autonomous growth, and self-repair, create an alternative paradigm requiring consideration of material properties over the entire building's life cycle, and accounting for the advantages of not relying in energy-intensive solutions to building adaptation problems (Almpani-Lekka et al., 2021; Mogas-Soldevila, 2021; Yadav et al., 2015).

In the next decade, industry will aim at high-quality, labor-friendly, minimum waste, low emission products demonstrating no harm to humans and the planet. Bio-based, biological, and living material systems in construction will certainly contribute to achieving these goals, and while doing so, will derive multifunctional, passive, adaptive, responsive, and aesthetically shifting solutions to old and new challenges.

References

Almpani-Lekka, D., et al. (2021). A review on architecture with fungal biomaterials: The desired and the feasible. *Fungal Biology and Biotechnology, 8*, 17. https://doi.org/10.1186/s40694-021-00124-5

Ashby, M. F. (2021). *Materials and the environment: eco-informed material choice*, 3rd edn. Elsevier Inc.

Attias, N., et al. (2019). Mycelium bio-composites in industrial design and architecture: Comparative review and experimental analysis. https://doi.org/10.1016/j.jclepro.2019.119037.

Aviv, D., Moradnejad, M., et al. (2020). Hydrogel-based evaporative and radiative prototype for hot-arid climates. In: *SimAUD Symposium Proceedings*, pp. 273–280.

Aviv, D., Wang, Z., et al. (2020). 'Surface generation of radiatively cooled building skin for desert climate. In: *ACADIA: Ecology and Ethics*, pp. 66–73.

Ball, P. (2012). Nature's Color Tricks. *Scientific American, 306*(5), 74–79. https://doi.org/10.1038/scientificamerican0512-74

Bayer, E., & McIntyre, G. (2012). *Method for making dehydrated mycelium elements and product made thereby*. USPTO.

Bechthold, M., & Weaver, J. C. (2017). Materials science and architecture. *Nature Reviews Materials, 2*(12), 17082. https://doi.org/10.1038/natrevmats.2017.82

Biobased Creations. (2019). *About The Growing Pavilion.*

Birch, A. (2016). *Reinventing the Green, BD Online.*

Burnard, M. D., & Kutnar, A. (2020). Human stress responses in office-like environments with wood furniture. *Building Research & Information, 48*(3), 316–330. https://doi.org/10.1080/09613218.2019.1660609

Correa, D., et al. (2015). 3D-printed wood: Programming hygroscopic material transformations. *3D Printing and Additive Manufacturing*, *2*(3), 106–116.

Cruz, M., & Beckett, R. (2016). Bioreceptive design: A novel approach to biodigital materiality. *Architectural Research Quarterly, 20*(1), 51–64. https://doi.org/10.1017/S1359135516000130

Dosier, G. (2011). *Methods for making construction material using enzyme producing bacteria.* USPTO.

Duro-Royo, J., et al. (2018). Designing a tree: Fabrication informed digital design and fabrication of hierarchical structures. In: *Proceedings of the IASS Annual Symposium: Construction-aware structural design.* Boston, pp. 1–7.

Elrayies, G. M. (2018). Microalgae: Prospects for greener future buildings. *Renewable and Sustainable Energy Reviews, 81*, 1175–1191. https://doi.org/10.1016/J.RSER.2017.08.032

Faircloth, B. (2015). *Plastics now: On architecture's relationship to a continuously emerging material*. Routledge.

Grönquist, P., & Bechert, S. (2020). From machine control to material programming : Self-shaping wood manufacturing of a high performance curved clt structure - urbach tower. *Fabricate 2020: Making Resilient Architecture.*

Haneef, M., et al. (2017). Advanced materials from fungal mycelium: Fabrication and tuning of physical properties. *Scientific Reports, 7*, 41292. https://doi.org/10.1038/srep41292

Heveran, C. M., et al. (2020). Biomineralization and successive regeneration of engineered living building materials. *Matter, 2*(2), 481–494. https://doi.org/10.1016/j.matt.2019.11.016

Holcim Foundation Awards. (2015). *Hy-Fi: Zero carbon emissions compostable structure, Sustainable Construction Global Finalist Certificate.*

Holt, G. A., et al. (2012). Fungal mycelium and cotton plant materials in the manufacture of biodegradable molded packaging material: Evaluation study of select blends of cotton byproducts. *Journal of Biobased Materials and Bioenergy, 6*(4), 431–439.

Jeronimidis, G. (2009). Biodynamics: Natural dynamic systems, material behaviour and adaptation in architecture and engineering. *Architectural Design, 74*(3), 90–95.

Joachim, M., & Silver, M. (2017). *XXL-XS : new directions in ecological design.* Actar D.

Jonkers, H. (2011) *Bacteria-based self-healing concrete.*

Jonkers, H. M. (2007) *Self healing concrete: A biological approach.* Springer, Dordrecht, pp. 195–204. https://doi.org/10.1007/978-1-4020-6250-6_9.

Krieg, O. D., & Menges, A. (2013). 'HygroSkin: A climate-responsive prototype project based on the elastic and hygroscopic properties of wood'. In: *Adaptive Architecture Proceedings of the 33rd Annual Conference of the Association for Computer Aided Design in Architecture (ACADIA).* Cambridge, Ontario, Canada, pp. 23–260.

Leboucq, P., et al. (2019) *Material atlas: The Growing Pavilion.* Amsterdam.

Lohmann, J. (2017). *The department of seeweed.*

Matzeu, G., et al. (2020). Large-scale patterning of reactive surfaces for wearable and environmentally deployable sensors. *Advanced Materials, 32*(28).

Mcsweeney, J., et al. (2015). Indoor nature exposure (INE): A health-promotion framework. *Health Promotion International, 30*(1), 126–139. https://doi.org/10.1093/heapro/dau081

Menges, A., & Reichert, S. (2015). Performative wood: Physically porgramming the responsive architecture of the HygroScope and HygroSkin Projects. *Architectural Design-Special Issue Material Synthesis*, 66–73.

Mogas-Soldevila, L. (2021). Material speculation: A science and craft. In: Jamelle, H. (ed.) *Under pressure: Essays on urban housing*. Routledge Taylor & Francis, 344p.

Mogas, L. (2018). Formalizing infinite properties: Beyond functionalism in design. *ACTAR urbanNEXT*, (Designing Matter).

Mogas, L., et al. (2021) Additively manufactured leather-like silk protein materials. *Materials & Design*, in review.

Mogas, L., Matzeu, G., & Omenetto, F. (2018) *Lachesis: Drawing the fabric of life.* Vimeo, USA. https://vimeo.com/304908228.

Munro, A., & Munro, G. (2006) *Full grown: Intersecting the boundaries between Nature and Art, Science and Horticulture, Manufacture and Design., Company Website.*

Niemeyer, C. M. (2001). Nanoparticles, proteins, and nucleic acids: Biotechnology meets materials science. *Angewandte Chemie International Edition, 40*(22), 4128–4158. https://doi.org/10.1002/1521-3773(20011119)40:22%3c4128::AID-ANIE4128%3e3.0.CO;2-S

Pawlyn, M. (2019). *Biomimicry in architecture* (2nd ed.). RIBA Publishing. https://doi.org/10.4324/9780429346774

Perini, K., et al. (2020). Experiencing innovative biomaterials for buildings: Potentialities of mosses. *Building and Environment, 172*, 106708. https://doi.org/10.1016/j.buildenv.2020.106708

Plank, J. (2005). Applications of biopolymers in construction engineering. In Steinbüchel, A. (ed.) *Biopolymers online*. Wiley. https://doi.org/10.1002/3527600035.bpola002.
Ratner, B. D. (Buddy, D. (2013) *Biomaterials science: An introduction to materials in medicine*. Academic Press.
Ravishankar, S., & Ji, S. (2021). Bio-architecture of living root bridges. in. Springer, Singapore, pp 881–887. https://doi.org/10.1007/978-981-16-0041-8_72.
Romano, R., et al. (2018). What is an adaptive façade? Analysis of recent terms and definitions from an international perspective. *Journal of Facade Design and Engineering, 6*(3), 65–76. https://doi.org/10.7480/JFDE.2018.3.2478
Roy, R., et al. (2020). Encapsulation techniques and test methods of evaluating the bacteria-based self-healing efficiency of concrete: A literature review. *Nordic Concrete Research, 62*(1). https://doi.org/10.2478/ncr-2020-0006.
Sandak, A., et al. (2019) *Bio-based building skin*. Springer Singapore (Environmental Footprints and Eco-design of Products and Processes), Singapore. https://doi.org/10.1007/978-981-13-3747-5.
Shtein, Z., & Shoseyov, O. (2017). When bottom-up meets top-down. *Proceedings of the National Academy of Sciences of the United States of America, 114*(3), 428–429. https://doi.org/10.1073/pnas.1619392114.
Srubar, W. V. (2021). Engineered living materials: Taxonomies and emerging trends. *Trends in Biotechnology, 39*(6), 574–583. https://doi.org/10.1016/J.TIBTECH.2020.10.009
Tai, Y. J., et al. (2018) Designing (for) decay: Parametric material distribution for controlled dissociation of water-based biopolymer composites. In: *Proceedings of the IASS Symposium 2018*.
The Living NY (2014) *Hy-Fi*.
Toivola, T. (2020). *A six-meter wide seaweed installation takes over Finnish shopping centre*. Aalto University, Alato University News.
Valenti, A., & Pasquero, C. (2021). 'The second life of micro-organisms: Bio-digital design for a new ecology of space and behavior. *AGATHÓN | International Journal of Architecture Art and Design, 9*, 42–53. https://doi.org/10.19229/2464-9309/942021
VenhoevenCS architecture+urbanism. (2006). *Sportplaza Mercator Building in Amsterdam (NL)*. https://venhoevencs.nl/projects/sportplaza-mercator/.
Vincent, J. (2012). *Structural biomaterials*. Princeton University Press. https://doi.org/10.1515/9781400842780/html.
Wegst, U. G. K., & Ashby, M. F. (2007). The mechanical efficiency of natural materials. *Philosophical Magazine*. https://doi.org/10.1080/14786430410001680935
Yadav, P., et al. (2015). Biomedical biopolymers, their origin and evolution in biomedical sciences: A systematic review. *Journal of Clinical and Diagnostic Research: JCDR, 9*(9), ZE21–5. https://doi.org/10.7860/JCDR/2015/13907.6565.

Chapter 3
Dynamic Electro-, Mechanochromic Materials and Structures for Multifunctional Smart Windows

Yao Zhao, Yanbin Li, and Jie Yin

Abstract As one of the key elements in building envelopes, smart windows that can adaptively block and transmit sunlight for energy saving are promising to construct energy-efficient buildings and reduce the greenhouse gas emissions. Most smart windows are based on electro-, thermo-, mechano-, and photochromics for single-purpose, passive energy saving. Considering the large surface area of windows in modern buildings, recent research advances witness the demand of embodying multifunctionality in smart windows for integrating additional beneficial functions, including energy storage and self-powering, self-cleaning, and even water harvesting to tackle the water scarcity challenge. These multifunctionality requires redesigning conventional smart windows in terms of new materials selection and synthesis, device fabrication, surface features, and structural designs. In this chapter, we will briefly discuss the recent advances in multifunctional smart windows, including harnessing electrochromism and thermochromism for energy storage and generation alongside integrated perovskite solar cells, surface wetting for self-cleaning, and combined wetting and optical properties in novel wrinkling and kirigami structures for water harvesting. The involved materials synthesis, device fabrication, characterization, mechanisms, and multifunctional performances are discussed. We hope it can provide constructive insights for designing next-generation multifunctional smart windows to make buildings more energy efficient and environmentally sustainable.

Keywords Multifunctional smart windows · Energy storage · Self-powering · Self-cleaning · Water harvesting · Wrinkling · Kirigami

---

Y. Zhao · Y. Li · J. Yin (✉)
Department of Mechanical and Aerospace Engineering, North Carolina State University, Raleigh, NC 27695, USA
e-mail: jyin8@ncsu.edu

Y. Zhao
e-mail: yzhao59@ncsu.edu

Y. Li
e-mail: yli255@ncsu.edu

J. Wang et al. (eds.), *Advanced Materials in Smart Building Skins for Sustainability*,
https://doi.org/10.1007/978-3-031-09695-2_3

3.1 Introduction

To date, buildings account for 30% of the global energy consumption (Isaac & van Vuuren, 2009; Pérez-Lombard et al., 2008; Vieira et al., 2019) To meet the 2050 Carbon Neutrality mission (Guterres, 2020; EU, 2020; DOE, 2021) advocated by United Nations, energy-efficient buildings are highly demanded to save energy and reduce the greenhouse gas emissions (Piccolo & Simone, 2015). Smart windows, as one of the key elements in building envelopes, can modulate the light transmittance into the buildings for reducing the electrical, cooling or heating energy consumptions for energy saving (Bechinger et al., 1996; Granqvist, 2000; Casini, 2018; Wang et al., 2019b). They can block the sunlight with low transmittance in hot and shiny days to maintain cool indoor environment and keep warm with high transmittance during cold days. By tuning solar irradiation and/or radiative cooling, it can effectively reduce the burden of heating, ventilation and air conditioning (HVAC) that account for about 50% of building energy consumption (Ke et al., 2018; Pérez-Lombard et al., 2008). Thus, smart window is one of the ideal candidates to reduce the building energy consumption (Dussault et al., 2012; Parkin & Manning, 2006; Wang et al., 2016b; Zhang et al., 2020).

Current state-of-the-art smart windows can be generally classified into four categories in terms of the working mechanisms for modulating the optical transmittance: thermochromic (Aburas et al., 2019; Kamalisarvestani et al., 2013), electrochromic (Azens & Granqvist, 2003; Macêdo et al., 1992), photochromic (Tällberg et al., 2019), and mechanochromic (Ge et al., 2015; Lin et al., 2017). Correspondingly, their optical transmittance can be tuned by means of temperature (Aburas et al., 2019; Kamalisarvestani et al., 2013), electric potentials (Azens & Granqvist, 2003; Macêdo et al., 1992), light intensity or light wavelengths (Chun et al., 2021; Tällberg et al., 2019; Wang et al., 2019a), and mechanical deformations (Ge et al., 2015; Lin et al., 2017). However, these smart windows are often designed for the single functionality of blocking or letting through sun light for passive energy saving.

Considering the large surface area of windows or façade in modern buildings, how to integrate large-area surface structures and novel materials for embodying multifunctionality into smart windows has recently attracted growing research interest. In addition to the energy saving, some representative strategies for achieving multifunctional performances include energy harvesting and energy storage, electricity generation for self-powering, surface wettability for self-cleaning, and water harvesting from environments for addressing the water scarcity issue in draught areas. In this chapter, we will focus on energy-efficient smart windows with such novel multifunctionalities, including energy storage (Sect. 3.2.1), electricity generation (Sect. 3.2.2), self-cleaning (Sect. 3.2.3), and water harvesting (Sect. 3.2.4). Materials synthesis, device fabrication, and some key features of these smart windows will be discussed based on several representative works. Lastly, perspectives for future developments are discussed in Sect. 3.3.

3.2 Multifunctional Smart Windows

3.2.1 Combined Energy Saving and Energy Storage

In addition to tuning optical transmittance for better indoor thermal management, smart windows are also desired to be energy-saving effectively (Baetens et al., 2010; Cai et al., 2016; Casini, 2018; Cao et al., 2019). Recently, electrochromic smart window that can also store energy is becoming an attractive topic. Electrochromic smart window is based on the electrochromism to change the materials' optical properties under a small electric potential (Rauh, 1999). Generally, electrochromic materials with large transmittance modulation for vis- and near-IR light are preferred (Kim et al., 2015a; Yang et al., 2012). Similar to batteries or capacitors, electric energy can also be stored or released by charging or discharging the electrochromic materials. The energy can be stored through pseudocapacitance, i.e., the reversible faradic reactions at vicinities of electrodes (Conway et al., 1997; Fleischmann et al., 2020; Wang et al., 2012). Coloration efficiency (CE) is one of the most important parameters to evaluate the electrochromic smart windows. CE is defined as the optical density change (ΔOD) enabled by per unit charge (ΔQ) transmitting into or extracted from the electrochromic smart windows (Conway et al., 1997; Fleischmann et al., 2020; Wang et al., 2012). Traditionally, lower charge density materials are preferred for higher CE. However, for energy-storage smart windows, large charge density materials should be selected given their large energy storage capacity (Cao et al., 2019). In the following, some representative works about energy-storage electrochromic smart windows will be discussed from the aspects of fabrication methods, mechanisms, performances, and potential advantages.

WO_x is one of the mainstream electrochromic materials that exhibit excellent electrochromic performances (Arakaki et al., 1995; Yu et al., 2019). Yang et al. (2014) utilized WO_3 for fabricating large-scale electrochromic smart windows. Firstly, the WO_3 film was composed by growing WO_3 nanosized thin layer on FTO glasses with physical vapor deposition method. A high current (100 A) was applied on the raw material WO_3 powder, and a low pressure (4×10^{-4} Pa) was maintained for the reaction chamber. Then two electrodes were sealed into a cell with H_2SO_4 electrolyte. Figure 3.1a shows the SEM image of the synthesized WO_3 on the FTO glass. Figure 3.1b demonstrates the thin thickness from the cross-section view. Figure 3.1c is the AFM image of the film. Figure 3.1d shows the color change with the applied voltage. It was found that the transmittance change could be 63.7% for 633 nm incident light. However, Cao et al. (2019) found that WO_x exhibits low charge density and poor energy storage performance. By using Ta-doped TiO_2 nanocrystals, they fabricated a new smart window with charge storage density of at least two times higher than WO_x. During fabrication, the colloidal Ta-doped TiO_2 nanocrystals were composed with a one-pot method. Figure 3.1e shows the TEM image of the as-synthesized Ta-doped TiO_2 nanocrystals. Then the working electrode was prepared. The TiO_2 hexane solution was spin-coated on an ITO glass, followed by annealing at

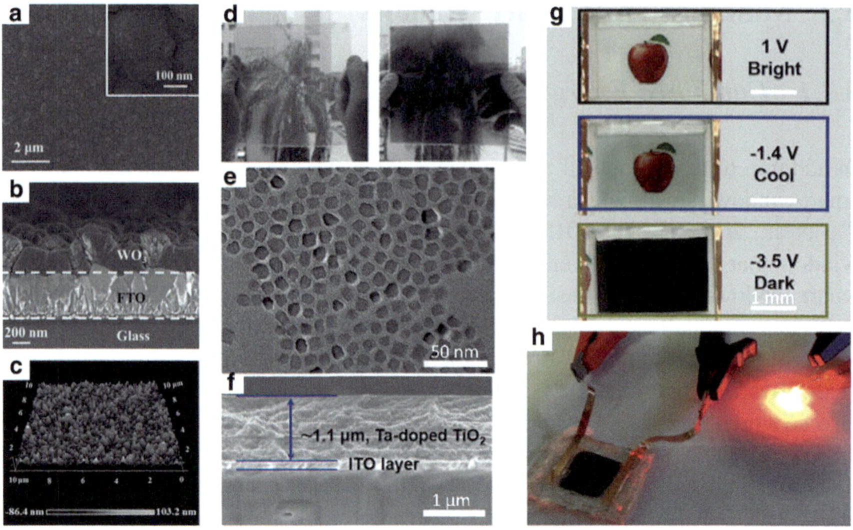

Fig. 3.1 The SEM images of **a** the synthesized WO_3 film (inset: the magnified SEM image) and **b** the cross-section of the layers of the film. **c** The AFM image of the WO_3 thin film. **d** The outdoor demonstration of the electrochromic behavior of the smart window. Reproduced with permission (Yang et al., 2014). Copyright by Wiley-VCH. **e** The TEM images of the as-prepared Ta-doped TiO_2 nanoparticles. **f** The SEM image of the cross-section of the Ta-doped TiO_2 film. **g** The digital images of the smart window at different voltages. **h** The smart window is powering a red LED. Reproduced with permission (Cao et al., 2019). Copyright by Cell Press

400 °C. Figure 3.1f shows the cross-section of the working electrode. Next, the NiO-based counter electrode was made by spin-coating a precursor solution containing $Ni(CH_3COO)_2{\cdot}4H_2O$, CH_3COOLi, and $Ti(OC_3H_7)_4$ in ethanol on an ITO glass, followed by annealing at 350 °C. Last, the working electrode and counter electrode were integrated into a cell with 0.5 mm spacing, and the electrolyte $LiClO_4$ was injected inside. This smart window can gradually change the color from transparent to dark under different voltages (see Fig. 3.1g, h). The outstanding energy storage of this smart window is demonstrated by easily lighting up a LED device.

Apart from the high performances, the fabrication technologies of smart windows are also important. Cai et al. (2017) fabricated large-area multifunctional smart windows using the inkjet printing method. Such a method has several advantages such as low cost, high-efficiency materials usage, and suitability for large-scale manufacturing (Corzo et al., 2019; Sundriyal & Bhattacharya, 2018; Xu et al., 2015). In their work, two CeO_2/TiO_2 and WO_3/PEDOT:PSS [poly(3,4-ethylenedioxythiophene)–poly(styrene sulfonate)] inks were developed. Both inks used the mixture of DI water, ethylene glycol, and diethylene glycol n-butyl ether as the solvent. Then the inks were printed on FTO-coated glasses, and dried at 60 °C for 2 h. The CeO_2/TiO_2 film needed additional 450 °C annealing for better adhesion. Finally, the smart window was fabricated by assembling the two films into cells with 2 mm spacing using H_2SO_4 as the electrolyte. Figure 3.2a–b show the SEM images of the printed CeO_2/TiO_2

and WO_3/PEDOT:PSS films. Figure 3.2c–d are the TEM images of the nanoparticles of the CeO_2/TiO_2 and WO_3/PEDOT:PSS. The reversible redox reactions on the electrodes are:

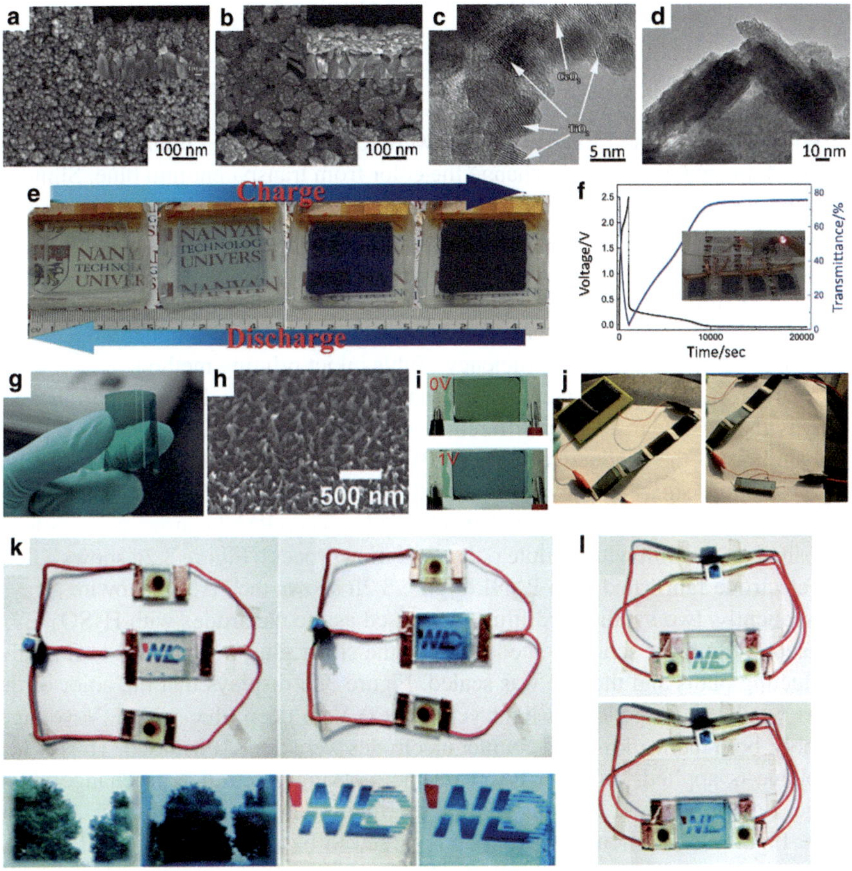

Fig. 3.2 The SEM images of printed **a** CeO_2/TiO_2 film and **b** WO_3/PEDOT:PSS film (insets: the magnified images). The TEM images of prepared **c** CeO_2/TiO_2 nanoparticles and WO_3/PEDOT:PSS nanoparticles. **e** The color change of the charging and discharging processes of the smart window. **f** The normal galvanostatic charging and slow discharging of the smart window and the corresponding optical responses (inset: a red LED can be lighted up with four devices connected in series). Reproduced with permission (Cai et al., 2017). Copyright by Wiley-VCH. **g** The digital image of a single electrode. **h** The morphology of the PANI nanowire array. **i** The images of the smart window at different voltages. **j** The smart windows are charging with a solar cell (left) and powering an LCD (right). Reproduced with permission (Wang et al., 2012). Copyright by Royal Society of Chemistry. The colored and bleached states of the Type I **k** and Type II **l** smart windows. Reproduced with permission (Xie et al., 2014). Copyright by Royal Society of Chemistry

$$WO_3 + xH^+ + xe^- \leftrightarrow H_xWO_3$$

$$PEDOT^+ \cdot PSS^- + H^+ + e^- \leftrightarrow PEDOT^0 \cdot PSS^- \cdot H^+$$

$$4CeO_2 + 4H^+ + 4e^- \leftrightarrow 2Ce_2O_3 + 2H_2O$$

$$TiO_2 + xH^+ + xe^- \leftrightarrow TiOOH$$

When the WO_3 and $PEDOT^+ \cdot PSS^-$ were reduced and Ce_2O_3 and TiOOH were oxidized, the materials would change the color from transparent into blue. Simultaneously, electric energy could be stored. Figure 3.2e demonstrates the color transition of the smart window during charging and discharging processes. The color of the device becomes darker from almost colorless with the applied electric potential. Figure 3.2f further displays the optical transmittance change for 633 nm incidental light. Notably, four devices connected in series can light up a commercial LED for two hours, demonstrating the efficiency of this inkjet printing method.

Apart from the optical performance and fabrication, excellent structural flexibility is also desired to adapt to different working environments. Using organic PANI (polyaniline) as the conductive material, Wang et al. (2012) developed a new flexible energy-storage smart window. In fabrication, a PEDOT:PSS in DMSO (dimethyl sulfoxide) solution was spin-coated on a PET film. Then PANI nanowire arrays were deposited on the film with a dilute polymerization process. Figure 3.2g shows a flexible electrode fabricated with PANI. Figure 3.2h shows the PANI nanowire arrays. Subsequently, two synthesized films were used as the electrodes with H_2SO_4-PVA gel electrolyte layer scrape-coated. Finally, the silver glue was applied to form the conducting paths and the cell was sealed. Figure 3.2i displays that the color of the smart window changed from light yellow at 0 V to deep blue with 1 V voltage. Notably, both the working and counter electrodes were the PANI-based. Thus, when a voltage is applied, only the PANI on the working electrode can change color. Large flexibility could be achieved by sacrificing certain transmittance modulations. Figure 3.2j demonstrates that commercial Si-based solar cells could be used to charge the smart window. Similarly, Xie et al. (2014) also integrated solar cells to power smart windows with self-fabricated dye-sensitized solar cell. The WO_3 electrochromic film was fabricated by electrochemically depositing on ITO glass with tungstic acid. Pt was deposited on ITO glasses as counter electrode. Then the smart window cells were fabricated by stacking and sealing the two electrodes with H_2SO_4-PVA gel as electrolyte. For solar cells, the TiO_2 electrode was manufactured by screen printing with commercial P25 TiO_2. After annealing at 500 °C and sensitized with ethanol solution, the TiO_2 electrode was integrated with Pt electrode into dye-sensitized solar cell using DMPII, LiI, I_2 and 4-TBP in methoxy propionitrile as electrolyte. Figure 3.2k–l show two designing strategies. For Type I (Fig. 3.2k), the smart window was controlled by a single-pole-double-throw switch to selectively change between coloration and bleaching states. For Type II (Fig. 3.21), the colorless and blue states can be reversibly changed under strong sunlight with a double-pole double-throw switch.

In addition to the electrochromic effect, new materials other than electrochromic materials can also be used for energy-storage smart windows. For example, Zhou et al. (2020) proposed a hydrogel-based smart window. The used hydrogel is a poly (N-isopropylacrylamide) (PNIPAm)-based network, which could reversibly change the transmittance at different temperatures. The mechanism was believed to be a hydrophilic to hydrophobic transition at lower critical solution temperature (LCST). The hydrogel can become transparent (high solar transmission) when the environmental temperature is lower than LCST and vice versa (Matsumoto et al., 2018; Wu et al., 2020). Moreover, the hydrogel can contain a large amount of water inside, showing outstanding thermal energy storage (~250 kJ kg^{-1}) and large specific heat capacity (4.2 kJ kg^{-1} K^{-1}). The hydrogel-based smart window can be manufactured by two steps. First, the hydrogel was synthesized with mixing NIPAm (N-isopropylacrylamide) monomer, the crosslinker acrylamide, the catalyst TEMED (N,N,N,N-tetramethylethylenediamine), and the initiator APS (ammonium peroxydisulfate). Second, the device can be assembled by pouring the hydrogel and DI water into a glass box and sealing. Figure 3.3 shows the samples at the meter scale. The window was transparent at lower temperatures and gradually changed into opaque at higher temperatures. Notably, this hydrogel-based smart window directly stored and released thermal energy instead of electric energy, which makes it tunable only by environmental temperature rather than electricity.

To conclude, smart windows can modulate indoor temperature by turning transmittance with the electrochromic effect. Specific materials such as WO_x and PANI could be used under different requirements. In addition to the color change under applied voltages, energy can also be stored on the electrodes through the electrochemical reactions, which enables more energy saving. Apart from the electrochromic

Fig. 3.3 The transmittance changes at different time of a day. The window is half filled with the hydrogel-based smart window. Reproduced with permission. (Zhou et al., 2020) Copyright by Cell Press

smart windows, other smart windows, such as hydrogel-based ones, also provide more choices. Instead of storing electricity, thermal energy from the environment is stored. This may provide some important insights since they do not need extra electricity input.

3.2.2 Combined Energy Saving and Self-powering

The electrochromic smart window is a good choice for energy-efficient buildings given its tunable transmittance by optical modulation and energy-storage capability. However, they usually need external power for charging and cannot intrinsically generate electricity. Here, in this section, we will discuss the strategies to construct self-powered smart windows.

The first method is by integrating a galvanic cell with the smart window. Wang et al. (2014) designed a self-rechargeable electrochromic smart window by integrating with aluminum electrodes. The electrochromic iron-based compound materials can be charged when the aluminum is oxidized. The benefit for the iron-based compound is that iron (II, III) hexacyanoferrate (II, III) (Prussian blue, PB) can be reduced to colorless Prussian white (PW), and the Prussian white can spontaneously oxidized with dissolved oxygen. The reactions are denoted as:

$$Al \leftrightarrow Al^{3+} + 3e^{-}$$

$$\mathrm{KFe}^{III}\left[\mathrm{Fe}^{II}(\mathrm{CN})_6\right](\mathrm{PB}) + \mathrm{K}^{+} + e^{-} \leftrightarrow \mathrm{K_2Fe}^{II}\left[\mathrm{Fe}^{II}(\mathrm{CN})_6\right](\mathrm{PW})$$

During fabrication, the PB films were first prepared by the electrochemical deposition method. Specifically, ITO glasses were coated in the mixture of K_3FeCN_6, $FeCl_3$, and KCl with current density 50 mA/cm^2. Then the cell was assembled by sealing the PB film and an ITO glass with an aluminum strip on one edge together and KCl as the electrolyte. Figure 3.4a demonstrates that the smart window can change the color with self-powering battery-like capabilities, as demonstrated by the LED lighting-up in Fig. 3.4a (i and ii) when two electrodes were connected at colored state. After changing into bleached state, the LED cannot be lighted (Fig. 3.4a, iii and iv). The cell can go back to the colored state when the electrodes were disconnected. Then the LED can be lighted again if the electrodes were connected at the colored state (Fig. 3.4a, v and vi). About 38.5% transmittance change could be achieved by this self-powered smart window.

Moreover, thermochromic smart window can tune optical transmittance by modulating the incident sunlight. Zhou et al. (2013) use the common VO_2 material (Cui et al., 2018, Chen et al., 2011, Li et al., 2017; Chen et al. 2019; Madida et al., 2014) and designed a thermochromic smart window that can generate electricity. The mechanism for the VO_2 thermochromic effect is attributed to the reversible metal-semiconductor phase transition (Chen et al., 2011; Cui et al., 2018; Golubev

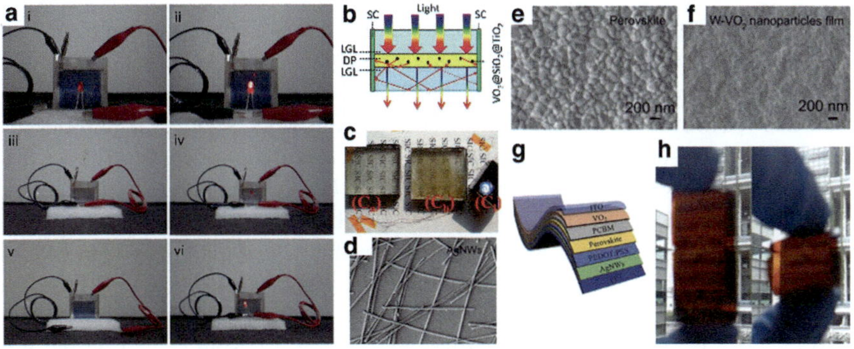

Fig. 3.4 **a** The performances of the smart window. (i) The two electrodes are disconnected at colored state. (ii) An LED is lighted up with electrodes connected. (iii) The bleached states with two electrodes connected. (iv) The LED is not powered with electrodes connected at bleached state. (v) The smart window is recovered with electrodes disconnected. (vi) The LED is lighted up. Reproduced with permission (Wang et al., 2014). Copyright by Nature Publishing Group. **b** The design strategy of the smart window. SC, LGL and DP denote the solar cell, light guider layer and low reflective index medium, respectively. **c** The smart window can power a 1.5 V LED. Reproduced with permission (Zhou et al., 2013). Copyright by Nature Publishing Group. The SEM images of **d** the AgNW arrays, **e** the perovskite film, and **f** the VO_2 film. **g** The schematic of the layers of the smart window. **h** The digital images of the perovskite/W-VO_2 film with different bending curvatures. Reproduced with permission (Meng et al., 2022). Copyright by Elsevier

et al., 2001). At a lower temperature, the VO_2 has a monoclinic crystalline structure and is transparent to the near-infrared light. At higher temperatures, the VO_2 turns into metallic state with a tetragonal crystal structure and reflects the near-infrared light (Miller & Wang, 2015; Wang et al., 2016a, 2018). The transition temperature is 68 °C (Hong-Chen et al., 2005; Wang et al., 2005, 2017). Therefore, the VO_2-based smart windows can automatically tune the indoor temperature by blocking/passing the incident light. To better use the scatted light, Zhou et al. (Zhou et al., 2013) assembled solar cells on both sides of the smart windows (Fig. 3.4b). The smart window consisted of light guider layers (LGL), low reflective index medium (DP), and solar cells on both sides. Thus, scattered lights can be picked by the solar cells without affecting the transmittance. To achieve better scattering effects and lower absorption by VO_2, a core-shell-shell $VO_2/SiO_2/TiO_2$ was synthesized. First, the core was synthesized by annealing V_2O_5 and oxalic acid at 240 °C. Then, the VO_2/SiO_2 was prepared with TEOS hydrolysis method. Third, the as-synthesized VO_2/SiO_2 were dispersed in ethanol, followed by adding tetrabutyl titanate and $NH_3(H_2O)$ for forming the $VO_2/SiO_2/TiO_2$. Next, the $VO_2/SiO_2/TiO_2$/PU composite film was prepared by mixing the particles, saline coupler KH-570 and polyurethane in DI water, and then casted on a PC plate. Finally, the smart window was assembled with the strategy shown in Fig. 3.4b. Thus, solar light could be scattered by this smart window and LED was lighted up at the same time (Fig. 3.4c).

In addition to embodying smart windows with energy-saving or energy-generating function, other solar devices can also be integrated for smart window functions. Meng et al. (2022) integrated a perovskite solar cell that can both generate electricity and tune transmittance. Perovskite solar cell is a newly emerging solar cell with a high power conversion efficiency (Ansari et al., 2018; Chen et al., 2018; Li et al., 2018; Mei et al., 2014; Yin et al., 2014; Zhang et al., 2021). In this work, Meng et al. introduced a layer of Tungsten-doped VO_2 (W-VO_2) nanoparticles for thermochromic effect. The perovskite photovoltaic materials used was $Cs_{0.05}FA_{0.85}MA_{0.10}Pb\ (I_{0.97}Br_{0.03})_3$ and the electron transport layer was [6,6]-Phenyl-C_{60}-butyric acid methyl ester (PCBM) for better energy-level matching. The device was fabricated mainly by spin-coating method. First, a silver nanowire (AgNW) network was deposited on a PET film. The SEM image is shown in Fig. 3.4d. Then a layer of PEDOT:PSS was spin-coated onto the AgNW. Subsequently, a perovskite solution was spin-coated on the film with N_2 protection. The perovskite solution was synthesized by mixing CsI, MABr, PbI_2, FAI, and MACl in DMF/DMSO solution. The SEM image of the perovskite layer is shown in Fig. 3.4e. After annealing at 100 °C, the PCBM chlorobenzene solution was spin-coated, and then annealed at 90 °C. Next, BCP saturated solution was spin-coated. The last spin-coated layer was a Si-Al gel containing W-VO_2 and ethanol. Figure 3.4f displays the morphology of the W-VO_2 layer. After all the spin-coated layers were finished, a nanosized ITO layer was deposited on the film with magnetron sputtering method. The overall stacking strategy was illustrated in Fig. 3.4g. Its power conversion efficiency is about 16.1%. Since the window was fabricated with PET substrates, it also exhibits certain flexibility. Figure 3.4h shows the fabricated perovskite/VO_2 film at different bending curvature. The solar cell performance did not compromise much at a bending radius of 5 mm. Nonetheless, the transmittance of this window was not high, showing an average about 25.5% visible transmittance.

With the higher energy-saving requirements for green buildings, it is beneficial to design smart windows that are capable of both saving energy and self-powering or outputting power to other appliances. It is ideal if the unwanted incident sunshine can be utilized to generate electricity. Although the performances of this smart windows are not perfect due to the comprised transmittance when integrated with solar cells, these issues could be solved by future advancements of materials and fabrication technologies.

3.2.3 Combined Energy Saving and Self-cleaning

In addition to the above functionalities of storing and generating energy, smart windows can also be designed with reversible light transmittance and tunable wetting properties. Smart windows with tunable optical properties play a significant role in buildings given their potentials in modulating solar irradiation to tune buildings internal thermal conditions (Granqvist, 2016, Llordés et al., 2013; Cai et al., 2016; Cui et al., 2018; Ke et al., 2018; Khandelwal et al., 2017). Moreover, smart windows with tunable surface wettability that can achieve self-cleaning capability can provide many other applications, such as for architectural/vehicle windows with better vision, microfluidic devices, and lab equipment surficial pollutant removement. In this section, multifunctional smart windows with switchable light transmittance and wetting tunability are discussed.

To date, reversible optical switching has been accomplished either by molecular arrangements change (Cupelli et al., 2009; Guo et al., 2017; Kim et al., 2015b), suspend particles (Vergaz et al., 2008), or by the oxidation-reduction reaction of chromogenic materials (Granqvist, 2014, 2000; Bechinger et al., 1996) stimulated by external light (Wu et al., 2017), or electrical (Dyer et al., 2007; Gesheva et al., 2012), thermal (Zheng et al., 2015), and chemical (Zhang et al., 2017a) factors. However, these smart windows are either chemically unstable during optical switch or difficult to prepare (Lampert, 2004). Thus, a rather simple and high efficient method based on mechanical wrinkling is proposed to fabricate multifunctional smart windows with integrated fast transparency switch and tunable surficial wettability. Surface wrinkling for smart window relies on the dynamical and reversible surface morphology manipulation. Bowden et al. (1998) first reported the fabrication of gold nanoscale surface patterning through wrinkling of gold nanofilms on soft poly (dimethylsiloxane) (PDMS) substrate. After coated on PDMS substrate at an elevated temperature, the gold nanofilm could be compressed due to the larger thermal expansion coefficient in PDMS substrates during cooling. When temperature goes beyond a critical value, spontaneous ordered and disordered wrinkling surface patterns occur at small scales. Thereafter, surface wrinkling with both flat surface and multileveled micro-structures have been extensively studied for dynamic scaffolds (Kim et al., 2010), tunable diffraction gratings (Harrison et al., 2004), microlens arrays (Chan & Crosby, 2006), and flexible electronic devices (Chandra et al., 2007), as well as multifunctional smart windows with transparency/wettability tunability (Jiang et al., 2018; Ke et al., 2019a; Khang et al., 2006; Kim et al., 2013, 2018; Lee et al., 2010; Li et al., 2021b; Lin et al., 2018; Tomholt et al., 2020; Zhang et al., 2017b).

The wrinkled surface morphology characterized by the wrinkle periodicity and height is mainly determined by the applied mechanical strain. Thus, it exhibits high controllability and dynamical tunability, making it promising for constructing mechanochromic (MC) smart windows. Here, we discuss several representative MC smart windows with different patterned (micro/nano-pillar (Lee et al., 2010), anisotropy (Kim et al., 2013), self-similar (Lin et al., 2018), additional delamination buckling (Zhang et al., 2017b) and cracks-based (Tomholt et al., 2020) surface morphologies.

Figure 3.5a (i) illustrates the procedures to pattern structured PDMS films with surface wrinkles. AAO (anodic aluminum oxide) templates with nanopore were first fabricated by a two-step anodization. The template surface was then treated with octadecyltrichlorosilane for easy releasing and preparation of multiple replicated PDMS structures. Uncured PDMS was poured over the patterned AAO template and cured. The PDMS film was then peeled off from the template. Next, the nanopillar structured PDMS film was uniaxially stretched and exposed to ultraviolet-ozone (UVO) radiation. (Heptadecafl uoro-1,1,2,2-tetrahydrodecyl) trichlorosilane is coated on the stretched PDMS film to achieve the enhanced tunable wettability. Finally, periodic wavy microstructures were generated on the PDMS substrate upon

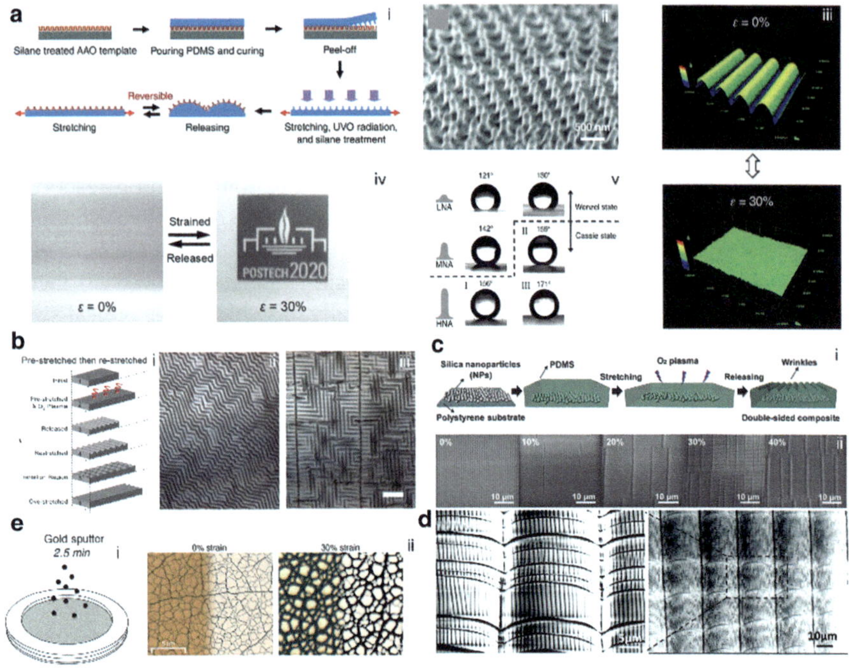

Fig. 3.5 Different fabrication strategies of MC smart windows. **a–d**, wrinkle based with nanopillar microstructure **a** (Lee et al., 2010), surficial anisotropy **b** (Kim et al., 2013), multiple states **c** (Kim et al., 2018) and self-similarity **d** (Lin et al., 2018). **e**, MC smart window with cracks (Tomholt et al., 2020). Reproduced with permissions. Copyright by Wiley Publishing Group

releasing the pre-stretched strain with preserved nanopillars. The nanopillar patterned wrinkling could be reversibly flattened upon stretching, but the nanopillar pattern remained. Figure 3.5a (ii) shows the SEM image on the features of the micro-wrinkling surface with nanopillars. It forms a very well-ordered and uniform microscopic pillar structures (with height ~300 nm and diameter ~150 nm) with hexagonal non-close-packed arrays on the micro-wrinkles (with 31 µm wavelength and 4.4 µm amplitude) over a large area without collapse. The confocal laser scanning microscopy (CLSM) images (Fig. 3.5a (iii)) demonstrated that the micro-wrinkle pattern morphs under the applied uniaxial mechanical strain. Gradually increasing the applied strain decreases the wrinkle amplitude and finally leaves only the nanopillar structures

Figure 3.5a (iv) shows that reversible opaque to transparent transition upon re-stretching the wrinkled sample, where the optical transmittance increases from 9.2 at 0% applied strain to 92 at 30% strain. The opacity is arising from the extensive scattering of light induced by the periodic micrometer-sized wrinkling surface. The tapered periodic nanopillar arrays produced the gradient effective refractive index that increases from the pillar peaks to the pillar troughs bypassing the pillar bodies. Thus, the entire visible spectrum is minimally reflected. For wettability of this wrinkle structured surface, the contact angle increases when the surface structure becomes rougher (i.e., micro-wrinkle appears or the aspect ratio of nanopillar increases). For strained or released films, the contact angle hysteresis increases with the surface roughness. Water droplets can easily penetrate the valleys between the low-aspect-ratio nanopillars with similar height. Then, a continuous stable solid-liquid-gas (three-phase) contact line would be created with high water adhesion, thus, water droplet could become pinned firmly on the surface without any sliding even though the samples were vertically titled. However, for either strained or released nanopillar arrays with relatively higher aspect ratios, the surface exhibits superhydrophobicity with water contact angle larger than 150° and contact angle hysteresis less than 10°. Water droplet on this surface was in Cassie state and can easily roll away.

In addition to uniaxial mechanical strain, sequential stretching methods have also been studied to generate multiple state and self-similar wrinkling microstructures. Figure 3.5b (Kim et al., 2013) shows that upon stretching the wrinkled PDMS film, three different types of wrinkling morphologies can be generated with changing wavy direction. Moreover, treating soft PDMS substrates with embedded stiff silica nanoparticles (Fig. 3.5c (i); Chandra et al., 2007), the modulus mismatch enables multiple micro-structured wrinkled surfaces, where the micro-wrinkles can change from a uniform to nonuniform form with studs upon increasing the stretching strain. Similarly, by following the two-stage stretching-releasing procedure, micro-wrinkled structures with self-similarity (Lin et al., 2018) could be achieved, which exhibits multistate optical transmittance (from totally bluish to intermediate state with structural color and finally transparency by increasing the stretching strain) and controllable droplet transport behaviors (pinning to rolling away). High strain sensitivity of transmittance with low mechanical strain and crack-free micro-wrinkle pattern can be achieved by applying biaxial mechanical strains (Ke et al., 2019a).

When the bonding between soft substrate and top stiff layer (for example embedded with metal/crystalline nanoparticles) is weak, buckling driven delamination or cracks in the stiff top layer can occur. Buckling delamination forms from the local blisters to larger periodic patterns with the increasing compression strain. Compared to wrinkling, buckling delamination can tolerate relatively large strains (Thomas et al., 2015; Zang et al., 2013, 2017b; Zhang & Yin, 2018), which makes it promising for smart windows with extreme tunable optical properties. Combined wrinkle-crack surface microstructures could enhance surface roughness and thus strengthen the optical scattering effect (Zeng et al., 2016). Figure 3.5d shows that wrinkled PDMS coated with gold-based nanoparticles could generate cracks after air inflation, resulting from the nonuniform strain field along the radial direction. After deflation, cracks would disappear. Thus, the coated gold layer can tune the near-infrared transmission to change the indoor thermal environment, while the combined wrinkles and cracks can tune the visual effect (Tomholt et al., 2020).

To date, most MC smart windows are developed by surface instability while more advancing methods are emerged with promising performances (Ke et al., 2019a). However, for practical commercialized applications, challenges still exist and need to be addressed. For example, the mechanical strains need to be smaller to generate a large transmittance modulation and the mechanical robustness of such smart windows need to be improved for long-time endurance.

3.2.4 Combined Energy Saving and Water Harvesting

Harvesting water from air is promising for solving the water crisis all around the world (Zhai et al., 2006; Fathieh et al. 2018; Lekouch et al., 2011). Water harvesting includes water/fog condensation, droplet transport and collection. The large-area windows in modern buildings become an ideal platform to not only save energy but also harvest and collect water from the environment. The tunable surface wettability of smart windows can control the water droplets' transport for collection and water harvesting functionality. In the following, different smart windows with water droplet movement control are discussed based on several representative works.

Achieving effective repulsion of low-surface-tension liquids on solid substrate is necessary for fine droplets control. Creating liquid-infused surface on porous film-based substrates provides an efficient method. Figure 3.6a illustrates the process to fabricate the temperature-responsive liquid infused porous surfaces (Zhai et al., 2006) for controlling water droplet movement at room temperature. The underlying mechanism is based on the paraffin that can easily change from a solid to liquid state upon temperature changes. Water droplets on this composite surface would be repelled or pinned responding to the environmental temperature. The left figure in Fig. 3.6a shows that the porous surfaces are composed of chitin nanofibers and poly(acrylic acid) and formed through the layer-by-layer self-assembly (see the SEM images in Fig. 3.6a, right), followed by introducing the mixed solidifiable and liquid paraffin into the hydrophobic nanofibrous surfaces as lubricant oil layer. Consequentially, the solidification of the paraffin changes the surface morphology and the solidifiable/liquid paraffin mixing control the droplets' movement. Figure 3.6a (right-bottom) shows that the contact angle of the droplets sharply decreased with the increasing temperature. For example, for a 1: 25 solidifiable/liquid paraffin surface, the droplet's contact angle can decrease rapidly from near 99–53° when increasing the temperature from 22 °C to the paraffin melting point near 28 °C. The sliding angle of the water droplet would decrease with the temperature and switched from an immobilized state to free sliding. This is attributed to the change of surface tension that improves the droplet mobility. At a low temperature (<22 °C), water droplets would be pinned even when the paraffin surface was titled to 90°. At a high temperature, water droplets will be cloaked by the lubricant fluid that is easy to evaporate, thus, the contact angle will increase to propel the droplets and roll away. However, at a low temperature, water droplets will not be cloaked and become pinned by the stable paraffin layer.

Compared to the chemical treatment, mechanical strains are preferred to tune the micro-surface wettability given its simplicity. As demonstrated by Lin et al. (Lin et al., 2017), self-similar hierarchical wrinkles (Fig. 3.6b, left) have been verified to exhibit excellent droplet movement control behavior. As shown in Fig. 3.6b (right),

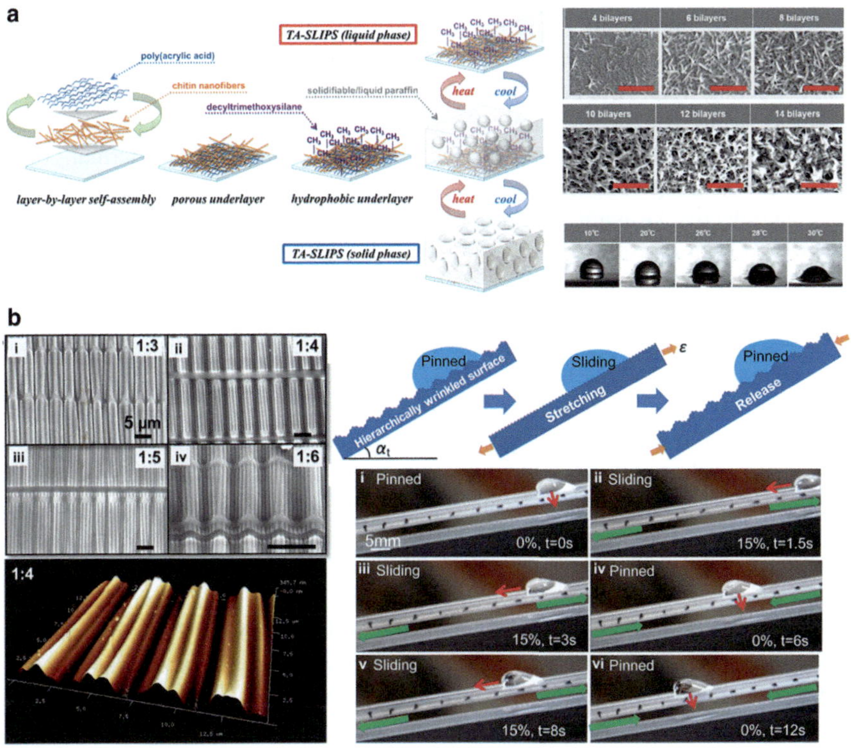

Fig. 3.6 Multifunctional smart windows for water droplet movement control. **a** Liquid-infused surface covered with paraffin. Reproduced with permission (Zhai et al., 2006). Copyright by Royal Society of Chemistry. **b** Self-similar hierarchical wrinkled smart window (Lin et al., 2017). Reproduced with permissions. Copyright by ACS Publications

the water droplets were initially pinned on the micro-wrinkled structure surface. After stretching, the pinned water droplet would slide. Due to the larger surface roughness, the droplet on the wrinkled surface would be in a Wenzel wetting state. When the wrinkled surface is stretched, its surface roughness decreases, which weakens the Wenzel effect and the water droplet loses its balance to slide.

Thus, by chemically or mechanically treating surface with different nano-/microstructures, smart windows could be effectively utilized to better collect water from the environmental air. However, water droplets are also highly dependent of the harvester's macroscopic geometry for high efficient delivery (Bai et al., 2020). Therefore, amounts of water harvesters with distinct geometrical forms have been proposed, such as cactus-like cone arrays (Ju et al., 2012; Lee et al., 2019) and gradient structural components (Xu et al., 2016). However, these water harvesters are challenging to be integrated with smart windows due to their unsuitable structural shapes. Recently, a highly stretchable kirigami structure has been proposed for multifunctional smart window applications that demonstrate efficient energy saving by dynamically blocking (Tang et al., 2017; Yi et al., 2018), modulating (Ke et al., 2019b), energy-transforming (Lamoureux et al., 2015) solar light and as well collecting water (Bai et al., 2020; Li et al., 2021a).

For kirigami structures with parallel slits, uniaxial stretching can simply buckle sectional strips and generate periodic arrays of pores to tune the thermal condition inside buildings. Figure 3.7a shows that Tang et al. (2017), Yi et al. (2018) used the reflective kirigami structure with periodically distributed parallel cuts as an integrated system for designing environmental adaptive building envelopes. When constructing the kirigami structure with temperature-responsive shape memory polymers (Fig. 3.7a, top), the integrated kirigami design could actively open or close its pores in response to environmental temperature change. Both simulation and experimental results show that the generated arrays of pores and adaptively tilted ribbons could collectively diffuse the sunlight more evenly for electricity saving. Ke et al. (2019b) have recently proposed a bi-functional kirigami smart windows by embedding the elastomer with plasmonic vanadium dioxide (VO_2) nanoparticles (Fig. 3.7b, left). It has similar working mechanisms as the VO_2 based thermochromic smart windows discussed in Sect. 3.2.3. The structural geometry change can tune the visible light while the temperature-dependent localized surface plasmon resonance (LSPR) can tune the ultraviolet, visible and near-infrared light regions. With the increase of the temperature, the transmittance in the UV-vis-NIR can reduce from 75% (opened state) to nearly 0% (close state). This work opens a new avenue by integrating both macroscopic structure and composition materials to facilitate enhanced solar energy modulation. Moreover, attributing to the dynamically tunable tilted strips, researchers also designed kirigami-based smart windows that are capable of generating energy by mounting with photovoltaic materials to transform solar right into electricity (Fig. 3.7c, (Lamoureux et al., 2015)).

Additionally, after coating with hydrophobic treatments, kirigami structures can be designed with unique local geometries for water harvesting. Kirigami sheets in triangular cut patterns coated with hydrophobic infused paraffin materials could directionally and continuously collect water droplets from the environment. Recently, by introducing origami folds to kirigami sheet, Li et al. (2021a) invented an aerodynamics-assisted kirigami water harvester that can be scalable to the meter scale. Figure 3.8b (right) shows that stretching the kirigami sheet assisted with folds can transform the sheet into a 3D structure with periodically distributed pyramidal local pop-ups. Thus, the fluid-structure interactions lead to the formation of tube-like vortices penetrating through the opening pores. Near the cortex core, droplets would be attracted and accumulated, finally getting ejected to the collectors with their growing size. Based on the 1 m^2 (Fig. 3.8b, right-bottom) sample, this design was demonstrated with highly feasible and excellent collection rates up to 3.5 kg/h, which is 2.71 times from wire mesh structure. The water harvesting behavior of this design is mainly determined by the macroscopic structure rather than the coating materials as demonstrated in (Li et al., 2021a) after testing large amounts of different hydrophobic materials (Fig. 3.8b (left)). It should be noted that in addition to the kirigami structures, structure-based smart windows can also be extended to other forms such as origami and rigid mechanisms (Ke et al., 2019a; Li et al., 2021b).

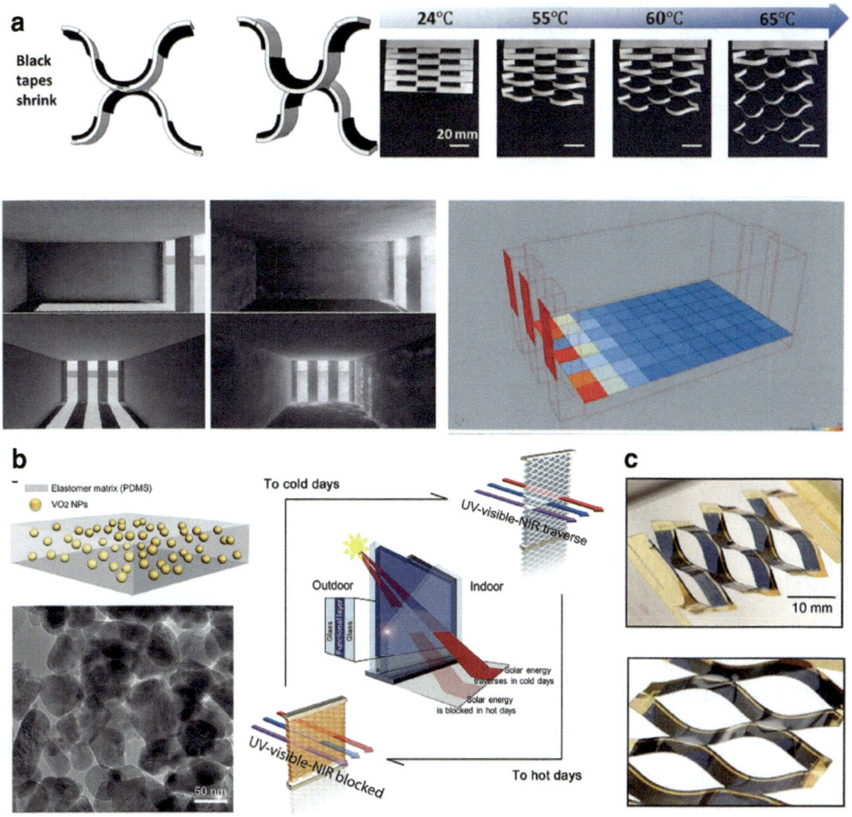

Fig. 3.7 Kirigami based smart window with dynamically tunable optical transmittance for energy saving and generation. **a** Thermally tunable kirigami façade with shape memory polymers (Tang et al., 2017; Yi et al., 2018). Reproduced with permissions. Copyright by Wiley Publishing Group and Elsevier. **b** Bifunctional kirigami based smart window fabricated by VO_2-embedded elastomers (Ke et al., 2019b). Reproduced with permission. Copyright by Cell Press. **c** Dynamic kirigami structure mounted with photovaltic material for energy generation (Lamoureux et al., 2015), Reproduced with permission. Copyright by Nature Publishing Group

3.3 Conclusion and Outlook

This chapter summarizes the advancement in smart windows with novel functionalities, such as energy storage, energy generation, self-cleaning, and water harvesting. We discussed several representative works from the perspectives of working mechanisms, materials synthesis, device fabrication, characterization, and functionality. Specifically, electricity can be stored by electrochemical reactions for the energy-storage electrochromic smart windows; some smart windows can utilize the scattered sunlight to generate electricity for self-powering during modulating the indoor light irrigation; by tuning the hydrophobicity, smart windows can self-clean by washing

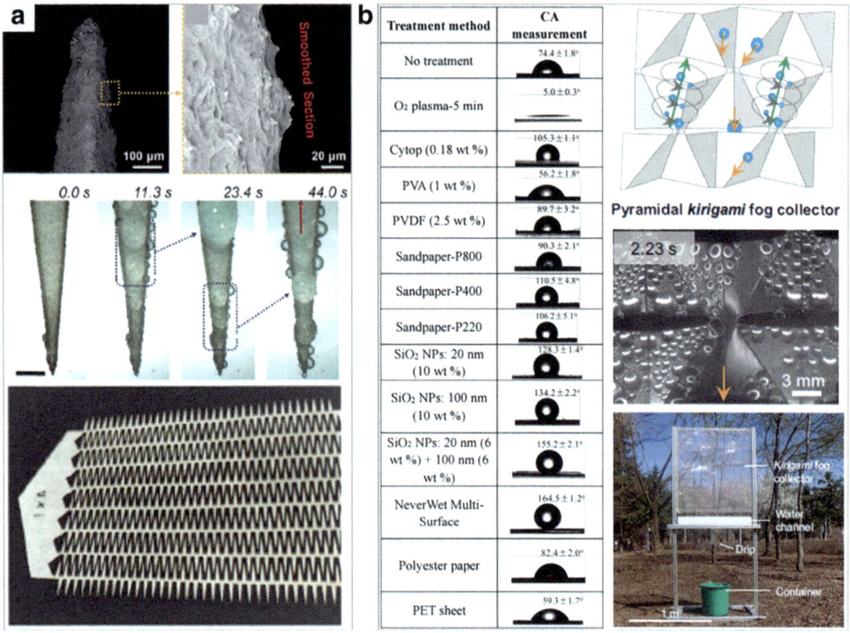

Fig. 3.8 Kirigami-based structures for water harvesting (Li et al., 2021a). Reproduced with permission. Copyright by Nature Publishing Group

away dusts to keep high performance; smart windows with special structures or by turning wettability can also harvest water, which is vital for energy and resource saving in drought areas.

Despite the advantages of the novel functionalities, limitations and challenges still remain for the current state-of-the-art smart windows. First, some smart windows do need energy input. For example, the electrochromic smart windows need charging, and the mechanochromic smart windows require mechanical loads for structure shifting. Second, the light transmittance modulations need improvements. For instance, some functionalities such as energy generation are achieved by sacrificing the light transmittance. The transmittance shifting temperature for VO_2 is about 68 °C, which is too high in terms of the practical use. Third, transmittance manipulation of some smart windows is fully passive.

Therefore, to overcome the issues, future studies are needed to improve the transmittance with lower or even zero energy consumption, harvesting energy from undesired sunlight without reducing the transmittance modulation, and better controllability by using high-performance materials and more novel designs. Furthermore, some new smart windows that harvest heat from the environment instead of sunlight, such as hydrogel-based smart windows, may also be a promising trend. Additionally, although adaptive façades have been developed for energy saving, such as the examples shown in Fig. 3.9, smart windows combined with adaptive façade are still little investigated. In terms of scalability, not only the performances of smart windows are

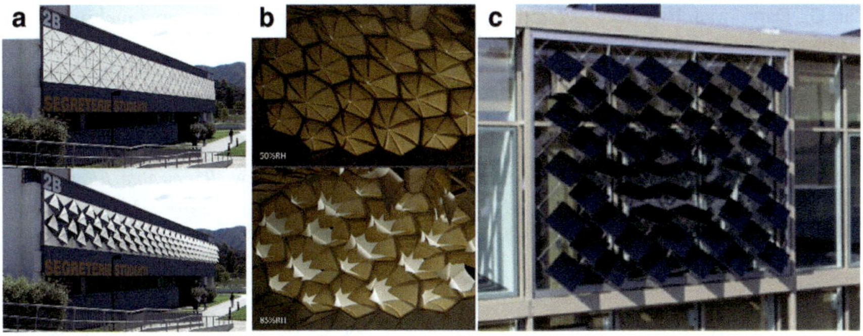

Fig. 3.9 **a** Fully-closed and open façades. Reproduced with permission (Cimmino et al., 2017) Copyright by Elsevier. **b** Adaptive façades actuated by moisture. Reproduced with permission (Reichert et al., 2015) Copyright by Elsevier. **c** The adaptive solar façades on the House of Natural Resources at ETH Zurich. Reproduced with permission (Powell et al., 2018) Copyright by Elsevier

needed to improve, but also simpler and more economic manufacturing methods are also preferred. Lastly, we hope this chapter can provide some insights for the future smart window designs for green building research and industry.

Acknowledgments The authors acknowledge the funding support from NSF (CMMI2005374) and NSF (CMMI-2013993), United States.

References

Aburas, M., Soebarto, V., Williamson, T., Liang, R., Ebendorff-Heidepriem, H., & Wu, Y. (2019). Thermochromic smart window technologies for building application: A review. *Applied Energy, 255*, 113522.

Ansari, M. I. H., Qurashi, A., & Nazeeruddin, M. K. (2018). Frontiers, opportunities, and challenges in perovskite solar cells: A critical review. *Journal of Photochemistry and Photobiology c: Photochemistry Reviews, 35*, 1–24.

Arakaki, J., Reyes, R., Horn, M., & Estrada, W. (1995). Electrochromism in NiOx and WOx obtained by spray pyrolysis. *Solar Energy Materials and Solar Cells, 37*, 33–41.

Azens, A., & Granqvist, C. (2003). Electrochromic smart windows: Energy efficiency and device aspects. *Journal of Solid State Electrochemistry, 7*, 64–68.

Baetens, R., Jelle, B. P., & Gustavsen, A. (2010). Properties, requirements and possibilities of smart windows for dynamic daylight and solar energy control in buildings: A state-of-the-art review. *Solar Energy Materials and Solar Cells, 94*, 87–105.

Bai, H., Zhao, T., Wang, X., Wu, Y., Li, K., Yu, C., Jiang, L., & Cao, M. (2020). Cactus kirigami for efficient fog harvesting: Simplifying a 3D cactus into 2D paper art. *Journal of Materials Chemistry A, 8*, 13452–13458.

Bechinger, C., Ferrere, S., Zaban, A., Sprague, J., & Gregg, B. A. (1996). Photoelectrochromic windows and displays. *Nature, 383*, 608–610.

Bowden, N., Brittain, S., Evans, A. G., Hutchinson, J. W., & Whitesides, G. M. (1998). Spontaneous formation of ordered structures in thin films of metals supported on an elastomeric polymer. *Nature, 393*, 146–149.

Cai, G., Darmawan, P., Cheng, X., & Lee, P. S. (2017). Inkjet printed large area multifunctional smart windows. *Advanced Energy Materials, 7*, 1602598.
Cai, G., Wang, J., & Lee, P. S. (2016). Next-generation multifunctional electrochromic devices. *Accounts of Chemical Research, 49*, 1469–1476.
Cao, S., Zhang, S., Zhang, T., Yao, Q., & Lee, J. Y. (2019). A visible light-near-infrared dual-band smart window with internal energy storage. *Joule, 3*, 1152–1162.
Casini, M. (2018). Active dynamic windows for buildings: A review. *Renewable Energy,* 119, 923–934.
Chan, E. P., & Crosby, A. J. (2006). Fabricating microlens arrays by surface wrinkling. *Advanced Materials, 18*, 3238–3242.
Chandra, D., Yang, S., & Lin, P.-C. (2007). Strain responsive concave and convex microlens arrays. *Applied Physics Letters, 91*, 251912.
Chen, C., Liu, D., Zhang, B., Bi, W., Li, H., Jin, J., Chen, X., Xu, L., Song, H., & Dai, Q. (2018). Carrier interfacial engineering by bismuth modification for efficient and thermoresistant perovskite solar cells. *Advanced Energy Materials, 8*, 1703659.
Chen, S., Wang, Z., Ren, H., Chen, Y., Yan, W., Wang, C., Li, B., Jiang, J., & Zou, C. (2019). Gate-controlled VO2 phase transition for high-performance smart windows. *Science Advances, 5*, eaav6815.
Chen, Z., Gao, Y., Kang, L., Du, J., Zhang, Z., Luo, H., Miao, H., & Tan, G. (2011). VO2-based double-layered films for smart windows: Optical design, all-solution preparation and improved properties. *Solar Energy Materials and Solar Cells, 95*, 2677–2684.
Chun, S. Y., Park, S., Lee, S. I., Nguyen, H. D., Lee, K.-K., Hong, S., Han, C.-H., Cho, M., Choi, H.-K., & Kwak, K. (2021). Operando Raman and UV-Vis spectroscopic investigation of the coloring and bleaching mechanism of self-powered photochromic devices for smart windows. *Nano Energy, 82*, 105721.
Cimmino, M. C., Miranda, R., Sicignano, E., Ferreira, A. J. M., Skelton, R. E., & Fraternali, F. (2017). Composite solar façades and wind generators with tensegrity architecture. *Composites Part b: Engineering, 115*, 275–281.
Conway, B. E., Birss, V., & Wojtowicz, J. (1997). The role and utilization of pseudocapacitance for energy storage by supercapacitors. *Journal of Power Sources, 66*, 1–14.
Corzo, D., Almasabi, K., Bihar, E., Macphee, S., Rosas-Villalva, D., Gasparini, N., Inal, S., & Baran, D. (2019). Digital inkjet printing of high-efficiency large-area nonfullerene organic solar cells. *Advanced Materials Technologies, 4*, 1900040.
Cui, Y., Ke, Y., Liu, C., Chen, Z., Wang, N., Zhang, L., Zhou, Y., Wang, S., Gao, Y., & Long, Y. (2018). Thermochromic VO2 for energy-efficient smart windows. *Joule, 2*, 1707–1746.
Cupelli, D., Pasquale Nicoletta, F., Manfredi, S., Vivacqua, M., Formoso, P., De FILPO, G., & Chidichimo, G. 2009. Self-adjusting smart windows based on polymer-dispersed liquid crystals. *Solar Energy Materials and Solar Cells, 93*, 2008–2012.
DOE. 2021. DOE Launches International Clean Energy Initiatives to Tackle Climate Crisis. In Energy, D. O. (Ed.). https://www.energy.gov/articles/doe-launches-international-clean-energy-initiatives-tackle-climate-crisis.
Dussault, J.-M., Gosselin, L., & Galstian, T. (2012). Integration of smart windows into building design for reduction of yearly overall energy consumption and peak loads. *Solar Energy, 86*, 3405–3416.
Dyer, A. L., Grenier, C. R. G., & Reynolds, J. R. (2007). A Poly(3,4-alkylenedioxythiophene) electrochromic variable optical attenuator with near-infrared reflectivity tuned independently of the visible region. *Advanced Functional Materials, 17*, 1480–1486.
EU. 2020. 2050 long-term strategy. In European Commission, E. U. (Ed.). https://ec.europa.eu/clima/eu-action/climate-strategies-targets/2050-long-term-strategy_en.
Fathieh, F., Kalmutzki Markus, J., Kapustin Eugene, A., Waller Peter, J., Yang, J., & Yaghi Omar, M. (2018). Practical water production from desert air. *Science Advances, 4*, eaat3198.

Fleischmann, S., Mitchell, J. B., Wang, R., Zhan, C., Jiang, D.-E., Presser, V., & Augustyn, V. (2020). Pseudocapacitance: From fundamental understanding to high power energy storage materials. *Chemical Reviews, 120*, 6738–6782.

Ge, D., Lee, E., Yang, L., Cho, Y., Li, M., Gianola, D. S., & Yang, S. (2015). A robust smart window: Reversibly switching from high transparency to angle-independent structural color display. *Advanced Materials, 27*, 2489–2495.

Gesheva, K. A., Ivanova, T. M., & Bodurov, G. (2012). Transition metal oxide films: Technology and "smart windows" electrochromic device performance. *Progress in Organic Coatings, 74*, 635–639.

Golubev, V. G., Davydov, V. Y., Kartenko, N. F., Kurdyukov, D. A., Medvedev, A. V., Pevtsov, A. B., Scherbakov, A. V., & Shadrin, E. B. (2001). Phase transition-governed opal–VO2 photonic crystal. *Applied Physics Letters, 79*, 2127–2129.

Granqvist, C. G. (2000). Electrochromic tungsten oxide films: Review of progress 1993–1998. *Solar Energy Materials and Solar Cells, 60*, 201–262.

Granqvist, C. G. (2014). Electrochromics for smart windows: Oxide-based thin films and devices. *Thin Solid Films, 564*, 1–38.

Granqvist, C. G. (2016). Recent progress in thermochromics and electrochromics: A brief survey. *Thin Solid Films, 614*, 90–96

Guo, S.-M., Liang, X., Zhang, C.-H., Chen, M., Shen, C., Zhang, L.-Y., Yuan, X., He, B.-F., & Yang, H. (2017). Preparation of a thermally light-transmittance-controllable film from a coexistent system of polymer-dispersed and polymer-stabilized liquid crystals. *ACS Applied Materials and Interfaces, 9*, 2942–2947.

Guterres, A. (2020). Carbon neutrality by 2050: the world's most urgent mission. In U. Nations (Ed.). https://www.un.org/sg/en/content/sg/articles/2020-12-11/carbon-neutrality-2050-the-world%E2%80%99s-most-urgent-mission.

Harrison, C., Stafford, C. M., Zhang, W., & Karim, A. (2004). Sinusoidal phase grating created by a tunably buckled surface. *Applied Physics Letters, 85*, 4016–4018.

Hong-Chen, W., Xin-Jian, Y., Jian-Jun, L., & Yi, L. (2005). Fabrication and characterization of nanocrystalline VO 2 thin films. *Chinese Physics Letters, 22*, 1746–1748.

Isaac, M., & van Vuuren, D. P. (2009). Modeling global residential sector energy demand for heating and air conditioning in the context of climate change. *Energy Policy, 37*, 507–521.

Jiang, B., Liu, L., Gao, Z., & Wang, W. (2018). A general and robust strategy for fabricating mechanoresponsive surface wrinkles with dynamic switchable transmittance. *Advanced Optical Materials, 6*, 1800195.

Ju, J., Bai, H., Zheng, Y., Zhao, T., Fang, R., & Jiang, L. (2012). A multi-structural and multi-functional integrated fog collection system in cactus. *Nature Communications, 3*, 1247.

Kamalisarvestani, M., Saidur, R., Mekhilef, S., & Javadi, F. S. (2013). Performance, materials and coating technologies of thermochromic thin films on smart windows. *Renewable and Sustainable Energy Reviews, 26*, 353–364.

Ke, Y., Chen, J., Lin, G., Wang, S., Zhou, Y., Yin, J., Lee, P. S., & Long, Y. (2019a). Smart windows: Electro-, thermo-, mechano-, photochromics, and beyond. *Advanced Energy Materials, 9*, 1902066.

Ke, Y., Yin, Y., Zhang, Q., Tan, Y., Hu, P., Wang, S., Tang, Y., Zhou, Y., Wen, X., Wu, S., White, T. J., Yin, J., Peng, J., Xiong, Q., Zhao, D., & Long, Y. (2019b). Adaptive thermochromic windows from active plasmonic elastomers. *Joule, 3*, 858–871.

Ke, Y., Zhou, C., Zhou, Y., Wang, S., Chan, S. H., & Long, Y. (2018). Emerging thermal-responsive materials and integrated techniques targeting the energy-efficient smart window application. *Advanced Functional Materials, 28*, 1800113.

Khandelwal, H., Schenning, A. P. H. J., & Debije, M. G. (2017). Infrared regulating smart window based on organic materials. *Advanced Energy Materials, 7*, 1602209.

Khang, D.-Y., Jiang, H., Huang, Y., & Rogers John, A. (2006). A stretchable form of single-crystal silicon for high-performance electronics on rubber substrates. *Science, 311*, 208–212.

Kim, H.-N., Ge, D., Lee, E., & Yang, S. (2018). Multistate and on-demand smart windows. *Advanced Materials, 30*, 1803847.

Kim, J., Ong, G. K., Wang, Y., Leblanc, G., Williams, T. E., Mattox, T. M., Helms, B. A., & Milliron, D. J. (2015a). Nanocomposite architecture for rapid, spectrally-selective electrochromic modulation of solar transmittance. *Nano Letters, 15*, 5574–5579.

Kim, J., Yoon, J., & Hayward, R. C. (2010). Dynamic display of biomolecular patterns through an elastic creasing instability of stimuli-responsive hydrogels. *Nature Materials, 9*, 159–164.

Kim, M., Park, K. J., Seok, S., Ok, J. M., Jung, H.-T., Choe, J., & Kim, D. H. (2015b). Fabrication of microcapsules for dye-doped polymer-dispersed liquid crystal-based smart windows. *ACS Applied Materials and Interfaces, 7*, 17904–17909.

Kim, P., Hu, Y., Alvarenga, J., Kolle, M., Suo, Z., & Aizenberg, J. (2013). Rational design of mechano-responsive optical materials by fine tuning the evolution of strain-dependent wrinkling patterns. *Advanced Optical Materials, 1*, 381–388.

Lamoureux, A., Lee, K., Shlian, M., Forrest, S. R., & Shtein, M. (2015). Dynamic kirigami structures for integrated solar tracking. *Nature Communications, 6*, 8092.

Lampert, C. M. (2004). Chromogenic smart materials. *Materials Today, 7*, 28–35.

Lee, S. G., Lee, D. Y., Lim, H. S., Lee, D. H., Lee, S., & Cho, K. (2010). Switchable transparency and wetting of elastomeric smart windows. *Advanced Materials, 22*, 5013–5017.

Lee, S. J., Ha, N., & Kim, H. (2019). Superhydrophilic-superhydrophobic water harvester inspired by wetting property of cactus stem. *ACS Sustainable Chemistry and Engineering, 7*, 10561–10569.

Lekouch, I., Muselli, M., Kabbachi, B., Ouazzani, J., Melnytchouk-Milimouk, I., & Beysens, D. (2011). Dew, fog, and rain as supplementary sources of water in south-western Morocco. *Energy, 36*, 2257–2265.

Li, H., Chen, C., Jin, J., Bi, W., Zhang, B., Chen, X., Xu, L., Liu, D., Dai, Q., & Song, H. (2018). Near-infrared and ultraviolet to visible photon conversion for full spectrum response perovskite solar cells. *Nano Energy, 50*, 699–709.

Li, J., Ran, R., Wang, H., Wang, Y., Chen, Y., Niu, S., Arratia, P. E., & Yang, S. (2021a). Aerodynamics-assisted, efficient and scalable kirigami fog collectors. *Nature Communications, 12*, 5484.

Li, Y., Zhao, Y., Chi, Y., Hong, Y., & Yin, J. (2021b). Shape-morphing materials and structures for energy-efficient building envelopes. *Materials Today Energy, 22*, 100874.

Li, M., Magdassi, S., Gao, Y., & Long, Y. (2017). Hydrothermal synthesis of VO2 polymorphs: Advantages, challenges and prospects for the application of energy efficient smart windows. *Small, 13*, 1701147.

Lin, G., Chandrasekaran, P., Lv, C., Zhang, Q., Tang, Y., Han, L., & Yin, J. (2017). Self-similar hierarchical wrinkles as a potential multifunctional smart window with simultaneously tunable transparency, structural color, and droplet transport. *ACS Applied Materials and Interfaces, 9*, 26510–26517.

Lin, G., Zhang, Q., Lv, C., Tang, Y., & Yin, J. (2018). Small degree of anisotropic wetting on self-similar hierarchical wrinkled surfaces. *Soft Matter, 14*, 1517–1529.

Llordés, A., Garcia, G., Gazquez, J., & Milliron, D. J. (2013). Tunable near-infrared and visible-light transmittance in nanocrystal-in-glass composites. *Nature, 500*, 323–326.

Macêdo, M. A., Dall'Antonia, L. H., Valla, B., & Aegerter, M. A. (1992). Electrochromic smart windows. *Journal of Non-Crystalline Solids, 147–148*, 792–798.

Madida, I. G., Simo, A., Sone, B., Maity, A., Kana Kana, J. B., Gibaud, A., Merad, G., Thema, F. T., & Maaza, M. (2014). Submicronic VO2–PVP composites coatings for smart windows applications and solar heat management. *Solar Energy, 107*, 758–769.

Matsumoto, K., Sakikawa, N., & Miyata, T. (2018). Thermo-responsive gels that absorb moisture and ooze water. *Nature Communications, 9*, 2315.

Mei, A., Li, X., Liu, L., Ku, Z., Liu, T., Rong, Y., Xu, M., Hu, M., Chen, J., Yang, Y., Grätzel, M., & Han, H. (2014). A hole-conductor–free, fully printable mesoscopic perovskite solar cell with high stability. *Science, 345*, 295–298.

Meng, Y., Li, X., Wang, S., Lau, C., Hu, H., Ke, Y., Tan, G., Yang, J., & Long, Y. (2022). Flexible smart photovoltaic foil for energy generation and conservation in buildings. *Nano Energy, 91*, 106632.

Miller, M. J., & Wang, J. (2015). Influence of grain size on transition temperature of thermochromic VO2. *Journal of Applied Physics, 117*, 034307.

Parkin, I. P., & Manning, T. D. (2006). Intelligent thermochromic windows. *Journal of Chemical Education, 83*, 393.

Pérez-Lombard, L., Ortiz, J., & Pout, C. (2008). A review on buildings energy consumption information. *Energy and Buildings, 40*, 394–398.

Piccolo, A., & Simone, F. (2015). Performance requirements for electrochromic smart window. *Journal of Building Engineering, 3*, 94–103.

Powell, D., Hischier, I., Jayathissa, P., Svetozarevic, B., & Schlüter, A. (2018). A reflective adaptive solar façade for multi-building energy and comfort management. *Energy and Buildings, 177*, 303–315.

Rauh, R. D. (1999). Electrochromic windows: An overview. *Electrochimica Acta, 44*, 3165–3176.

Reichert, S., Menges, A., & Correa, D. (2015). Meteorosensitive architecture: Biomimetic building skins based on materially embedded and hygroscopically enabled responsiveness. *Computer-Aided Design, 60*, 50–69.

Sundriyal, P., & Bhattacharya, S. (2018). Inkjet-printed sensors on flexible substrates. In S. Bhattacharya, A. K. Agarwal, N. Chanda, A. Pandey & A. K. Sen (Eds.), *Environmental, chemical and medical sensors*. Singapore: Springer Singapore.

Tällberg, R., Jelle, B. P., Loonen, R., Gao, T., & Hamdy, M. (2019). Comparison of the energy saving potential of adaptive and controllable smart windows: A state-of-the-art review and simulation studies of thermochromic, photochromic and electrochromic technologies. *Solar Energy Materials and Solar Cells, 200*, 109828.

Tang, Y., Lin, G., Yang, S., Yi, Y. K., Kamien, R. D., & Yin, J. (2017). Programmable kiri-kirigami metamaterials. *Advanced Materials, 29*, 1604262.

Thomas, A. V., Andow, B. C., Suresh, S., Eksik, O., Yin, J., Dyson, A. H., & Koratkar, N. (2015). Controlled crumpling of graphene oxide films for tunable optical transmittance. *Advanced Materials, 27*, 3256–3265.

Tomholt, L., Geletina, O., Alvarenga, J., Shneidman, A. V., Weaver, J. C., Fernandes, M. C., Mota, S. A., Bechthold, M., & Aizenberg, J. (2020). Tunable infrared transmission for energy-efficient pneumatic building façades. *Energy and Buildings, 226*, 110377.

Vergaz, R., Sánchez-Pena, J.-M., Barrios, D., Vázquez, C., & Contreras-Lallana, P. (2008). Modelling and electro-optical testing of suspended particle devices. *Solar Energy Materials and Solar Cells, 92*, 1483–1487.

Vieira, F., Sarmento, B., Reis-Machado, A. S., Facão, J., Carvalho, M. J., Mendes, M. J., Fortunato, E., & Martins, R. (2019). Prediction of sunlight-driven CO2 conversion: Producing methane from photovoltaics, and full system design for single-house application. *Materials Today Energy, 14*, 100333.

Wang, H., Yi, X., & Li, Y. (2005). Fabrication of VO2 films with low transition temperature for optical switching applications. *Optics Communications, 256*, 305–309.

Wang, J., Zhang, L., Yu, L., Jiao, Z., Xie, H., Lou, X. W., & Wei Sun, X. (2014). A bi-functional device for self-powered electrochromic window and self-rechargeable transparent battery applications. *Nature Communications, 5*, 4921.

Wang, K., Wu, H., Meng, Y., Zhang, Y., & Wei, Z. (2012). Integrated energy storage and electrochromic function in one flexible device: An energy storage smart window. *Energy and Environmental Science, 5*, 8384–8389.

Wang, L., Liu, Y., Zhan, X., Luo, D., & Sun, X. (2019a). Photochromic transparent wood for photo-switchable smart window applications. *Journal of Materials Chemistry C, 7*, 8649–8654.

Wang, N., Duchamp, M., Dunin-Borkowski, R. E., Liu, S., Zeng, X., Cao, X., & Long, Y. (2016a). Terbium-doped VO2 thin films: Reduced phase transition temperature and largely enhanced luminous transmittance. *Langmuir, 32*, 759–764.

Wang, N., Goh, Q. S., Lee, P. L., Magdassi, S., & Long, Y. (2017). One-step hydrothermal synthesis of rare earth/W-codoped VO2 nanoparticles: Reduced phase transition temperature and improved thermochromic properties. *Journal of Alloys and Compounds, 711*, 222–228.
Wang, S., Gao, W., Hu, X.-Y., Shen, Y.-Z., & Wang, L. (2019b). Supramolecular strategy for smart windows. *Chemical Communications, 55*, 4137–4149.
Wang, S., Liu, G., Hu, P., Zhou, Y., Ke, Y., Li, C., Chen, J., Cao, T., & Long, Y. (2018). Largely lowered transition temperature of a VO2/carbon hybrid phase change material with high thermal emissivity switching ability and near infrared regulations. *Advanced Materials Interfaces, 5*, 1801063.
Wang, Y., Runnerstrom, E. L., & Milliron, D. J. (2016b). Switchable materials for smart windows. *Annual Review of Chemical and Biomolecular Engineering, 7*, 283–304.
Wu, L. Y. L., Zhao, Q., Huang, H., & Lim, R. J. (2017). Sol-gel based photochromic coating for solar responsive smart window. *Surface and Coatings Technology, 320*, 601–607.
Wu, S., Zhang, Q., Deng, Y., Li, X., Luo, Z., Zheng, B., & Dong, S. (2020). Assembly pattern of supramolecular hydrogel induced by lower critical solution temperature behavior of low-molecular-weight gelator. *Journal of the American Chemical Society, 142*, 448–455.
Xie, Z., Jin, X., Chen, G., Xu, J., Chen, D., & Shen, G. (2014). Integrated smart electrochromic windows for energy saving and storage applications. *Chemical Communications, 50*, 608–610.
Xu, C., Cai, L., Zhong, M., & Zheng, S. (2015). Low-cost and rapid prototyping of microfluidic paper-based analytical devices by inkjet printing of permanent marker ink. *RSC Advances, 5*, 4770–4773.
Xu, T., Lin, Y., Zhang, M., Shi, W., & Zheng, Y. (2016). High-efficiency fog collector: water unidirectional transport on heterogeneous rough conical wires. *ACS Nano, 10*, 10681–10688.
Yang, L., Ge, D., Zhao, J., Ding, Y., Kong, X., & Li, Y. (2012). Improved electrochromic performance of ordered macroporous tungsten oxide films for IR electrochromic device. *Solar Energy Materials and Solar Cells, 100*, 251–257.
Yang, P., Sun, P., Chai, Z., Huang, L., Cai, X., Tan, S., Song, J., & Mai, W. (2014). Large-scale fabrication of pseudocapacitive glass windows that combine electrochromism and energy storage. *Angewandte Chemie International Edition, 53*, 11935–11939.
Yi, Y. K., Yin, J., & Tang, Y. (2018). Developing an advanced daylight model for building energy tool to simulate dynamic shading device. *Solar Energy, 163*, 140–149.
Yin, W.-J., Shi, T., & Yan, Y. (2014). Unusual defect physics in CH3NH3PbI3 perovskite solar cell absorber. *Applied Physics Letters, 104*, 063903.
Yu, H., Guo, J., Wang, C., Zhang, J., Liu, J., Zhong, X., Dong, G., & Diao, X. (2019). High performance in electrochromic amorphous WOx film with long-term stability and tunable switching times via Al/Li-ions intercalation/deintercalation. *Electrochimica Acta, 318*, 644–650.
Zang, J., Ryu, S., Pugno, N., Wang, Q., Tu, Q., Buehler, M. J., & Zhao, X. (2013). Multifunctionality and control of the crumpling and unfolding of large-area graphene. *Nature Materials, 12*, 321–325.
Zeng, S., Zhang, D., Huang, W., Wang, Z., Freire, S. G., Yu, X., Smith, A. T., Huang, E. Y., Nguon, H., & Sun, L. (2016). Bio-inspired sensitive and reversible mechanochromisms via strain-dependent cracks and folds. *Nature Communications, 7*, 11802.
Zhai, L., Berg, M. C., Cebeci, F. Ç., Kim, Y., Milwid, J. M., Rubner, M. F., & Cohen, R. E. (2006). Patterned superhydrophobic surfaces: Toward a synthetic mimic of the Namib desert beetle. *Nano Letters, 6*, 1213–1217.
Zhang, K., Zhang, M., Feng, X., Hempenius, M. A., & Vancso, G. J. (2017a). Switching light transmittance by responsive organometallic poly(ionic liquid)s: Control by cross talk of thermal and redox stimuli. *Advanced Functional Materials, 27*, 1702784.
Zhang, L., Zhao, Y., & Dai, Q. (2021). Recent progress in perovskite solar cell: Fabrication, Efficiency, and stability. In J. K. Roy, S. Kar, & J. Leszczynski (Eds.), *Development of solar cells: Theory and experiment*. Cham: Springer International Publishing.
Zhang, Q., Tang, Y., Hajfathalian, M., Chen, C., Turner, K. T., Dikin, D. A., Lin, G., & Yin, J. (2017b). Spontaneous periodic delamination of thin films to form crack-free metal and silicon ribbons with high stretchability. *ACS Applied Materials and Interfaces, 9*, 44938–44947.

Zhang, Q., & Yin, J. (2018). Spontaneous buckling-driven periodic delamination of thin films on soft substrates under large compression. *Journal of the Mechanics and Physics of Solids, 118*, 40–57.

Zhang, S., Cao, S., Zhang, T., & Lee, J. Y. (2020). Plasmonic oxygen-deficient TiO2-x nanocrystals for dual-band electrochromic smart windows with efficient energy recycling. *Advanced Materials, 32*, 2004686.

Zheng, S., Xu, Y., Shen, Q., & Yang, H. (2015). Preparation of thermochromic coatings and their energy saving analysis. *Solar Energy, 112*, 263–271.

Zhou, J., Gao, Y., Zhang, Z., Luo, H., Cao, C., Chen, Z., Dai, L., & Liu, X. (2013). VO2 thermochromic smart window for energy savings and generation. *Scientific Reports, 3*, 3029.

Zhou, Y., Wang, S., Peng, J., Tan, Y., Li, C., Boey, F. Y. C., & Long, Y. (2020). Liquid thermo-responsive smart window derived from hydrogel. *Joule, 4*, 2458–2474.

Chapter 4
Material Programming for Bio-inspired and Bio-based Hygromorphic Building Envelopes

Dylan Wood, Tiffany Cheng, Yasaman Tahouni, and Achim Menges

Abstract Building skins play a decisive role in maintaining occupant comfort. Adaptive building skins have been proposed to adjust to the weather, with mechanically complex multi-component solutions that require operating energy. Nature and its materials exhibit a fundamentally different strategy for environmental responsiveness; motile plant systems show entirely passive, integrative, hygroscopic actuation due to their cellulose-based material structure. Through a design and fabrication process we refer to as material programming, a bio-inspired and bio-based functional integration of actuator, sensor, and controller can be achieved. We present an overview of related research on weather responsive building components. Wood-based composite elements that respond to relative humidity without operational energy have been demonstrated at architectural-scale. This research was recently expanded through the additive manufacturing of custom-made natural fiber composites, allowing 4D-printed self-shaping compliant mechanisms based on highly differentiated and multifunctional plant movements with varying mechanical stiffnesses and actuation speeds. The application of 4D-printing to weather responsive shading systems still necessitates the codesign of materials, mechanism, and façade system as well as matching stimuli-responsiveness to ambient weather conditions and mass production at the scale of buildings. Overcoming these challenges will enable a more reliable, sustainable, and zero-energy solution for regulating comfort in the built environment.

Keywords Computational design · Additive manufacturing · Responsive materials · Hygroscopic actuation · Adaptive architecture · Building Skins · Hygromorphic · 4D-printing · Bio-based materials

D. Wood (✉) · T. Cheng · Y. Tahouni · A. Menges
Institute for Computational Design and Construction, University of Stuttgart, Keplerstr. 11, 70174 Stuttgart, Germany
e-mail: dylan.wood@icd.uni-stuttgart.de

J. Wang et al. (eds.), *Advanced Materials in Smart Building Skins for Sustainability*,
https://doi.org/10.1007/978-3-031-09695-2_4

4.1 Introduction

In nature, many materials exhibit shape-changing characteristics in response to stimuli such as temperature or moisture. In many plants, shape-change is used as functional, motile movements without living biological or electrical energy. The awns of the Erodium and Pelargonium seeds, for example, self-drill into the ground by reversibly coiling and uncoiling in response to the fluctuations of environmental humidity (Burgert & Fratzl, 2009; Jung et al., 2014). Another example, the standard spruce cone, can open up and release its seeds in response to a loss of moisture and resultant drying of the fibrous composite material in each and every scale in mass quantity (Dawson et al., 1997; Harlow et al., 1964; Reyssat & Mahadevan, 2009). Such a transformation is remarkable, given that it occurs passively as the material equalizes with the humidity of the surrounding air, after the cones are separated from the living biological functions of the tree (Poppinga et al., 2017). In the spruce cones, the variations in the morphology of cellular structure and layout of fibers within the scales have evolved over time to suit the specialized function of each type of cone, the environment in which the species grows, and the kinds of seeds they release.

Similar to the tissue of the spruce cones, a handful of well-known natural fibrous materials exhibit hygromorphic behavior, in that they dramatically expand or contract as they absorb and desorb moisture from the surrounding environment and can be engineered into a responsive system (Carneiro et al., 2013; Le Duigou et al., 2020). Common examples that can be found in our daily lives include cellulose-based materials such cotton, paper, and wood. In the context of adaptable building systems, three aspects of these natural hygromorphic materials are particularly interesting. First, many of these materials can naturally actuate with stimuli ranges found in daily and seasonal weather patterns. Second, the shape-changing characteristics are inherent or well-integrated into the material structure, giving them unmatched hygromorphic performance with a fraction of the processing and engineering required for synthetic "smart materials (Erb et al., 2013). Third, the materials are available in large quantities, at low costs, and in regenerative life cycles that can be considered viable and sustainable solutions at the scale of building facades. These qualities are advantageous for the development of ecologically constructed building envelopes explicitly designed to adapt with the natural ebbs and flows of weather on Earth.

Working with natural materials comes with unique challenges and opportunities concerning design, engineering, and manufacturing. We develop new types of bio-inspired and bio-based prototypes of passively adaptive building envelopes through careful decoding, re-programming, and re-packaging these natural materials—a design and fabrication process that we refer to as *material programming*. In this context, computational design and fabrication provide the key to the critical understanding and unitizing of the unique functional aspects of natural materials and systems, from the analysis of movement patterns to the physical programming of simple yet novel shape-changing structures. More specifically, material programming can be viewed here as arranging natural materials within larger systems and in higher function hygromorphic structures. The approach can be applied at various scales, resolutions, and specific material combinations but fundamentally involves developing a technical understanding of the materials themselves as well as a precise method of their arrangement in physical space.

Considering that building systems and construction have a substantial impact globally in terms of both carbon footprint and energy consumption, the question of how to more effectively regulate the interior climate of buildings is of increasing relevance across a range of cultures and climates (United Nations Environment Programme, 2020). Contrary to the pursuit of increased efficiency often at the expense of complexity in active systems for heating and cooling, passively adaptive building envelopes offer a potentially low-cost, reliable solution for regulating human comfort in both interior conditions and exterior public spaces (Grondzik & Kwok, 2020; Poppinga et al., 2018; Tabadkani et al., 2021). Starting with wood as well-known hygromorphic building material, we have developed principles for material programming with natural, plant-based materials with existing shapeshifting behaviors. From these principles, we explore this concept through the combination of additive manufacturing and the development of custom cellulose-based printing filaments that enable the 3D internal arrangement of material strands to design and fabricate intricate mesotructuring within each piece. Paired with computational methods for designing high resolution structuring, this approach to material programming opens up new functionalities including the choreography of actuation speed, tuning of response ranges and architecting of movement patterns.

4.2 Understanding and Deploying Wood as Pre-constructed Natural Hygromorphic Smart Material

Wood in its living and harvested state is inherently a hygromorphic material, adapting both its shape and its mechanical properties in response to changes in moisture. The hygroscopicity of wood comes from the complex material arrangement of cellulose microfibrils inside the tube-shaped cell walls that make up its primary structure. When wet, the cell walls expand volumetrically in the defined anisotropic coordinates of the material and simultaneously soften. When dried, the same wood material naturally contracts in volume and increases in stiffness—a phenomenon known for causing unwanted geometric deformations in wood boards as a result of variation in the natural structuring. As a living organism, the cellular structure of wood serves many purposes, one of which is the transport of water, which means the material by design is able to maintain its overall structure even when the cells are in a fully saturated state. Once harvested, wood materials continuously desorb and absorb moisture to maintain an equilibrium with the relative humidity of the surrounding air (Hoadley & Barbara, 2000). This constant, passive equalization and resulting form adaptation makes it a near-ideal material for the development of adaptive building systems intended by design to fluctuate with changes in the environment.

To implement wood in adaptive systems, first, the hygromorphic behaviors must be understood and quantified, an aspect which has for centuries been studied and quantified scientifically for the purpose of preventing unwanted movements in fields ranging from timber construction to art preservation. Beyond understanding the basic shape-changing principle, the directional change in volume can be translated into a change in curvature using a bilayer mechanism, a principle in which one layer of active material is connected to a perpendicular, resistive layer. Changes in the volume of the active layer results in a translational bending of the bilayer, a geometric movement significantly larger than the volume change of the active layer alone. Bilayer mechanisms have been studied extensively in both structural mechanics and in thermally responsive material in which isotropic shape change further limits the possible movement patterns in larger arrangements. Critically, the curvature of a bilayer system can be predicted with relatively high accuracy using an analytical model developed by Timoshenko in 1925 and more recently adapted for use with the anisotropic properties of wood (Grönquist et al., 2018; Rüggeberg & Burgert, 2015; Timoshenko, 1925).

Through a careful selection of the wood species and adjustment of controllable design parameters, the relatively untamed deformation in wood can be programmed into reliable movement patterns that respond to daily and seasonal fluctuations in relative humidity (Fig. 4.1). More specifically, European Maple (*Acer platanoides*) is chosen as it combines good values for the shrinking and swelling coefficient, structural integrity, and can be sourced with straight, even grains. Additionally, maple is known to exhibit a natural resistance to fungus and weather-related deterioration due to water and UV exposure. While the swelling coefficient and stiffness are ingrained within the material, the thickness of the veneers and the orientations of the fibers/grains can be used to further design the direction and magnitude of curvature within the bilayer build-up. In parallel, the moisture content of the active boards is used to embed the geometric configuration assigned to a specific corresponding relative humidity. Similarly, the restrictive bending layer made from wood can be used to further tune the curvature through a selection of the species and thickness. Alternatively, composite materials such as glass fiber reinforced plastic (GFRP) or natural fiber reinforced plastic textiles can be implemented for different mechanical properties. In the demonstrator projects HygroScope and HygroSkin from the ICD (Fig. 4.2), a thin 0.5 mm maple veneer is combined with lightweight GFRP textile to give an added mechanical spring back in the restrictive layer and counteract the long term effects of creep found in wood (Menges & Reichert, 2015; Reichert et al., 2015). Through the design of these simple parameters, responsive wood and wood composite parts can be tuned to curve cyclically, from flat to curved and vice versa, in a matter of seconds, in a relative humidity range of 20–95% or through direct exposure to water.

Despite the promising material qualities and relatively easy to construct mechanisms, hygromorphic climate-responsive wood systems have been implemented only in limited architectural scenarios. Weather responsive wood composite bilayers were designed and tested by the ICD prototypically in the project HygroScope, which used artificially accelerated simulation of real-world fluctuations in relative

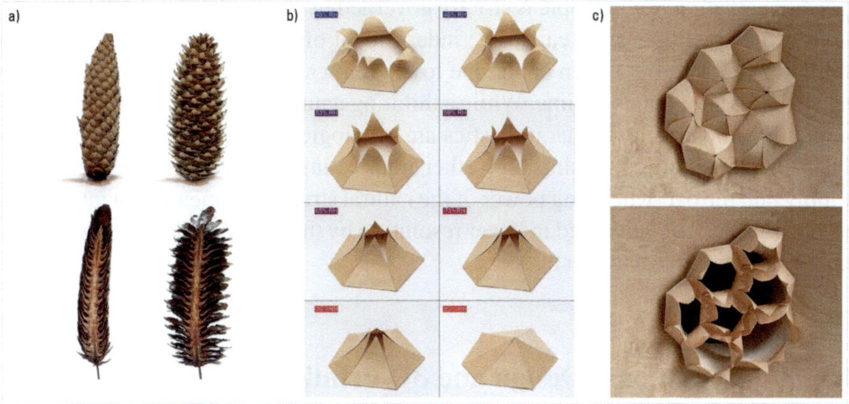

Fig. 4.1 Research at the ICD investigated the transfer of the pine scale actuation (**a**) to a hygroscopic veneer-composite bilayer system (**b**) for humidity-responsive apertures for architectural applications (**c**)

Fig. 4.2 The wood veneer-based bilayer system has been instrumentalized in full-scale architectural prototypes, demonstrating the hygroscopic response in outdoor conditions in HygroSkin Pavilion (**a**) as well as in controlled and accelerated humidity cycles from past weather data (**b**)

humidity to demonstrate the concept. Similar responsive parts are implemented as weather responsive apertures in the HygoSkin pavilion, designed to open to light and air in hot, dry, sunny conditions and close up in humid, rainy weather (Krieg et al., 2017). Wood bilayers have also been studied extensively for adaptive shading devices such as horizontal louvers that use a change in curvature to adjust the angle between their surface and the sun (Holstov et al., 2017; Vailati et al., 2018). When cleverly implemented, even single monolayers of wood can be used as self-ventilating adaptive building systems similar to the wood shingle cladding used historically in the Nordic regions and in recent studies (Davidova, 2016). While wood is unique in its ability to change shape naturally within a suitable range of relative humidity on earth and requires little to no extra process to implement in responsive systems, the scale of its structuring presents limitations in the fine-tuning and design. The speed of actuation, for example is strictly governed by the thickness and depth/width of the active boards, which cannot be independent of the mechanical performance. Similarly, wood suffers from creep due to the realignment of the microfibers when bent over longer periods of time. While exhibiting natural variation, both the bending orientation and structural characteristics are typologically hardcoded into the material itself, presenting constraints for further architecting the material at the mesoscale. Simply put, the overwhelming power of the natural structuring means that intentional tuning is practically governed in lower resolution by the width of the board (typically 25–50% of the tree diameter).

4.3 Computational Design and 3D Printing as Tools for Constructing Natural Material Systems

In contrast to the material constraints of solid wood materials, advances in computational design and digital manufacturing enable the modeling and production of features in increasingly fine resolution and accuracy. Hierarchical structuring of materials for higher-level functionality is common in many adaptive motile plant structures but has historically been difficult to emulate in man-made engineered structures, even at an abstracted level. Digital design of material gradients and differentiation has advanced considerably faster than the physical production methods, allowing designers to turn structures at the incredible resolution but typically abstract from the resolution allowed by the physical production methods (Duro-Royo et al., 2015; Oxman, 2010). 3D printing, on the other hand, enables the fabrication of intricate structures through a differentiated layering of material to form larger three-dimensional objects.

More specifically, Fused Filament Fabrication (FFF) is an ideal method for producing macro-scale parts (ca. 10+ cm) with mesoscale variation (0.5 m–1 mm resolution) from combinations of filament-based materials (Correa & Menges, 2017). Using FFF, strand based, multi-layer mesostructures can be constructed using low-cost, generic, 3D printing technology (Correa & Menges, 2017). An integrated computational design framework has been developed at the ICD for FFF-based 4D printing taking into account the material programming and processing parameters specific to the printing method, such as the anisotropy and stiffness that results from the sequence and quality of material deposition (Cheng et al., 2020). The developed design approach simplifies the complexity of considering highly differentiated structures into a networked assembly of functional regions defined using the intuitive geometric descriptors from existing CAD workflows. This workflow allows the assignment and tuning of material properties and bending behavior—including the direction, magnitude, and orientation or curvature—to be directly translated into the necessary fabrication data for producing the desired hygromorphic response (Fig. 4.3). Parameters in the strand spacing, strand thickness, layer height, and functional patterning can additionally be defined in each sub-millimeter printing layer for constructing varying levels of directional stiffness, time scale of hygro-responsiveness, and elasticity in the overall composite build-ups.

Incorporating this functional mesostructure allows for extending and augmenting the hygromorphic characteristics of wood materials using both wood bilayer actuators and hygromorphic printing filaments. The project HygroFold demonstrates the use of 3D printing through the development of a custom mesostructure with multi-directional bending and an integrated folding hinge integrated with a wood bilayer actuator (Fig. 4.4). A parallel approach displays how meso structural design principles can be used to emulate the sophisticated structuring and transformations found in biological examples of passive movements such as the pine cone scales (Fig. 4.5). Here, variants of the basic bilayer mechanism are built up using standard thermoplastics for the restrictive layers and commercially available wood fiber-based plastic filaments that respond to moisture change through water submersions and drying (Correa et al., 2015, 2020).

Fundamentally, mesostructural tailoring allows for the decoupling of parameters that are inherently linked in solid wood material, opening up the potential to further architect and choreograph movement. Speed of actuation for example can be designed separately from the desired curvature through adjustment of the spacing and resulting porosity of the structure while at the same time counter adjusting the spacing in the restrictive layer to maintain the magnitude of curvature (Fig. 4.6). Through this approach, similar curvatures can be reached on dramatically different time scales, under the same actuation condition (Tahouni et al., 2021) At the level of the mechanism, 3D printing also enables the design of in-plane pattern variation to combine multiple areas of bilayer effect with alternating directions of curvature, variable orientations of curvature, and compliant hinges. This is exemplified by the construction of self-shaping curved crease folding surfaces in which the actuation comes from inside the curved surfaces rather than an external or localized actuation at the edges or hinge

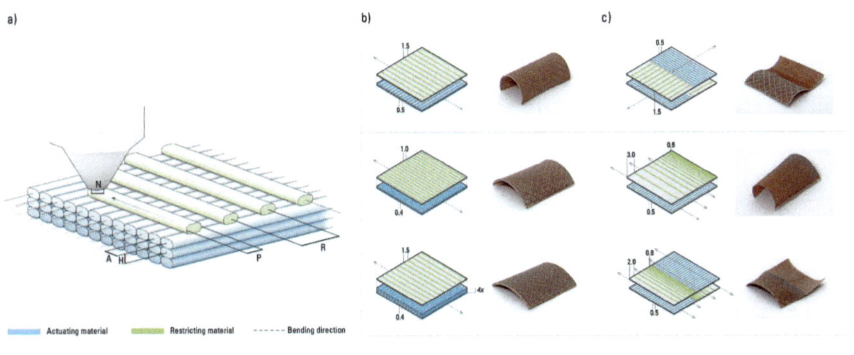

Fig. 4.3 The material programming of functional mesostructures is based on the fabrication parameters of FFF 3D printing (**a**) including the nozzle size N, layer height H, toolpath spacing of the active A and restrictive R filaments, as well as the resulting extrusion paths P. Through the mesostructure design, bending can be tuned, for example in (**b**) the amount of curvature and (**c**) even combined together in different configurations. (Figure adapted with permission from Cheng et al. 2021; p. 5)

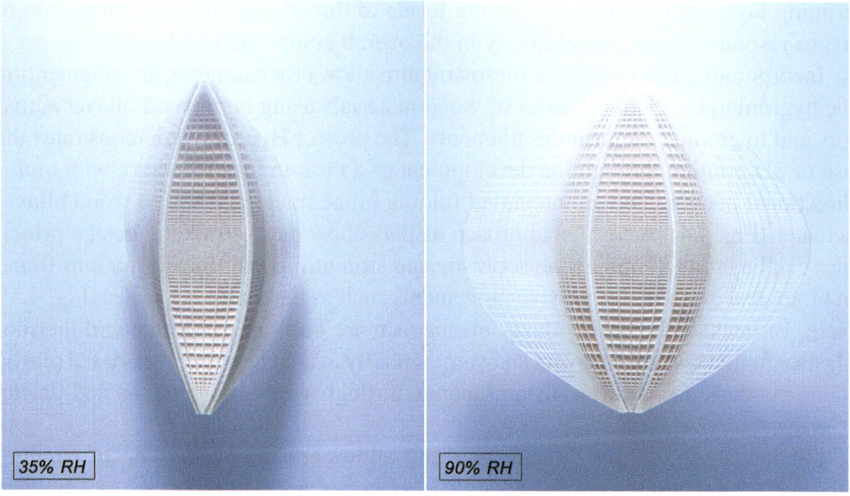

Fig. 4.4 The Hygrofold is a curved folding mechanism combining bio-based 3D-printing and wood actuators, which self-shapes and folds at low RH (left, shown at 35%) and unfolds at high RH (right, shown at 90%)

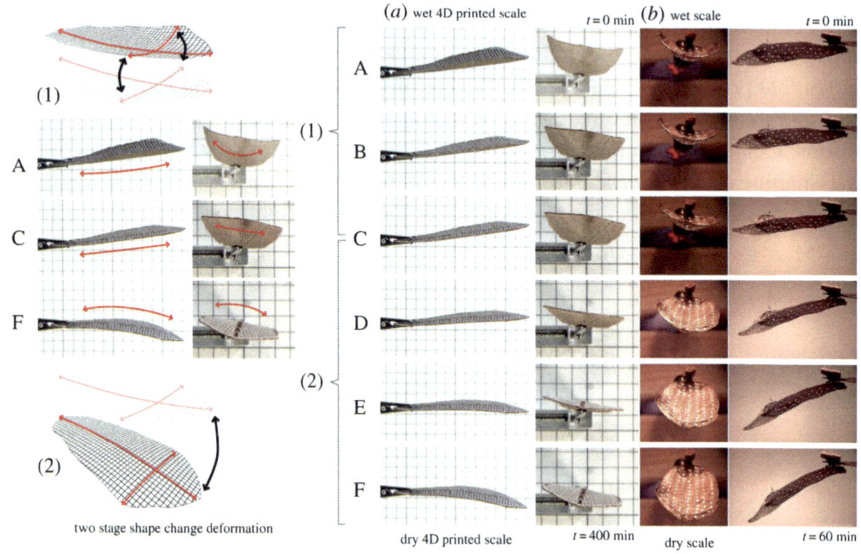

Fig. 4.5 A 4D printed artificial scale with two-phase shape deformation compared with the two-phase movement of the Bhutan pine scale. Starting fully wet (top) both scales show predominantly transversal curvature. The transversal curvature decreases (**a**–**c**) in the first stage of actuation, then the longitudinal curvature increases (**c**–**f**) until the maximum bending angle is reached when fully dry (bottom). (Figure adapted with permission from Correa et al., 2020; p. 13)

zones (Tahouni et al., 2020).Complex folding patterns can be used to geometrical amplify the actuation and/or to dramatically enhance geometric stiffness and depth in larger tessellations (Fig. 4.7), thereby demonstrating the complex movement patterns and functionalities enabled by distributing the actuation throughout the material via 3D printing.

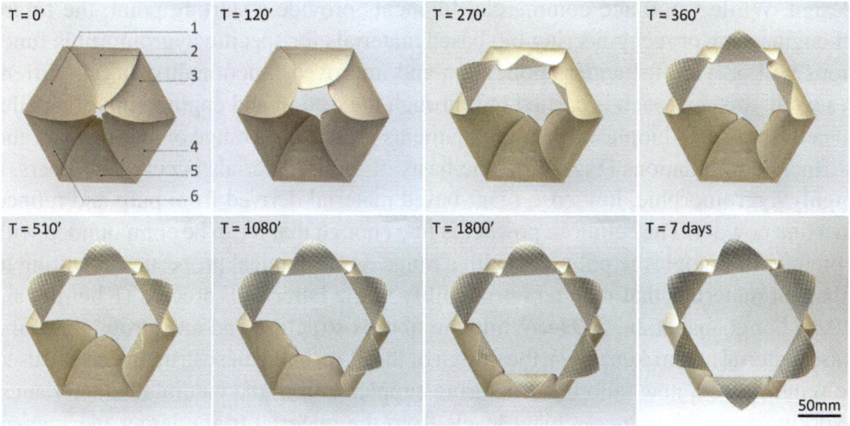

Fig. 4.6 Sequential shape-change in a 4D printed aperture. The timelapse shows six flaps taking turns to bend until the successful opening upon drying in 40% RH. (Figure adapted with permission from Tahouni et al., 2021; p. 16)

Fig. 4.7 Self-shaping curved folding mechanism (**a**) and lens tessellation design (**b**) and their transformation from flat (left) to folded (right) state. (Figure adapted with permission from Tahouni et al., 2020; p. 1)

4.4 Material Co-design for Bio Based, Hygromorphic Materials and Next-Generation 4D Printed Smart Structures

The next generation of hygromorphic smart structures is based on cooperative development of both material, digital manufacturing, and computational mechanism design. While wood and commercial filaments provide a starting point, the future of engineering or reengineering bio based materials for specific hygromorphic functions will provide for wider application and improved functionality. In our current research, this approach is carried out through the design and engineering of a pallet of cellulose-filled bioplastic printing filaments presenting a range of hygroscopic and stiffness combinations (Fig. 4.8). The basis of these materials is cellulose fibers, a highly hygromorphic, low-cost, plant-based material derived from pulp and refined to a fine powder. The cellulose powder is fine enough that it can be compounded with common thermoplastic polymers with a range of mechanical properties, resulting in filament materials that maintain printability in the latter FFF process (Kliem et al., 2020; Langhansl et al., 2021). While the fibrous structure and anisotropicity of the base material are broken down, the design of the 3D printed mesostructure can be used to reintroduce a new tailored structuring supplementing the natural characteristics. Working at multiple hierarchical levels from the material to the larger mechanism the approach maximizes the potential of the material through targeted design moves at each level. While the properties of the cellulose base material are fixed, the overall hygro response for example can be adjusted by the percentage of cellulose to matrix material in compounding, the stiffness of the matrix material as well as in the design of the bilayer mesostructure. The engineering of the custom filaments is also valuable in ensuring compatibility with materials used in other parts of the mesostructure with no active roles.

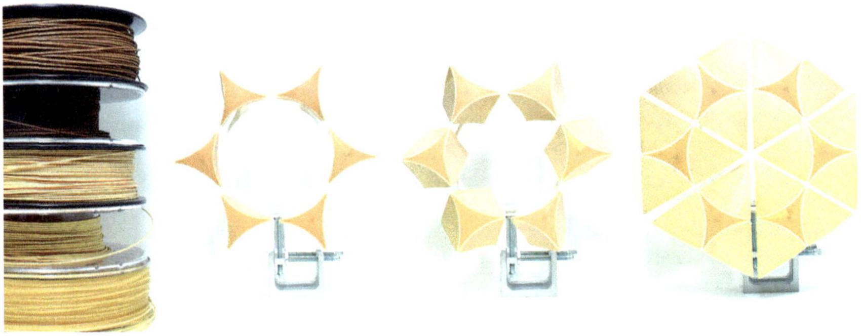

Fig. 4.8 Development of custom hygroscopic printing filaments composed of biobased matrix polymers and cellulose powder for humidity responsive smart structures based on the curved folding mechanism, transforming from fully folded (left) to nearly flat (right)

Through such multi-scale co-development approach, highly functional smart structures and weather-adaptive building envelopes can be created (Fig. 4.9). In the material level, the use of pure cellulose powder and selective matrix polymers results in high responsiveness and large shape-transformations in the printed structures. As a result, the smart structures can fully transform, for example open and close, in response to naturally occurring ranges of relative humidity. An optimized arrangement of mesostructures allows for this transformation to be fast, responding to weather changes in real-time. Furthermore, this shape-change is fully reversible and repeatable in many cycles, allowing for long-term use of such structures in real-world applications. On the mechanism level, the use of bilayer structures and curved folding geometries provides a motion amplification effect that further enhances the shape-change of the smart structures. Such shape-change can be utilized for different functionalities in building skins, such weather-adaptive ventilation or shading. As a result, the zero-energy system can dramatically enhance the performance of the building system, reducing the energy consumption related to heating, cooling and ventilation through utilizing the environment and natural resources.

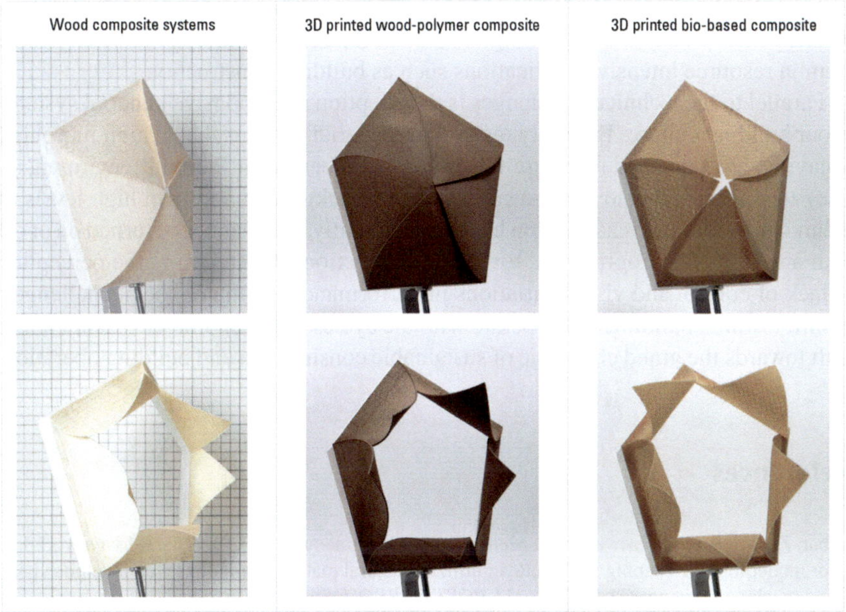

Fig. 4.9 The ICD has developed a series of bio-inspired weather-responsive apertures made from natural materials such as wood composites as well as newly developed bio-based materials for 3D printing

4.5 Future Perspectives—Learning to Build and Live with Biobased Materials for Sustainable Building Systems

Our work shows the value of deploying biobased hygromorphic materials in prototypical, passively responsive systems. However, working with natural materials presents challenges in contemporary design and manufacturing where precision, predictability and repeatability is highly valued. Future work will need to balance the unique natural functionality of these materials with the standards of engineering and design, to best utilize the range of sustainably harvested materials we have to work with (Sanandiya et al., 2020). One approach is to use data driven methods for optimizing the use of materials that come with inherent variation in material properties and structure and consider them advantageously during manufacturing (Morin et al., 2020; Sanandiya et al., 2020; Tamke et al., 2021). Combining advances in machine vision and applied machine learning is especially promising for utilizing hygromorphic functions which are sensitive to variation in natural material structuring and are challenging to solve using classical mechanical or material models (Akbar et al., 2022). The low cost, high availability, and regenerative aspects of naturally responsive materials favors further development in digital design and manufacturing technology to better utilize them in resource intensive applications such as building construction.

Parallel to the technical challenges is the adoption of responsive material systems in our building culture. Passively responsive material systems that exhibit morphing behaviors are common in nature but can be unexpected in the built environment. They tend to operate slowly and silently in the background and with high levels of redundancy. While this adaptation happens efficiently, they cannot be turned on or off with a switch or reprogrammed for different functions. Overcoming the perception of lack of control and visual variations in performance are major social challenges. A shift towards buildings that operate with the cycles of the natural environment is a shift towards the grand challenge of sustainable construction and building operation.

References

Akbar, Z., Wood, D., Kiesewetter, L., Menges, A., & Wortmann, T. (2022). A data-driven workflow for modelling self-shaping wood bilayer utilizing natural material variations with machine vision and machine learning. *CAADRIA 2022-POST-CARBON*. Sydney, Australia.

Burgert, I., & Fratzl, P. (2009). Actuation systems in plants as prototypes for bioinspired devices. *Philosophical Transactions of the Royal Society A, 367*(1893), 1541–1557.

Carneiro, V. H., Meireles, J., & Puga, H. (2013). Auxetic materials – A review. *Materials Science-Poland, 31*(4), 561–571.

Cheng, T., Tahouni, Y., Wood, D., Stolz, B., Mülhaupt, R., & Menges, A. (2020). Multifunctional mesostructures: Design and material programming for 4D-printing. In Symposium on Computational Fabrication (pp. 1–10). New York, NY, USA: ACM.

Cheng, T., Thielen, M., Poppinga, S., Tahouni, Y., Wood, D., Steinberg, T., Menges, A., & Speck, T. (2021). Bio-inspired motion mechanisms: Computational design and material programming of self-adjusting 4D-printed wearable systems. *Advanced Science, 8*(13), 2100411.

Correa, D., & Menges, A. (2017). Fused filament fabrication for multi-kinematic-state climate-responsive aperture. In A. Menges, B. O. Sheil, R. Glynn & M. Skavara (Eds.), *Fabricate* (pp. 190–195). UCL Press.

Correa, D., Papadopoulou, A., Guberan, C., Jhaveri, N., Reichert, S., Menges, A., & Tibbits, S. (2015). 3D-Printed Wood: Programming Hygroscopic Material Transformations. *3D Printing and Additive Manufacturing, 2*(3), 106–116.

Correa, D., Poppinga, S., Mylo, M. D., Westermeier, A. S., Bruchmann, B., Menges, A., & Speck, T. (2020). 4D pine scale: Biomimetic 4D printed autonomous scale and flap structures capable of multi-phase movement, *Philosophical transactions. Series A, Mathematical, Physical, and Engineering Sciences, 378*(2167), 20190445.

Davidova, M. (2016). *Wood as a primary medium to architectural performance: A case study in performance oriented architecture approached through systems oriented design*. Technical University of Liberec.

Dawson, C., Vincent, J. F. V., & Rocca, A.-M. (1997). How pine cones open. *Nature, 390*(6661), 668.

Duro-Royo, J., Mogas-Soldevila, L., & Oxman, N. (2015). Flow-based fabrication: An integrated computational workflow for design and digital additive manufacturing of multifunctional heterogeneously structured objects. *Computer-Aided Design, 69*, 143–154.

Erb, R. M., Sander, J. S., Grisch, R., & Studart, A. R. (2013). Self-shaping composites with programmable bioinspired microstructures. *Nature Communications, 4*(1), 1–8.

Grondzik, W. T., & Kwok, A. G. (2020). *Mechanical and electrical equipment for buildings*. Hoboken, New Jersey: Wiley.

Grönquist, P., Wittel, F. K., & Rüggeberg, M. (2018). Modeling and design of thin bending wooden bilayers. *PLoS ONE, 13*(10), e0205607.

Harlow, W. M., Côté, W. A., & Day, A. C. (1964). The opening mechanism of pine cone scales. *Journal of Forestry, 62*(8), 538–540.

Hoadley, R. B., & Barbara, L. H, E. o. (2000). *Understanding wood: A Craftsman's guide to wood technology*. Taunton Press.

Holstov, A., Farmer, G., & Bridgens, B. (2017). Sustainable Materialisation of Responsive Architecture. *Sustainability, 9*(3), 435 [Online]. https://doi.org/10.3390/su9030435.

Jung, W., Kim, W., & Kim, H.-Y. (2014). Self-burial mechanics of hygroscopically responsive awns. *Integrative and Comparative Biology, 54*(6), 1034–1042.

Kliem, S., Tahouni, Y., Cheng, T., Menges, A., & Bonten, C. (2020). Biobased smart materials for processing via fused layer modeling. *Fracture and damage mechanics: Theory, simulation and experiment* (p. 20034). Mallorca, Spain: AIP Publishing.

Krieg, O. D., Christian, Z., Correa, D., Menges, A., Reichert, S., Rinderspacher, K. and Schwinn, T. (2017). Hygroskin: Meteorosensitive Pavilion. In F. Gramazio, M. Kohler & S. Langenberg (Eds.). *Fabricate* (pp. 272–279) UCL Press.

Langhansl, M., Dörrstein, J., Hornberger, P., & Zollfrank, C. (2021). Fabrication of 3D-printed hygromorphs based on different cellulosic fillers. *Functional Composite Materials, 2*(1), 1–8.

Le Duigou, A., Correa, D., Ueda, M., Matsuzaki, R., & Castro, M. (2020). A review of 3D and 4D printing of natural fibre biocomposites. *Materials and Design, 194*, 108911.

Menges, A., & Reichert, S. (2015). Performative wood: Physically programming the responsive architecture of the hygroscope and hygroskin projects. *Architectural Design, 85*(5), 66–73.

Morin, M., Gaudreault, J., Brotherton, E., Paradis, F., Rolland, A., Wery, J., & Laviolette, F. (2020). Machine learning-based models of sawmills for better wood allocation planning. *International Journal of Production Economics, 222*, 107508.

Oxman, N. (2010). Structuring materiality: Design fabrication of heterogeneous materials. *Architectural Design, 80*(4), 78–85.

Poppinga, S., Nestle, N., Šandor, A., Reible, B., Masselter, T., Bruchmann, B., & Speck, T. (2017). Hygroscopic motions of fossil conifer cones. *Scientific Reports, 7*, 40302.

Poppinga, S., Zollfrank, C., Prucker, O., Rühe, J., Menges, A., Cheng, T., & Speck, T. (2018). Toward a new generation of smart biomimetic actuators for architecture. *Advanced Materials, 30*(19), e1703653.

Reichert, S., Menges, A., & Correa, D. (2015). Meteorosensitive architecture: Biomimetic building skins based on materially embedded and hygroscopically enabled responsiveness. *Computer-Aided Design, 60*, 50–69.

Reyssat, E., & Mahadevan, L. (2009). Hygromorphs: From pine cones to biomimetic bilayers. *Journal of the Royal Society, Interface, 6*(39), 951–957.

Rüggeberg, M., & Burgert, I. (2015). Bio-inspired wooden actuators for large scale applications. *PLOS ONE, 10*(3), e0120718 [Online]. https://doi.org/10.1371/journal.pone.0120718.

Sanandiya, N. D., Ottenheim, C., Phua, J. W., Caligiani, A., Dritsas, S., & Fernandez, J. G. (2020). Circular manufacturing of chitinous bio-composites via bioconversion of urban refuse. *Scientific Reports, 10*(1), 4632.

Tabadkani, A., Roetzel, A., Li, H. X., & Tsangrassoulis, A. (2021). Design approaches and typologies of adaptive facades: A review. *Automation in Construction, 121*, 103450.

Tahouni, Y., Cheng, T., Wood, D., Sachse, R., Thierer, R., Bischoff, M., & Menges, A. (2020). Self-shaping curved folding. In *Symposium on Computational Fabrication* (pp. 1–11). New York, NY, USA: ACM.

Tahouni, Y., Krüger, F., Poppinga, S., Wood, D., Pfaff, M., Rühe, J., Speck, T., & Menges, A. (2021). Programming sequential motion steps in 4D-printed hygromorphs by architected mesostructure and differential hygro-responsiveness. *Bioinspiration and biomimetics, 16*(5).

Tamke, M., Gatz, S., Svilans, T., & Ramsgard Thomsen, M. (2021). *Tree to Product-Prototypical workflow connecting Data from tree with fabrication of engineered wood structure-RawLam, WCTE 2021*. Santiago.

Timoshenko, S. (1925). Analysis of bi-metal thermostats. *Journal of the Optical Society of America, 11*(3), 233.

United Nations Environment Programme. (2020). *Global status report for buildings and construction: Towards a zero-emission, efficient and resilient buildings and construction sector*.

Vailati, C., Bachtiar, E., Hass, P., Burgert, I., & Rüggeberg, M. (2018). An autonomous shading system based on coupled wood bilayer elements. *Energy and Buildings, 158*(9), 1013–1022.

Chapter 5
Solar-Thermal Conversion in Envelope Materials for Energy Savings

Mohammad Elmi and Julian Wang

Abstract Designing and incorporating solar-absorptive materials into building envelopes has been conventionally and widely adopted in the passive solar design paradigm. With the emergence of newly discovered spectrally selective materials, studies of and discussions about solar-thermal conversion have gained renewed research interest. This chapter introduces the fundamental processes and mechanisms of solar-thermal conversion in typical building materials and the application of such conversion in building envelopes. Both conventional and emerging application methods and principles are discussed. Potential future directions are then identified regarding the use of photothermal materials in architecture.

Keywords Photothermal conversion · Solar absorption · Solar radiation · Envelope materials · Passive solar design

5.1 Introduction

Building energy consumption accounts for a substantial share of the total energy demand, and it is continuing to expand rapidly due to the world's rising population and drive for urbanization (Wu et al., 2018). Numerous studies have been conducted with the goal of achieving sustainable buildings. Sustainability in building design can be studied using different approaches ranging from sustainability during construction (Syed et al., 2021) to reduce operational energy consumption in buildings (Chel & Kaushik, 2018). One of the major challenges for sustainable development is reducing energy consumption while also preserving occupants' comfort (Ke et al., 2018). The first and most important aspect of energy-efficient construction is passive building design (Chel & Kaushik, 2018). The passive solar design includes ways of maintaining buildings' thermal comfort without using mechanical or electrical equipment, helping to reduce the energy consumption of buildings (Stevanović, 2013). Attaching sunspaces to building façades to reduce energy consumption is a basic passive use

M. Elmi · J. Wang (✉)
Department of Architectural Engineering and Department of Architecture, Pennsylvania State University, University Park, PA, USA
e-mail: julian.wang@psu.edu

J. Wang et al. (eds.), *Advanced Materials in Smart Building Skins for Sustainability*,
https://doi.org/10.1007/978-3-031-09695-2_5

of solar energy that has been employed since the 1970s (Gainza-Barrencua et al., 2021). New materials and discoveries in nano-/micro-scale light-to-heat conversion have paved novel ways of using solar energy in buildings. Light-to-heat conversion has been studied extensively for biomedical applications (Chen et al., 2017; Kim et al., 2019; Li et al., 2016), but there is a lack of research on the various possible applications of the photothermal effect in energy-efficient buildings. The purpose of this chapter is to review prior research using light-to-heat energy conversion to make buildings more energy-efficient and study different photothermal applications in buildings.

In this chapter, each light-to-heat energy conversion mechanism is described in detail, followed by light-to-heat conversion applications in buildings. This chapter will help researchers become familiar with various applications of the photothermal effect in buildings, as well as the specific materials used in such applications. Moreover, beyond applications, a detailed study of the mechanisms of photothermal conversion will help researchers improve their design efficiency.

5.2 Photoactivation Modes

Solar radiation is an electromagnetic wave consisting of photons with different frequencies. Depending on its properties, when light strikes matter, photons can be reflected, absorbed, transmitted/refracted, or scattered. The absorbed photons can cause various energy conversion processes (Wu et al., 2019). Eventually, the energy that has been absorbed is released. Depending on the substance, the release of that energy can take place in one of many closely similar but mechanistically distinct modes, including those that are photoconductive, photoelectrochemical, photovoltaic, photothermal, and fluorescent. In theory, distinct applications may each have specific photoexcitation modes; however, most devices concurrently photoactivate in various modes, resulting in these modes in some cases being interchangeable. Photoconductive modes occur when a substance becomes more electrically conductive as a result of the absorption of light. In this step, photoexcited electrons are transferred from the atoms or molecules with which they are linked. If an electric field is supplied, this causes the electron to travel towards the cathode and an electrical current is generated. In metallic conductors, the current created is proportional to the surface area accessible on the conductor and the intensity of the light. In photoelectrochemical photoactivation, a high-energy electron causes a redox reaction at the interface between the electrode and electrolyte. As a consequence, the material may corrode, resulting in changes to the electrical properties of the electrode/electrolyte interface. In a solid-state device, the photovoltaic mode occurs when an electron is excited by a photon, without the presence of an electric field. However, the electron does not become separated from its hole, resulting in the photocurrent being restricted from leaving the device. In contrast to photoelectrochemical activation, photovoltaic activation does not result in the production of a chemical reaction. The photothermal mode is characterized by the transfer of energy to an electron by a photon. In this mode, the absorbed energy is not converted into the formation of another photon, a release of electrons, or an increase in the electron's energy level, but rather the absorbed or surplus energy is released as heat or kinetic energy, which can cause the material's temperature to increase. The last mode is the photoelectromagnetic mode. In this mode, absorbed energy in a photoexcited electron is released as a formation of a fluorescent photon that has less energy and, consequently, a longer wavelength than the incident photon (Archer, 2002; Kozai & Vazquez, 2015).

5.3 Photothermal Mechanisms

Based on a study by Chen et al., there are three photothermal mechanisms: (1) plasmonic localized heating, (2) electron/hole generation and relaxation, and (3) thermal vibration of molecules (Chen et al., 2019). Different photothermal materials exhibit a variety of photothermal mechanisms. The electromagnetic radiation absorption of metallic materials, good solar absorbers, can lead to the localized surface plasmon resonance effect. Metallic materials such as gold, silver, copper, palladium, and aluminum have been widely produced, modified, and explored for use in photothermal applications, either as nanoparticles (NPs) or as composites. Au NPs have demonstrated outstanding plasmonic resonance in the visible-near infrared spectrum, as well as chemical stability. Because of its remarkable efficiency of light-to-heat energy conversion and flexibility of control in terms of size and structure, gold has received a great deal of attention as a high-performance photothermal metal. Semiconductor photothermal materials mostly have electron/hole generation and relaxation mechanisms. A semiconductor is a material that conducts electricity between the conductor and insulator. It is usually opaque to visible light and transparent to infrared radiation. Semiconductors' absorption in the visible region is largely reliant on the bandgap. Organic semiconductors' band gap is generally 2.5–4 eV, whereas the band gap for inorganic semiconductors is 1–2 eV. The band gap can be modified to decrease the band gap or produce energy levels within the band gap in various ways, including: (1) fine-tuning the size, structure, and composition, (2) self-doping and creating disorder, and (3) changing the position of conduction and valence band by doping or vacancy creation (Li et al., 2021; Tao et al., 2019). Semiconductors with smaller bandgaps tend to have a broad absorption spectrum and can harvest more photons, increasing their photothermal conversion ability (Su et al., 2020). Reduced TiO_2 NPs are mostly used by researchers as common photothermal materials because of their broad UV-Vis spectrum of light absorption, efficient heat energy conversion, chemical stability, and low toxicity and cost (Fuzil et al., 2021). The photothermal conversion mechanism of carbon-based nanomaterials is the thermal vibration of molecules. A photothermal material capable of absorbing the entire spectrum of light and reflecting nothing is an excellent photothermal converter. Although for this purpose a perfectly black body is a desirable material, such bodies can be impractical because they also feature high reflection (Wu et al., 2019). Polymers have a photothermal conversion mechanism similar to that of carbon materials, also related to the thermal vibration of molecules. Polymeric materials offer the benefits of being sufficiently flexible and easily moldable, and have broad and high solar absorption (Tao et al., 2019). Numerous polymers exist, including polydopamine, polyaniline (PAN), and PPy, which have significant π electron delocalization structures and hence exhibit distinctive optical characteristics. PAN is an excellent photothermal material, owing to its high absorption of light, which can be converted to heat with minimal energy loss. PPy offers superior photostability, low light reflection, and solution-based fabrication, making it easier to use when managing film thickness (Wu et al., 2019).

In general, each photothermal material has a single dominant photothermal mechanism. However, in some instances, several photothermal mechanisms may be at work together. This is especially the case for hybrid photothermal materials, which are composed of two or more components with distinct photothermal mechanisms (Chen et al., 2019; Fuzil et al., 2021). Each photothermal mechanism is discussed in greater detail in the following subsections. In the next section, a time-scale overview of each mechanism is given.

5.3.1 Plasmonic Localized Heating

Plasmonic localized heating resulting from surface plasmon resonance occurs when the frequency of the incident light matches the natural frequency of electrons on the metal's surface. This match causes electrons to excite and, consequently, hot electrons are generated. The excited hot charge carriers oscillate collectively with the incident electromagnetic field, causing heat generation via the joule mechanism. The hot electrons quickly redistribute their energy via the electron–electron scattering process, which generates heat in the plasmonic element (Anis et al., 2019; Chen et al., 2019). The schematic of this mechanism is shown in Fig. 5.1.

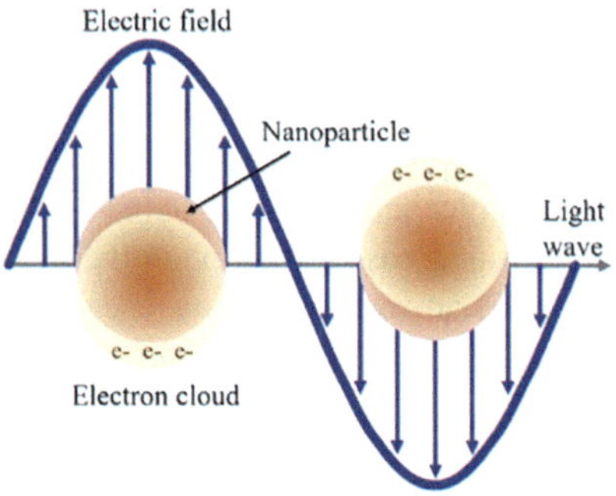

Fig. 5.1 Schematic of the plasmonic localized heating mechanism

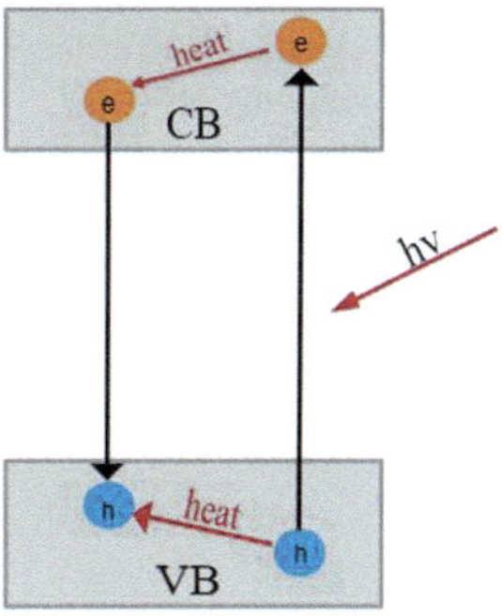

Fig. 5.2 Electron/hole generation and relaxation mechanism

5.3.2 Electron/Hole Generation and Relaxation

When a semiconductor is exposed to sunlight with an energy similar to or higher than its bandgap, electron/hole pairs are generated. Because in narrow bandgap semiconductors the energy of the majority of photons from incoming solar light is greater than the bandgap, above-bandgap electron/hole pairs are formed. These electron/hole pairs relax to the band edges and their surplus energy is released by radiative or nonradiative phonon relaxation. When nonradiative relaxation occurs and the energy is released by phonons, the lattice is locally driven to heat up and light energy is converted to heat (see Fig. 5.2). The temperature distribution in the lattice is established based on the material's recombination and optical absorption properties. Conversely, in broad bandgap semiconductors, most of the absorbed light energy is re-emitted as photons when the electron/hole pair recombines at the bandgap edge, leading to a significantly lower light-to-heat conversion efficiency (Almond et al., 1996; Fuzil et al., 2021; Gao et al., 2019; Zhu et al., 2018).

5.3.3 Thermal Vibration of Molecules

In the thermal vibration of molecules, low energy photons can excite weak electrons from the π orbital to π^* orbital. The excited electrons are raised from the ground state, the highest occupied molecular orbital, to an orbital with higher energy, the lowest unoccupied molecular orbital. Then, when the excited electron relaxes back to the ground state, the heat is released and the conversion of light-to-heat energy is completed. Figure 5.3 demonstrates the schematic of this mechanism. Carbon materials and certain polymers are among the materials engaging in light-to-heat energy conversion through this mechanism (Gao et al., 2019; Vélez-Cordero & Hernández-Cordero, 2015).

There are several materials that have multiple forms of light-to-energy conversion. For example, some narrow bandgap semiconductors such as Ti_2O_3 have a high photothermal conversion efficiency, as well as a photovoltaic effect (Zhu et al.,

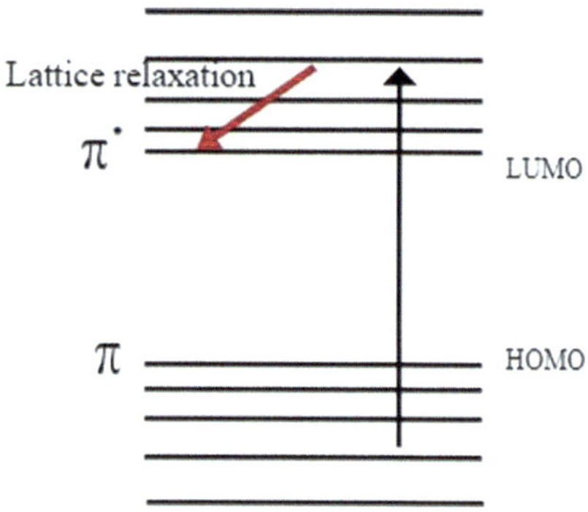

Fig. 5.3 Mechanism for the thermal vibration of molecules

2018). Other semiconductors (e.g., TiO_2) are widely utilized for photoelectrochemical purposes (Elmi et al., 2020) and also demonstrate a photovoltaic effect (Bai & Zhou, 2014). Furthermore, when light strikes the surface of a metal, a photothermal effect can occur; additionally, hot photoexcited electrons can interact with molecules on the metal's surface and induce photochemistry (Brongersma et al., 2015). Moreover, there are compound materials that demonstrate multiple forms of light-to-energy conversion. Substances with switchable absorption properties are among these materials and have been widely used as an absorber layer in smart windows, offering practical applications such as reducing energy consumption and improving the built environment. Upon illumination, the photothermal effect switches the absorber layer from a transparent state to an absorbing photovoltaic state. After cooling, the absorber layer is returned to the transparent state and the device acts as a window to visible light (Wheeler et al., 2017).

5.4 Timescales of Photothermal Mechanisms

As mentioned above, there are different mechanisms for light-to-heat energy conversion. This conversion process can be summarized from the timescale view as follows. A Jablonski diagram is helpful for studying the thermal vibrations of molecules' timescale mechanisms. Based on this diagram, a molecule is promoted from its ground state to a higher state by an absorbed photon on a timescale of the order of 10^{-15} s (1 fs). Then, vibrational relaxation and, consequently, heat generation occur on a timescale of 10^{-12}–10^{-10} s (1–100 ps) (Lichtman & Conchello, 2005). In the plasmonic localized heating mechanism, electron/hole excitation occurs on a timescale of 10^{-15}–10^{-13} s (1–100 fs). The electron–electron scattering and relaxation of hot electrons occur around 10^{-13}–10^{-12} s (100 fs–1 ps). Then, equilibration with a lattice occurs on a timescale of several picoseconds. The dynamics of this process are well described by the two-temperature model, where the electron's temperature and the temperature of the lattice are eventually equal. Heat is transferred to the surroundings in the final step of this process, which can take from 10^{-10} to 10^{-8} s (100 ps–10 ns) (Brongersma et al., 2015).

5.5 Performance and Applications of Building Photothermal Materials

In this section, some important characteristics of photothermal materials are discussed, as well as applications of these photothermal effects in buildings envelopes. One of the major factors in photothermal materials' performance is how well they can absorb solar radiation. In other words, a good photothermal material has a high rate of solar absorption. In this context, solar absorptance is typically calculated as the ratio of total absorbed solar radiation to the total incident solar radiation, as in the equation below (Boström et al., 2003; Gao et al., 2021):

$$\alpha = \frac{\int_{300}^{2500} |1 - R(\lambda)| I(\lambda) d\lambda}{\int_{300}^{2500} I(\lambda) d\lambda}$$

where $I(\lambda)$ is the solar spectral irradiance and $R(\lambda)$ is the reflectance at wavelength λ.

Another important factor in photothermal materials' performance is light-to-heat energy conversion efficiency. This factor shows how efficiently the material can convert absorbed energy to heat. Methods for calculating light-to-heat energy conversion efficiency differ based on the material and photothermal mechanisms. One experimental method for determining light-to-heat energy conversion efficiency involves measuring the matter's temperature rise and, consequently, its specific heat capacity. The heat energy generated by incident light is then compared to the energy from absorbed light (Gao et al., 2019; Jiang et al., 2013; Li et al., 2017).

Thermal management plays a crucial role in photothermal materials' performance. When the desired heat is produced by the material, it should be transferred efficiently to meet users' needs, with minimum heat loss (Fuzil et al., 2021; Gao et al., 2021).

Photothermal materials have numerous applications in solar energy systems, including saving energy and reducing energy consumption. Employing photothermal materials in building envelopes can reduce building energy consumption and help achieve the goal of green buildings. Recently, a number of studies have explored applications of light-to-heat energy conversion in building envelopes. Based on this prior work, we categorized these applications into three main groups: (1) employing photothermal materials as a coating on window surfaces to improve their thermal performance, (2) applications of the photothermal effect in the photo-thermochromic and phase change materials used in building envelopes, and (3) solar-thermal conversion in conventional passive solar designs. Below, each application category is discussed, followed by a review of related research.

5.5.1 Photothermal Materials Applied to Improve Windows' Thermal Performance

When light strikes a photothermal material, a large part of its energy is converted to heat. After coating the surface of the window with a photothermal material, the photothermal effect heats the window's surface and causes a temperature increase. Then, the produced heat is transferred to the inside room by radiation heat transfer. This process is schematically illustrated in Fig. 5.4.

The solar heat gain coefficient (SHGC) of a window can be calculated as (Zhang et al., 2021):

$$\mathrm{SHGC} = \tau_s + N_i \alpha_s$$

where τ_s is the solar transmittance, N_i is the inward-flowing fraction of absorbed radiation, and α_s is the solar absorptance of the window.

Zhang et al. performed a study showing that photothermal materials coated on a window's surface can individually increase the SHGC (Zhang et al., 2021). They compared two different windows, one with a photothermal coating and one with a low-e coating. Both windows had the same transmittance, τ_s. They demonstrated that the window with a photothermal coating had a higher SHGC. This result confirmed that the photothermal material had a direct impact on increasing the SHGC of the

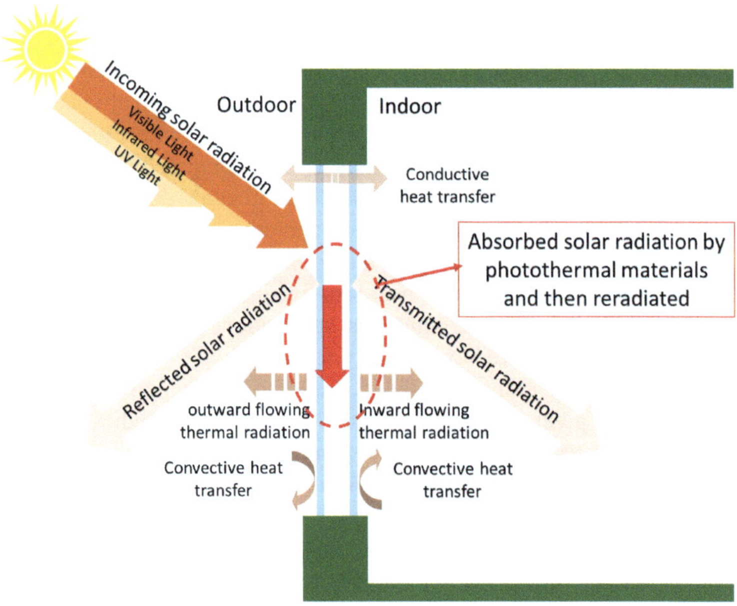

Fig. 5.4 Windows' thermal performance improvement from photothermal materials

window. Based on the similar mechanism, Md Jahid et al. developed a reversible window structure by embedding the photothermal materials into the double-pane window systems, which could achieve 0.2 to 0.6 solar heat gain modulations and a stable visible transmittance of 0.32 (Md Jahid et al., 2022). The building energy simulation presents the energy savings could be about 18% in mixed climates in relative to the baseline models built using the most recent energy codes.

Guo et al. constructed a transparent solar thermal surface on a window glazing. The surfaces constructed included gold nanoparticles with linear morphologies. The researchers studied the photothermal conversion effect on the window's surface and reported that as a result of employing such surfaces, the window glazing temperature increased up to 9.8 °C under solar irradiation while retaining a high transmittance in the visible range (Guo et al., 2021).

5.5.2 Photothermal Effect in Photo-Thermochromic and Phase Change Materials Used in Buildings Envelopes

The optical properties of photo-thermochromic smart windows change in response to changes in temperature, a fully passive method of light modulation with a broad variety of potential applications in energy-efficient buildings. However, one issue with using thermochromic windows is their high transition temperature, meaning that the optical properties of the window are changed at temperatures higher than room temperature, so they are not practical for use in building applications. For example, VO_2, which is commonly used as thermochromic material, has a transition temperature of 68 °C. To address this issue, a significant amount of research has been conducted that leverages the idea that the optical properties of thermochromic materials can be changed at room temperature by the photothermal effect. In the other words, the photothermal effect accelerates the phase transition of thermochromic materials and causes optical changes to occur in lower environmental temperatures (Zou et al., 2021). Hao et al. hybridized VO_2 with TiN nanoparticles to apply the photothermal effect and make thermochromic VO_2 applicable at room temperature (Hao et al., 2018). Ji prepared a composite of VO_2 and PbS for this same purpose (Ji, 2014). The schematic of this mechanism for VO_2/TiN glass coating is demonstrated in Fig. 5.5.

Hydrogels are also materials commonly appearing in photo-thermochromic windows. A number of studies have combined hydrogels with various photothermal materials to make them usable in window applications at room temperature. Tian et al. studied poly(*N*-isopropylacrylamide) hydrogel combined with polydopamine particles as a photothermal material and employed them for energy-efficient smart windows (Tian, 2021). Also, the combination of hydrogels with other materials such as graphene oxide, Cs_xWO_3, and antimony-doped tin oxide have been studied for use in smart window applications (Chou et al., 2017; Huang et al., 2015; Kim et al.,

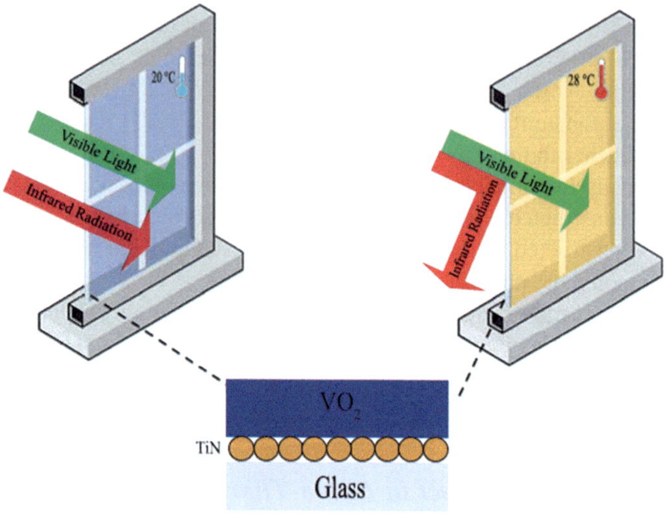

Fig. 5.5 Schematic of the VO_2/TiN glass coating used in smart windows

2014, 2015; Lee et al., 2016; Wu et al., 2018; Xu et al., 2018; Yeong Lee et al., 2017). Zou et al. prepared a research review of photo- and electro-driven thermochromic windows and studied VO_2-based composites, liquid crystals, and hydrogels (Zou et al., 2021).

The concept of applying the photothermal effect seen in phase change materials is the same as in photo-thermochromic windows, meaning that photothermal materials are applied to heat up another material to help its properties change in response to shifts in temperature. The difference is in the type of material with which the photothermal material is combined. Photothermal materials can be hybridized with the phase change materials used as filler in windows. In phase change materials, latent heat is stored when the heat is excessive and released when it is required. Research has pursued improving the photothermal properties of different phase change materials to make them usable in windows. For instance, Ma et al. studied photothermal property improvement for paraffin via binary Zn–ZnO nanoparticles (Ma et al., 2022). The researchers simulated double-layer glazing units with binary Zn–ZnO/paraffin nanofluids serving as a filler, and reported less energy consumption in buildings and more comfortable indoor environments.

Although most research in this category has used photothermal materials in glazing systems, Shen et al. studied the effects on the inside temperature of buildings produced by photo-thermochromic smart coatings applied to sun-facing walls and roofs (Shen et al., 2021). The authors reported that employing these photo-thermochromic coatings could help manage the indoor temperature.

5.5.3 Solar-Thermal Conversion in Conventional Passive Solar Designs

Another way that solar-thermal conversion is used in buildings to reduce energy consumption is by employing conventional passive solar designs such as Trombe walls. Trombe walls, or solar heating walls, come in several different types, but a classic Trombe wall consists of a wall, glass, and the air space in between (see Fig. 5.6a). The materials used in Trombe walls have a high heat-storage capacity, and in order to increase the surface absorption, the external surface is colored black (Agrawal & Tiwari, 2010; Gan, 2006; Hestnes et al., 2013; Mekhilef et al., 2011; Saadatian et al., 2012). The mechanism for using solar energy via Trombe walls is that direct and diffuse solar radiation are absorbed by the wall during the day and converted into heat. This heat is transferred to the interior by radiation and convection during the night (Agrawal & Tiwari, 2010). There is a ventilated Trombe wall that features free or forced airflow between the glass and wall. When the temperature of the wall increases, the air in front is also warmed. This warm air then directly transfers heat to the interior (Szyszka et al., 2017) (see Fig. 5.6b).

Much research has sought ways of efficiently employing Trombe walls in buildings. Yu et al. studied purified Trombe walls with ventilation blinds (Yu et al., 2019),

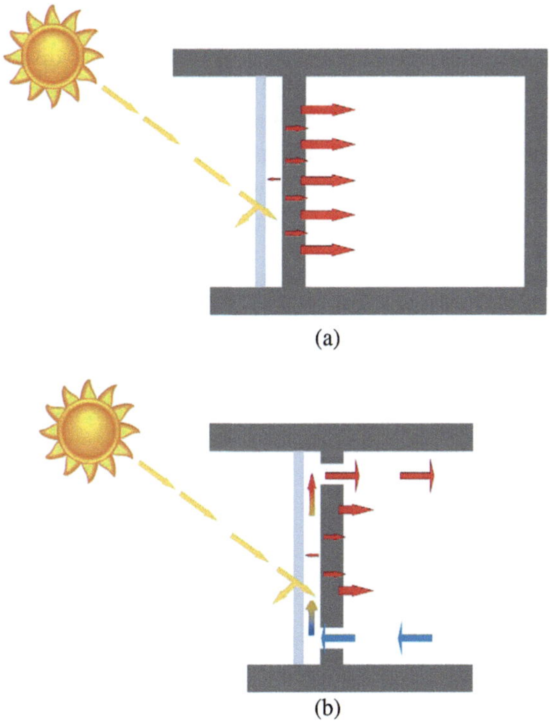

Fig. 5.6 A schematic of a classic Trombe wall: **a** non-ventilated and **b** ventilated

using TiO_2 and MnO_x–CeO_2 layers on both sides of the blinds for the summer and winter modes, respectively. Jie et al. modeled a Trombe wall with PV cells (Jie et al., 2007), establishing both electrical performance and temperature distribution. Szyszka et al. provided an overview of the uses of Trombe walls, based on the climate conditions of Central Europe (Szyszka et al., 2017). Saadatian et al. published a comprehensive review of research and developments related to Trombe walls (Saadatian et al., 2012), discussing different types and analyzing their efficiency in detail.

5.6 Future Research Directions for Using Photothermal Materials in Buildings Envelopes

With recent discoveries of photothermal materials, especially spectrally-selective nanoscale materials, solar-thermal conversion is now exhibiting great potential for building energy savings and environmental sustainability. Unique phenomena produced by photothermal materials under solar radiation have recently been reported that seem to challenge the conventional notion of macro-scale solar-thermal conversion features. However, fundamental explanations of the mechanism operating, derived from experimental and simulational works alike, have been ignored. Thus, continuous investigation of the physics and materials operating are needed. Furthermore, most recent research considering newly-discovered nanoscale solar-thermal conversion has been limited to applications for building windows. Although solar-thermal conversion has historically been applied in passive solar design and demonstrated strong energy-savings potential, very few studies have explored the potential application of nanoscale photothermal materials or structures in opaque building envelopes.

Acknowledgements This research is supported by NSF 2001207—CAREER: Understanding the Thermal and Optical Behaviors of the Near Infrared (NIR)-Selective Dynamic Glazing Structures.

References

Agrawal, B., & Tiwari, G. (2010). *Building integrated photovoltaic thermal systems: For sustainable developments.*

Almond, D., Patel, P., & Patel, P. (1996). *Photothermal science and techniques.*

Anis, S. F., Hashaikeh, R., & Hilal, N. (2019). Functional materials in desalination: A review. *Desalination, 468*, 114077. https://doi.org/10.1016/j.desal.2019.114077

Archer, M. D. (2002). Photovoltaics and photoelectrochemistry: Similarities and differences. *Physica E: Low-Dimensional Systems and Nanostructures, 14*(1–2), 61–64. https://doi.org/10.1016/S1386-9477(02)00450-2

Bai, J., & Zhou, B. (2014). Titanium dioxide nanomaterials for sensor applications. *Chemical Reviews, 114*(19), 10131–10176. https://doi.org/10.1021/cr400625j

Boström, T., Wäckelgård, E., & Westin, G. (2003). Solution-chemical derived nickel–alumina coatings for thermal solar absorbers. *Solar Energy, 74*(6), 497–503. https://doi.org/10.1016/S0038-092X(03)00199-3

Brongersma, M. L., Halas, N. J., & Nordlander, P. (2015). Plasmon-induced hot carrier science and technology. *Nature Nanotechnology, 10*(1), 25–34. https://doi.org/10.1038/nnano.2014.311

Chel, A., & Kaushik, G. (2018). Renewable energy technologies for sustainable development of energy efficient building. *Alexandria Engineering Journal, 57*(2), 655–669. https://doi.org/10.1016/J.AEJ.2017.02.027

Chen, C., Kuang, Y., & Hu, L. (2019). Challenges and opportunities for solar evaporation. *Joule, 3*(3), 683–718. https://doi.org/10.1016/j.joule.2018.12.023

Chen, W., et al. (2017). Black phosphorus nanosheet-based drug delivery system for synergistic photodynamic/photothermal/chemotherapy of cancer. *Advanced Materials, 29*(5), 1603864. https://doi.org/10.1002/adma.201603864

Chou, H. T., Chen, Y. C., Lee, C. Y., Chang, H. Y., & Tai, N. H. (2017). Switchable transparency of dual-controlled smart glass prepared with hydrogel-containing graphene oxide for energy efficiency. *Solar Energy Materials and Solar Cells, 166*, 45–51. https://doi.org/10.1016/J.SOLMAT.2017.01.025

Elmi, M., et al. (2020). Structural optimization of hematite nanorod array as water oxidation photoanodes.

Fuzil, N. S., et al. (2021). A review on photothermal material and its usage in the development of photothermal membrane for sustainable clean water production. *Desalination, 517*, 115259. https://doi.org/10.1016/j.desal.2021.115259

Gainza-Barrencua, J., Odriozola-Maritorena, M., Hernandez_Minguillon, R., & Gomez-Arriaran, I. (2021). Energy savings using sunspaces to preheat ventilation intake air: Experimental and simulation study. *Journal of Building Engineering, 40*, 102343. https://doi.org/10.1016/J.JOBE.2021.102343

Gan, G. (2006). Simulation of buoyancy-induced flow in open cavities for natural ventilation. *Energy and Buildings, 38*(5), 410–420. https://doi.org/10.1016/J.ENBUILD.2005.08.002

Gao, M., Peh, C. K., Meng, F. L., & Ho, G. W. (2021). Photothermal membrane distillation toward solar water production. *Small Methods, 5*(5), 2001200. https://doi.org/10.1002/smtd.202001200

Gao, M., Zhu, L., Peh, C. K., & Ho, G. W. (2019). Solar absorber material and system designs for photothermal water vaporization towards clean water and energy production. *Energy & Environmental Science, 12*(3), 841–864. https://doi.org/10.1039/c8ee01146j

Guo, M., Gao, L., Wei, Y., Ma, Y., Jianyong, Y., & Ding, B. (2021). Solar transparent radiators based on in-plane worm-like assemblies of metal nanoparticles. *Solar Energy Materials and Solar Cells, 219*. https://doi.org/10.1016/J.SOLMAT.2020.110796

Hao, Q., et al. (2018). VO_2/TiN plasmonic thermochromic smart coatings for room-temperature applications. *Advanced Materials, 30*(10), 1705421. https://doi.org/10.1002/adma.201705421

Hestnes, A. G., Hastings, R., & Saxhof, B. (2013). Solar energy houses: Strategies, technologies, examples, second edition. In *Solar energy houses: Strategies, technologies, examples* (2nd ed., Vol. 9781315067148, pp. 1–202). https://doi.org/10.4324/9781315067148

Huang, H., Ng, M., Wu, Y., & Kong, L. (2015). Solvothermal synthesis of Sb:SnO_2 nanoparticles and IR shielding coating for smart window. *Materials and Design, 88*, 384–389. https://doi.org/10.1016/J.MATDES.2015.09.013

Ji, Q. (2014). *Design of intelligent infrared control nanomaterial system*. Zhengzhou University.

Jiang, K., Smith, D. A., & Pinchuk, A. (2013). Size-dependent photothermal conversion efficiencies of plasmonically heated gold nanoparticles. *Journal of Physical Chemistry C, 117*(51), 27073–27080. https://doi.org/10.1021/jp409067h

Jie, J., Hua, Y., Wei, H., Gang, P., Jianping, L., & Bin, J. (2007). Modeling of a novel Trombe wall with PV cells. *Building and Environment, 42*(3), 1544–1552. https://doi.org/10.1016/J.BUILDENV.2006.01.005

Ke, Y., Zhou, C., Zhou, Y., Wang, S., Chan, S. H., & Long, Y. (2018). Emerging thermal-responsive materials and integrated techniques targeting the energy-efficient smart window application. *Advanced Functional Materials, 28*(22), 1800113. https://doi.org/10.1002/adfm.201800113

Kim, D., Lee, H., Yoon, J. (2014). Remote control of volume phase transition of hydrogels containing graphene oxide by visible light irradiation. *RSC Advances*. https://doi.org/10.1039/c4ra01537a

Kim, D., Lee, E., Lee, H. S., & Yoon, J. (2015). Energy efficient glazing for adaptive solar control fabricated with photothermotropic hydrogels containing graphene oxide. *Scientific Reports, 5*(1), 1–6. https://doi.org/10.1038/srep07646

Kim, M., Lee, J.-H., Nam, J.-M., Kim, M., Nam, J.-M., & Lee, J.-H. (2019). Plasmonic photothermal nanoparticles for biomedical applications. *Advancement of Science, 6*(17), 1900471. https://doi.org/10.1002/ADVS.201900471

Kozai, T. D. Y., & Vazquez, A. L. (2015). Photoelectric artefact from optogenetics and imaging on microelectrodes and bioelectronics: New challenges and opportunities. *Journal of Materials Chemistry B, 3*(25), 4965–4978. https://doi.org/10.1039/C5TB00108K

Lee, E., Kim, D., & Yoon, J. (2016). Stepwise activation of switchable glazing by compositional gradient of copolymers. *ACS Applied Materials & Interfaces, 8*(39), 26359–26364. https://doi.org/10.1021/ACSAMI.6B10091

Li, J., et al. (2021). Evaporation efficiency monitoring device based on biomass photothermal material for salt-resistant solar-driven interfacial evaporation. *Solar Energy Materials and Solar Cells, 222*, 110941. https://doi.org/10.1016/j.solmat.2020.110941

Li, R., Zhang, L., Shi, L., & Wang, P. (2017). MXene Ti_3C_2: An effective 2D light-to-heat conversion material. *ACS Nano, 11*(4), 3752–3759. https://doi.org/10.1021/acsnano.6b08415

Li, Z., et al. (2016). Multimodal imaging-guided antitumor photothermal therapy and drug delivery using bismuth selenide spherical sponge. *ACS Nano, 10*(10), 9646–9658. https://doi.org/10.1021/ACSNANO.6B05427

Lichtman, J. W., & Conchello, J. A. (2005). Fluorescence microscopy. *Nature Methods, 2*(12), 910–919. https://doi.org/10.1038/nmeth817

Ma, M., Xie, M., & Ai, Q. (2022). Study on photothermal properties of Zn–ZnO/paraffin binary nanofluids as a filler for double glazing unit. *International Journal of Heat and Mass Transfer, 183*, 122173. https://doi.org/10.1016/j.ijheatmasstransfer.2021.122173

Md Jahid, A., Wang, J., Zhang, E., Duan, Q., & Feng, Y. (2022) Energy savings potential of reversible photothermal windows with near infrared-selective plasmonic nanofilms. Energy Conversion and Management 263115705-S0196890422005015 115705 10.1016/j.enconman.2022.115705

Mekhilef, S., Saidur, R., & Safari, A. (2011). A review on solar energy use in industries. *Renewable and Sustainable Energy Reviews (Elsevier)*, Retrieved 22 March, 2022, from https://www.sciencedirect.com/science/article/pii/S1364032110004533

Saadatian, O., Sopian, K., Lim, C. H., Asim, N., & Sulaiman, M. Y. (2012). Trombe walls: A review of opportunities and challenges in research and development. *Renewable and Sustainable Energy Reviews, 16*(8), 6340–6351. https://doi.org/10.1016/J.RSER.2012.06.032

Shen, S., et al. (2021). A smart material built upon the photo-thermochromic effect and its use for managing indoor temperature. *Chemical Communications, 57*(69), 8628–8631. https://doi.org/10.1039/D1CC03379D

Stevanović, S. (2013). Optimization of passive solar design strategies: A review. *Renewable and Sustainable Energy Reviews, 25*, 177–196. https://doi.org/10.1016/J.RSER.2013.04.028

Su, L., et al. (2020). Synthesis of hollow copper sulfide nanocubes with low emissivity for highly efficient solar steam generation. *Solar Energy Materials and Solar Cells, 210*, 110484. https://doi.org/10.1016/j.solmat.2020.110484

Syed, M., Moeini, M., Okumus, P., Elhami-Khorasani, N., Ross, B. E., & Kleiss, M. C. B. (2021). Analytical study of tessellated structural-architectural reinforced concrete shear walls. *Engineering Structures, 244*. https://doi.org/10.1016/J.ENGSTRUCT.2021.112768

Szyszka, J., Kogut, J., Skrzypczak, I., & Kokoszka, W. (2017). Selective internal heat distribution in modified trombe wall. In *IOP Conference Series: Earth and Environmental Science* (Vol. 95, No. 4, p. 42018). https://doi.org/10.1088/1755-1315/95/4/042018

Tao, F., et al. (2019). Recent progress of nanostructured interfacial solar vapor generators. *Applied Materials Today, 17*, 45–84. https://doi.org/10.1016/j.apmt.2019.07.011

Tian, J., et al. (2021). Sunlight-driven photo-thermochromic hybrid hydrogel with fast responsiveness and durability for energy efficient smart windows. *Composites Part A: Applied Science and Manufacturing, 149*. https://doi.org/10.1016/J.COMPOSITESA.2021.106538

Vélez-Cordero, J. R., & Hernández-Cordero, J. (2015). Heat generation and conduction in PDMS-carbon nanoparticle membranes irradiated with optical fibers. *International Journal of Thermal Sciences, 96*, 12–22. https://doi.org/10.1016/J.IJTHERMALSCI.2015.04.009

Wheeler, L. M., et al. (2017). Switchable photovoltaic windows enabled by reversible photothermal complex dissociation from methylammonium lead iodide. *Nature Communications, 8*(1). https://doi.org/10.1038/s41467-017-01842-4

Wu, M., Shi, Y., Li, R., & Wang, P. (2018). Spectrally selective smart window with high near-infrared light shielding and controllable visible light transmittance. *ACS Applied Materials & Interfaces, 10*(46), 39819–39827. https://doi.org/10.1021/ACSAMI.8B15574

Wu, X., Chen, G. Y., Owens, G., Chu, D., & Xu, H. (2019). Photothermal materials: A key platform enabling highly efficient water evaporation driven by solar energy. *Materials Today Energy, 12*, 277–296. https://doi.org/10.1016/j.mtener.2019.02.001

Xu, Z., et al. (2018). Sunlight-induced photo-thermochromic supramolecular nanocomposite hydrogel film for energy-saving smart window. *Solar RRL, 2*(11). https://doi.org/10.1002/solr.201800204

Yeong Lee, H., Cai, Y., Bi, S., Nan Liang, Y., Song, Y., & Matthew Hu, X. (2017). A dual-responsive nanocomposite toward climate-adaptable solar modulation for energy-saving smart windows. *ACS Applied Materials & Interfaces, 9*(7), 6054–6063. https://doi.org/10.1021/acsami.6b15065

Yu, B., Li, N., & Ji, J. (2019). Performance analysis of a purified Trombe wall with ventilation blinds based on photo-thermal driven purification. *Applied Energy, 255*. https://doi.org/10.1016/J.APENERGY.2019.113846

Zhang, E., Duan, Q., Wang, J., Zhao, Y., & Feng, Y. (2021). Experimental and numerical analysis of the energy performance of building windows with solar NIR-driven plasmonic photothermal effects. *Energy Conversion and Management, 245*, 114594. https://doi.org/10.1016/J.ENCONMAN.2021.114594

Zhu, L., Gao, M., Peh, C. K. N., & Ho, G. W. (2018). Solar-driven photothermal nanostructured materials designs and prerequisites for evaporation and catalysis applications. *Materials Horizons, 5*(3), 323–343. https://doi.org/10.1039/c7mh01064h

Zou, X., et al. (2021). Research progress of photo-/electro-driven thermochromic smart windows. *Nanomaterials, 11*(12), 3335. https://doi.org/10.3390/NANO11123335

Chapter 6
Thermally Responsive Building Envelopes from Materials to Engineering

Hongyu Zhou and Yawen He

Abstract As the interface mediates between the interior space and outdoor environment, building envelope plays a crucial role in determining building systems' energy efficiency as well as structural safety. However, the static nature of current envelope design and operation is diametric to the mutable and transient forces and energies acting on our building stocks. Substantial energy-saving potential resides in more pervasive solutions that transcend the envelope's role from a passive barrier to a responsive functional assembly attuned to energy optimizations. This chapter provides an overview of technological developments that enable such responsive/or interactive building envelope concepts. First, a general classification of current responsive building envelope (RBE) technologies is discussed together with their sensing, actuation mechanisms, and application spaces. Emerging materials and future development outlook of RBE technologies are discussed.

Keywords Responsive building envelope · Environmental adaptivity · Materials · Mechanisms

6.1 Responsive Building Envelope: An Evolving Paradigm

Over the past decade, significant progress has been made to reduce building energy consumption including the innovations in high insulation façade, daylighting, and passive ventilation. However, despite the highly mutable and dynamic forces acting on our building stocks—i.e., from the changing climates and energies, to the occupant behaviors, current building practices still operate the envelope as a static system. Substantial energy-saving potential may reside in solutions that transcend the building envelope's role from a simple protective space divider to a more intelligent and responsive functional assembly attuned to climate and energy optimizations.

H. Zhou (✉) · Y. He
University of Tennessee, Knoxville, USA
e-mail: hzhou8@utk.edu

Y. He
e-mail: yhe43@vols.utk.edu

J. Wang et al. (eds.), *Advanced Materials in Smart Building Skins for Sustainability*,
https://doi.org/10.1007/978-3-031-09695-2_6

The dynamic nature of forces and energies acting on the building structure has prompted the emergence of high-performance building skins that are interactive and responsive to the environment (Wigginton & Harris, 2002; Stec & Paassen, 2005; Biloria & Sumini, 2007; Joe et al., 2013; Loonen et al., 2013; Loonen, 2015). Increasingly, building façades are being developed as complex systems of material assemblies attuned to climate and energy optimization—they are equipped with new performative materials, sensors, actuators, and artificial intelligence that support automated and dynamic functionalities of buildings, such as regulating natural daylighting, air, and sound transmission, thermal transfer, and interior air quality (Velikov & Thün, 2013). This paradigm shift from a static building envelope to a more intelligent 'building skin' that can sense and respond to environmental changes provides opportunities for energy saving, improving occupant comfort, and enabling adaptations to the changing climate.

To date, several approaches have been taken to achieve the "responsive building envelope (RBE)" concept including the utilization of intrinsic material properties such as the thermoresponse of bimetals and shape-memory polymers for self-ventilation and day-lighting control (Brigham, 2015; Rybkowski et al., 2015; Sung, 2010). In addition, recent research and development efforts have advanced materials and devices to achieve switchable or variable thermal properties in building envelope assemblies. Inspired by the response of animal skins to thermal environment variations, these variable insulations are designed to selectively transfer heat across the envelope, making it possible to insulate heat flux and dissipate/absorb heat on demand (Cui & Overend, 2019).

6.2 Classification of RBE

Responsive building envelopes (or RBEs) can be classified into one of the following general types, or a combination thereof, based on their thermal characteristics: (1) variable thermal insulations, (2) dynamic shading, (3) adaptive ventilation, and (4) variable thermal masses, as shown in Fig. 6.1.

6.2.1 Variable Thermal Insulations

Variable (or 'switchable') thermal insulation, in the form of an opaque panel that alternates between thermally conductive and insulating states, can be an effective means of regulating buildings' thermal environment by selectively transferring heat between the indoor and outdoor environments. To date, some pioneering work has been undertaken by researchers to develop variable insulation technologies intended for applications in the automotive and aerospace industries (Cui & Overend, 2019). The exploration of variable insulation technologies for building envelope applications is in early-stage development.

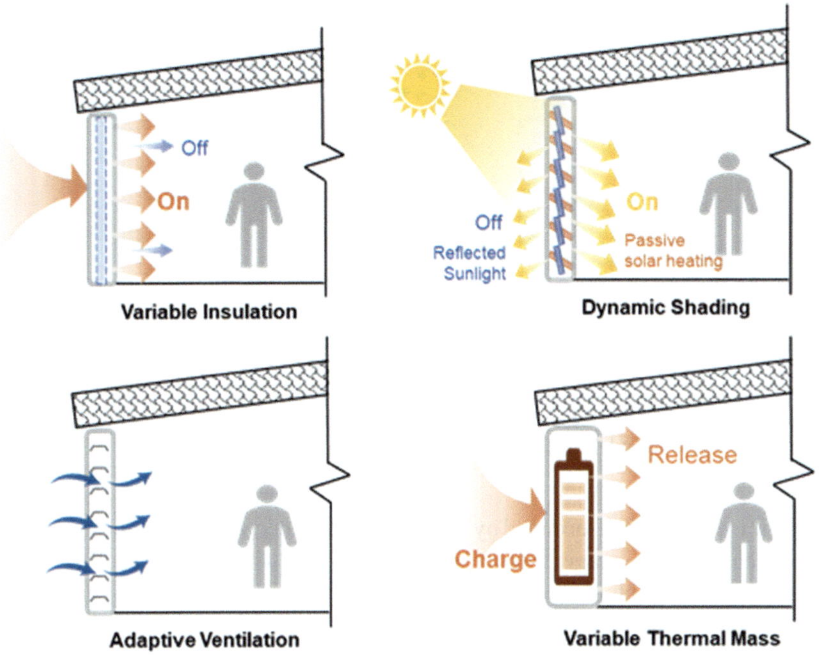

Fig. 6.1 Categories of responsive building envelope technologies

The variable thermal insulations (VTIs) act as 'heat valves or switches' to regulate heat flow between the outdoor environment and indoor spaces. They are intended to provide beneficial heat flow into (or out of) the indoor space when desired. Based on their actuation mechanisms, VTIs may be achieved through the density transition of its thermal carrier medium, the creation of an alternative heat flow path (i.e., via mechanical contact or suspended particles), or material phase change (Cui & Overend, 2019). Some VTIs behave like a 'binary switch' with an "on" and "off" modes—i.e., their thermal conductivities can change between two discrete states; whereas some other types of VTIs can gradually regulate their thermal properties based on the desired insulation levels.

6.2.1.1 Active Vacuum Insulation

Conductive heat transfer in the insulating material is dictated by the collision between gas molecules and solid surfaces. Depending on the mobility of gas molecules and the size of the cavity that encloses the gas phase, the heat transfer falls into three regimes: (i) the viscous regime, where heat transfer is dominated by the collisions between gas molecules, (ii) the molecular regime, where heat transfer is governed by the direct collision between the gas molecules and solid surfaces, and the transitional regime in between the first two (Yan Feng, 2005). The transition between one heat transfer

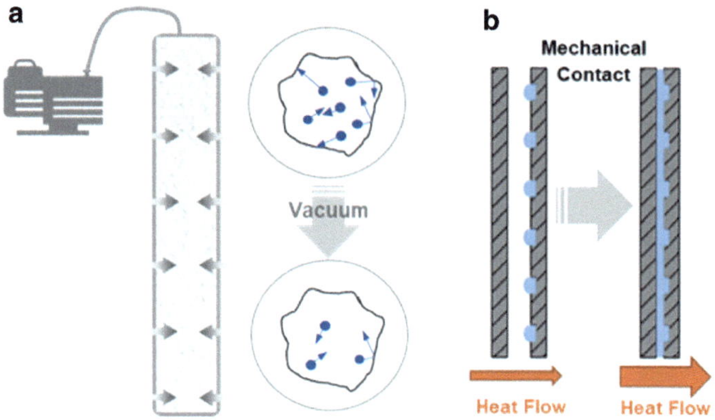

Fig. 6.2 Variable thermal insulations (VTIs) based on: **a** active/variable vacuum; and **b** mechanical contact (Cui & Overend, 2019)

regime to another is determined by the Knudsen number, which is the ratio of the mean free path (i.e. the average distance traveled between two successive collisions) to the characteristic length of the enclosure. Vacuum-based switchable thermal insulation works by alternating the heat transfer regime between the conductive viscous regime and the insulated molecular regime. This is achieved by reversibly pressurizing or evacuating the enclosure with a thermally conductive gas. The vacuum-based thermal switch typically consists of a micro-/or nano-porous core, an impermeable seal, and a (vacuum) pump controlling the pressure inside the sealed core, see Fig. 6.2a. In order to reduce the local Knudsen number and make it easier to achieve the pressure-sensitive molecular regime, porous filling materials such as aerogel or fumed silica (Berge et al., 2015) or fabricating microscale narrow gaps (Krielaart et al., 2015) are often used to artificially create small pores or gaps in the vacuum volume.

6.2.1.2 VTIs Based on Mechanical Contact

Mechanical contact is perhaps one of the most intuitive means to control the heat flux across insulation panels. The switching mechanism works by bringing/or separating two solid surfaces through mechanical contact to create conductive pathways. The presence/absence of mechanical contact causes a change in the dominant thermal transfer mechanism from gaseous convection/conduction to solid thermal conduction. As a result, rapid alternation in the overall heat transfer rate by orders of magnitude can be achieved. A mechanical thermal switch normally consists of three basic elements: (1) two solid surfaces, (2) an actuator which separates two surfaces or brings two surfaces into contact, and (3) an enclosed cavity filled with insulative gas or compressible porous insulative materials. The movement of two surfaces can be actuated by several mechanisms including actively controlled mechanical actuators (Pflug et al., 2015) and passive thermoresponsive materials (Cui & Overend, 2019).

Research efforts to improve the performance of mechanical-contact based VTIs have mainly focused on the heat transfer efficiency (both insulation and conduction capability to increase the 'switching ratio') and the actuation mechanisms.

6.2.1.3 VTIs Based on Suspended-Particles

The alternation in thermal conductivities of insulation panels can also be achieved through liquid-based suspensions—the heat transfer rate in fluidic media can be altered by introducing thermally conductive insoluble (typically solid) particles (Cui & Overend, 2019). Colloids, which consist of a liquid phase and microscopically dispersed suspended particles, have been used to create VTIs (Eapen et al., 2010). As the size of suspended particles in colloids is reduced to the nanoscale, the colloidal fluid is known as a nanofluid. The dispersed nanoparticles in nanofluids are subject to vigorous Brownian motion to form well-dispersed and stable two-phase (binary) or three-phase (trinary) mixtures. The effective thermal conductivities of these nanofluidic mixtures can be described by Fourier's law of heat conduction, which is a function of the thermal conductivities of the constituent phases, their volume fractions, and the spatial distributions. In nanofluids, microscopic heat transfer mechanisms such as microconvection on the particle-liquid interface (Jang & Choi, 2004), the ballistic phonon interactions between nanoparticles (Das et al., 2007), and the clustering of nanoparticles (Angayarkanni & Philip, 2015), also contribute to the heat transfer in the mixtures. Among these mechanisms, the clustering effect—i.e. the connectivity of particles, causes significant changes in the mixtures' thermal concavity by creating locally packed clusters with high internal thermal conductivity (Philip et al., 2008).

Tunable thermal properties can be achieved by modulating the connectivity and spatial distribution of clustered nanoparticles in the base fluid, leading to alternation in effective thermal conductivity. The formation of such clustering processes can be triggered by external fields such as temperature and magnetic fields (Eapen et al., 2010)—e.g., magnetite nanoparticles suspended in the fluid can align into chain-like structures when magnetic fields are applied. This process is reversible when the external magnetic field is removed, as illustrated in Fig. 6.3.

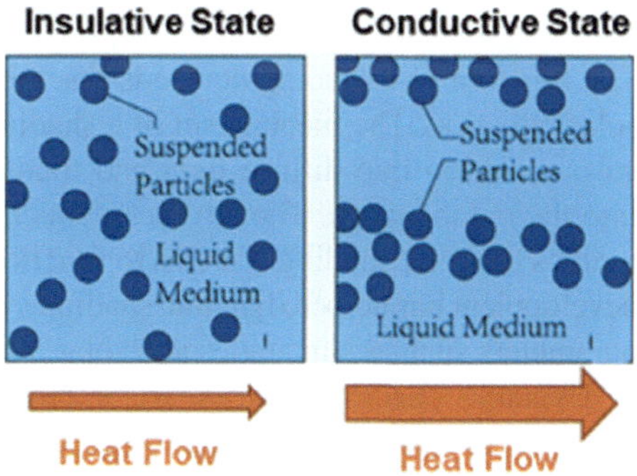

Fig. 6.3 Schematic showing the principle of suspended particle devices (Cui & Overend, 2019)

6.2.1.4 VTIs Based on Material Phase Change

Another pathway to achieve VTI is through the phase change of thermal carrier materials. The phase of substances undergoes physical transitions most commonly between solid, liquid and gaseous states triggered by condition changes such as temperature and pressure. During phase transitions, part of the potential energy within the materials will be converted into kinetic energy, or vice versa. Such reversible conversion in the evaporation-condensation cycle enables fluids/gases to transfer much more energy during the heat transfer process, leading to very high effective thermal conductance (Zohuri, 2011). One way to achieve VTI through material phase-change is to intermittently interrupt the evaporation-condensation cycle, which will lead to an alternation in heat transfer between the conductive state where heat is mostly transported by evaporation and condensation processes, and the insulated state where the gaseous conduction or single-phase natural convection is dominant. Based on this principle, technologies known as the variable conductance heat pipe (VCHP) are developed for VTIs. A typical VCHP consists of four elements: a working fluid which undergoes phase transition during heat transfer, a sealed container consisting of evaporator, condenser and transport section, a wick structure, and a control unit which regulates the evaporation-condensation cycle (Faghri, 1995).

Table 6.1 summarizes several prior research on variable thermal insulation technologies (Benson et al., 1994; Varga et al., 2002; Kimber et al., 2014; Loonen et al., 2014; Wu et al., 2014; Park et al., 2015; Pflug et al., 2015, 2018; Tomko et al., 2018). In these examples, "on-and-off" thermal switch was achieved by changing material thermal conductivity through hydration/dehydration (Tomko et al., 2018); changing porosity; changing materials orientation (Wu et al., 2014), or alternating the overall insulation through convective heat transfer (Pflug et al., 2015). Continuous thermal conductivity change was also achieved by varying the vacuum pressure of vacuum insulated panels (VIP) (Berge et al., 2015). Table 6.1 also lists the thermal conductivity (or R-value) range of the VTIs achieved through different mechanisms along with a brief description of the triggering of the heat transfer mechanisms.

6.2.2 *Dynamic Shading*

The function of a shading device is essential to shelter a building from undesirable solar radiation. The mechanism of a shading system functions on converting direct solar radiation into diffuse light and modulating the amounts of light penetration into the indoor space. The advancements in software and hardware provide opportunities for solar shading systems to function dynamically within their context. This development has helped dynamic shading systems respond to variable environmental parameters such as sun angles and solar insolation.

Dynamic shading devices are intelligent systems automatically operated in response to outdoor or indoor parameters in favor of a high comfort level and improved energy performance. The automatic response presents the key feature of

Table 6.1 Summary of technologies for variable thermal insulation

Technology	Triggering mechanism	Description	Thermal conductivity—W/mK (R-value Range—$m^2 \cdot K/W$)	
			λ_{min} (R_{min})	λ_{max} (R_{max})
Active vacuum	Pressure/evacuate the filling material (gas) or fabricated narrow gaps	Variable pressure VIP (Berge et al., 2015)	0.007	0.019
		Variable pressure aerogel blanket (Berge et al., 2015)	0.011	0.017
		Variable conductance insulation (Benson et al., 1994)	0.025(0.13)	0.2(1)
Thermodiode	Control direction of the heat flow through the diodes	Bi-directional thermo diode (Varga et al., 2002)	0.07	0.35
Mechanical contact	Bring two panels into mechanical contact or separation	Translucent element switchable u-value (Pflug et al., 2015)	0.0175 (0.33)	0.075 (1.43)
	Change thickness of air layer	Adaptive multilayer wall air layer (Kimber et al., 2014)	0.005 (0.13)	0.2 (5)
	Roller system	Movable insulation (Pflug et al., 2017)	(0.35)	(4)
Suspended particles	Form/break chain-like structures by external fields	Suspended particle device (Cui & Overend, 2019)	0.14–0.17	0.31–0.56
	Direction of nanotubes	Carbon nanotubes suspension in liquid (Wu et al., 2014)	0.4 (0.02)	1.2
Phase-Change Technologies	Leading to an alternation in heat transfer between (1) evaporation and condensation processes and (2) natural convection	Variable conductance heat pipe (VCHP) (Wu et al., 2014)	N.A	N.A
		Integrated flat-plate heat pipe (Wu et al., 2014)	N.A	N.A

dynamic shading systems—e.g., mechanically actuated shading devices have three main components: a sensor network to obtain data, a controller to determine the suitable action, and mechanical actuators to execute the response. Over the recent years, a number of semi-active and passive mechanisms have also been developed for dynamic shading.

The thermal and energy performance of shading systems involves two main aspects: (i) light regulation which includes preventing glare and maintaining adequate indoor illuminance, and (ii) providing acceptable thermal environments by controlling solar heat gains accompanied by direct and diffused solar irradiations (Al-Masrani & Al-Obaidi, 2019). The modulation of solar radiation can be achieved through mechanical actuation or morphing materials or kinetic devices. Some technologies studied in prior research are summarized in Table 6.2, which include dynamic louvers, adaptive kiri-kirigami structures (Tang et al., 2017) that can respond to irradiance changes, kinetic shading systems, etc. (Al-Masrani & Al-Obaidi, 2019). Some of the systems are self-adaptive to temperature changes (Dewidar et al., 2013) while others are triggered by electric (Velikov & Thün, 2013) or shape memory alloy (SMA) actuators (Fraternali et al., 2015; Pesenti et al., 2015). Solar shading systems have been proved to be effective in shedding intensive solar radiation to reduce building energy consumption, especially during peak hours. In addition to energy saving, passive shading systems are also effective bioclimatic means to maintain the balance between visual and thermal demands.

6.2.3 Adaptive Ventilation

Controllable ventilation systems (Pujadas-Gispert et al., 2020; Sung, 2016) can also behave as heat switches by changing the amount of air flow, and thus convective heat flow, into and out of building indoor spaces. Examples to implement adaptive ventilation as a form of responsive building envelope include the bio-inspired ventilated facade developed by Doris Sung and Ester Pujadas-Gispert et al. (2020), Sung (2016), which provides building surface cooling through natural ventilation, see Fig. 6.4a. Inspired by the opening and closing behavior of pine cones in response to hygroscopic conditions, Reichert et al. (Reichert et al., 2015) developed a responsive façade component, named Meteorosensitive Architecture, which harnesses the elastic deformation of veneer composite materials to make a tunable, humidity-responsive façade opening system, see Fig. 6.4b. An interesting feature of these concepts (or similar) is that the material acts as a sensor and actuator at the same time, which embeds the adaptive capacity as an inherent feature of the building shell. Another example of kinetic ventilation systems includes the mechanically actuated adaptive façade developed by Biloria et al. (2009), see Fig. 6.4c.

Table 6.2 Typical techniques for kinetic shading system

Technology	Motion pattern	Schematic	Description
Electric/mechanical actuation	Rotation	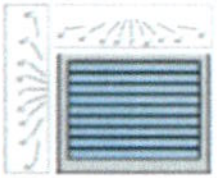	Louvres (Al-Masrani & Al-Obaidi, 2019)
	Translation		Roller shade (Al-Masrani & Al-Obaidi, 2019)
	Rotation + translation/folding		Origami structure (Tang et al., 2017)
	Rotation + translation/revolving/folding/sliding		Kinetic device (Biloria & Sumini, 2009; Dewidar et al., 2013; Velikov & Thün, 2013; Fraternali et al., 2015; Pesenti et al., 2015; Zawidzki, 2015)

(continued)

Table 6.2 (continued)

Technology	Motion pattern	Schematic	Description
	Deformation of dielectric elastomers	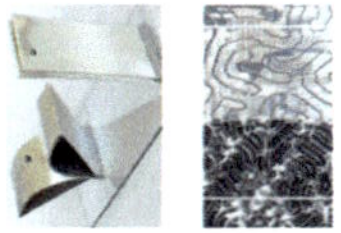	Dielectric system composed of Electro-Active Polymers with electric conduction features (Decker, 2013)
Deformation actuated device	Rotation + translation/revolving/folding/sliding	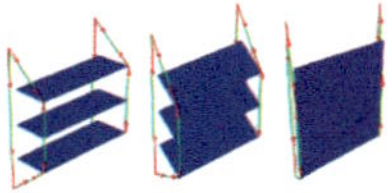	Shape memory alloy (SMA) actuated device (Fraternali et al., 2015; Pesenti et al., 2015)
Temperature actuated device	Rotation + translation/folding	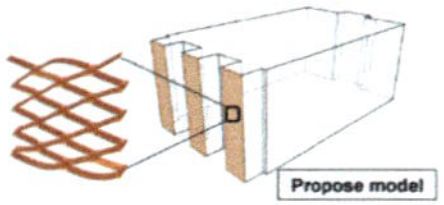	Self-adaptive kirigami metastructures (Tang et al., 2017)

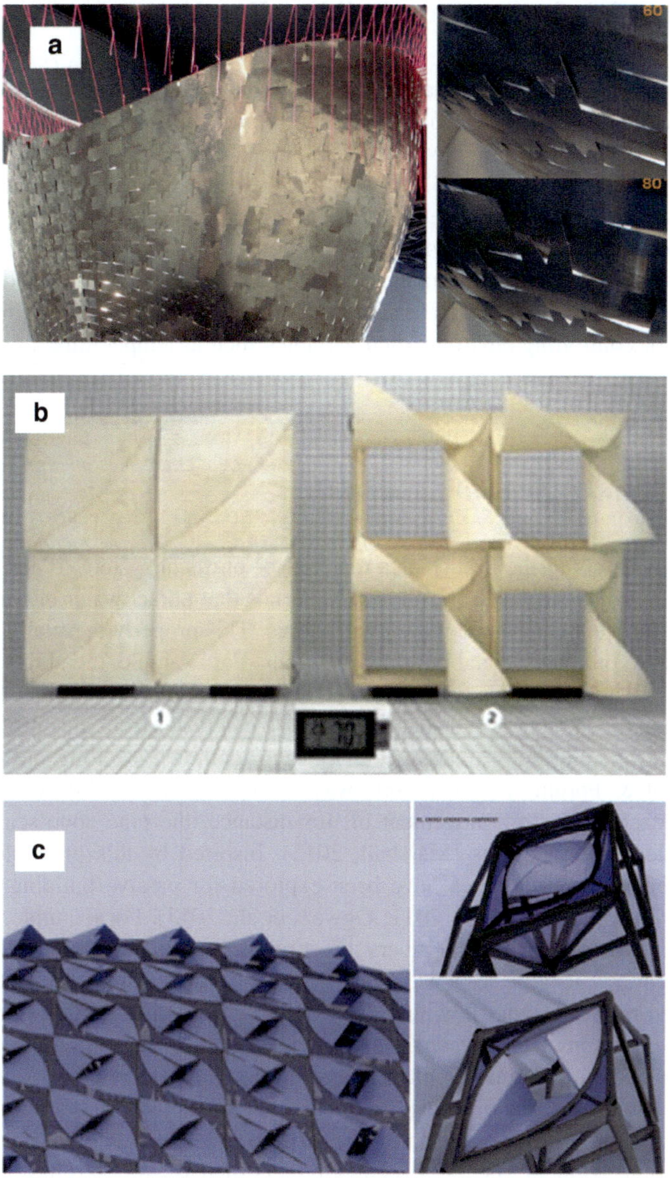

Fig. 6.4 Examples of adaptive ventilations: **a** the ventilated façade by Sung (2016); **b** meteorosensitive architectural project by Reichert et al. (2015); and **c** mechanical actuated façade by Biloria et al. (2009)

6.3 Materials for Adaptive Building Envelopes

Passive systems of dynamic activation that works with intrinsic properties of environmentally responsive smart materials represent a promising pathway for the development of adaptive building envelope technologies. Building performance can be improved with the use of passive systems to reduce energy consumption, and greenhouse gas emissions (Villegas et al., 2020), and provide opportunities to interact with outdoor environments. Environmentally responsive, or 'intelligent', materials play an essential role in achieving these goals. A brief discussion is provided herein to introduce some emerging materials used in architectural skins that have been explored for controlling environmental variables such as temperature, lighting, and humidity.

6.3.1 Humidity Sensitive Materials

Hygroscopy refers to the ability of a material or physical system to absorb water from the environment or release it back. Materials that attract water or water vapor from their ambient environment are hygroscopic. This property is usually seen as a disadvantage for conventional building materials. For example, the ability of wood to absorb moisture often leads to unwanted deformation of the materials. However, expansion and contraction in a specific direction can result from anisotropic deformation controlled by cell wall architecture through cellulose swelling and shrinkage (Ministerial & Forum, 2011). In this way, changes in the cellulose volume by humidity exchange allow movement of, for instance, the pine cone scales under hydrated/dehydration cycles (Marshall, 2015). Inspired by this principle, hygroscopic properties of materials have been explored for passive building envelope control (Ministerial & Forum, 2011; Ogwezi et al., 2013). For example, laminate composite systems with specific fibers direction were reported as principal components to achieve moisture response with autonomous movement. Architectural skin with closed modules under low humidity conditions and open modules under high humidity conditions was reported in (Augustin, 2018). Chen et al. (Villegas et al., 2020) fabricated a dynamic architectural surface made of a matrix of tiles elaborated with lime veneer, nylon, and plastic with hydro-sensitive capabilities due to the different porous densities of the composing materials.

In addition, synthetic superabsorbent polymers such as hydrogels have been used in dynamic envelope systems with two main approaches. The first focuses on the ability of the material to retain large amounts of water, and the second path leverages the volume change of super-absorbent polymers after hydration/desiccation. An example of the first approach is the hydroceramics proposed by Mitrofanova et al. (2014), which consist of a multi-cavity system that catches rainwater. The panels are made of clay and filled with hydrogel spheres. The system provides passive cooling to buildings through rainwater harvesting by the envelope module, and the long-term storage of water and moisture released throughout the day improves thermal exchanges between the building and the external environment as shown in Fig. 6.5. The second approach consists of a force-generating systems (Ayala Castro et al., 2017). The devices are based on an acrylic piece attached to hydrophobic fabric pockets filled with sodium polyacrylate spheres with a mesh in contact with it. When the humidity goes through the mesh, the volume change of the pockets is triggered to generate movement from one side to another to allow air flow. A different system which uses the force-generating hydrogels was reported (Roth, 2015), where a surface made of a matrix of silicone scales was fixed by a polyacrylates composite. When the surface is in contact with water, the composite net points swell, and the architectural scales can open and close when it shrinks.

The hydration/dehydration of certain materials has also been explored to create tunable materials that have reversibly variable thermal conductivities. To that end, Hopkins et al. (Tomko et al., 2018) developed topologically networked bio-inspired materials for reversible thermal conductivity switching. They propose that by varying the network topology, or crosslinked structure, of squid ring teeth-based bio-polymers through tandem-repetition of DNA sequences, the thermal switching ratio can be directly programmed. This on/off ratio in thermal conductivity switching is promising

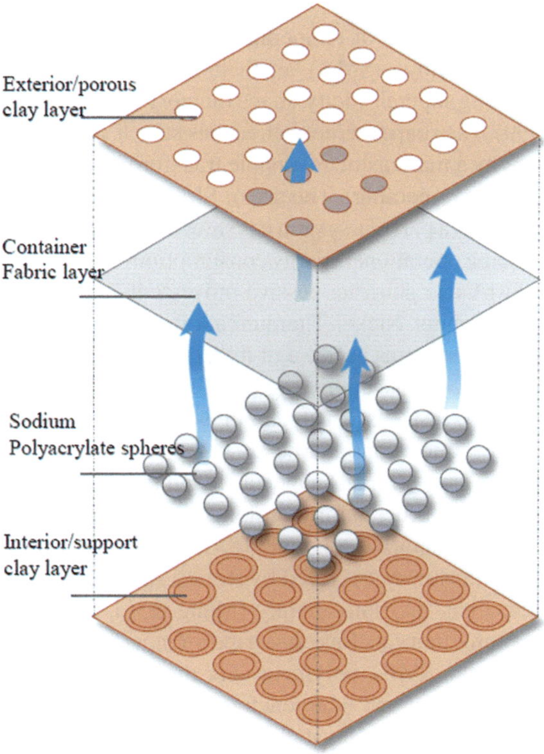

Fig. 6.5 The 'hydroceramic' concept developed by Mitrofanova et al. (2014)

in dynamic building envelopes where reversible thermal conductivity switching is desired as introduced in previous sections.

6.3.2 Temperature-Responsive Materials

Since temperature control is one of the main functional objectives of responsive building envelopes, 'temperature-responsive materials' have been widely explored for RBE applications. Hybrid shape memory materials are a class of stimulus-responsive components made of two different materials which do not have shape memory capabilities (Sun et al., 2012). For example, bi-metallic shape memory strips are made of two metallic pieces with different thermal expansion coefficients bonded by an adhesive. The system operation is based on the asymmetric stress distribution within the two laminated layers due to the expansion/contraction of each strip caused by thermal gradient differences. This phenomenon allows shape changes such as bending, triggered by direct or indirect heating. This principle has been applied to architectural surfaces with active thermal features triggered by solar irradiation as well as ambient temperature changes. An early report that utilized the behavior of thermobimetals in architecture was the stimulus-responsive skins by Sun et al. (2012), see Fig. 6.6a.

Another class of temperature-responsive materials is based on the use of shape memory alloys (SMA) as their active components. SMA has several features including superelasticity, shape memory effect, and high damping capacity (Liu et al., 1999). The shape memory effect has been leveraged to achieve a bi-directional movement by a martensitic reversible transformation because of temperature change (i.e., warming or cooling). The use of SMA in architecture either as a dynamic system by itself or as a part of a system has shown an important improvement in performance and reducing operational energy consumption—i.e., the temperature-responsive behavior of SMA can generate passive movements that do not rely on sensors and external energy inputs. Nickel-Titanium alloy (NiTi) is the most reported SMA in responsive envelope systems because of its reliable mechanical performance (Sun et al., 2012). Several responsive building envelopes which focused on heat and lighting control used NiTi SMA as an actuator or as components of the actuation mechanisms. These systems' operations are mostly based on prestressed SMA springs or wires trying to recover their original shape due to temperature change, which generates mechanical force in the process for morphing or actuation.

Shape memory polymers (SMPs) are stable polymer networks with reversible shape transitions triggered by stimuli such as temperature, pH, electricity, magnetic field, light, and ions mainly (Meng & Li, 2013). There are numerous molecular structures that can drive the shape memory effect in polymers. In the case of thermal-responsive SMPs, polymers with molecular entanglement, chemical crosslinking, crystallization and interpenetrated network may be used. The reversible switching transitions can be crystallization and vitrification as shown in Fig. 6.6b. During this process, shape change or mechanical force can be generated when the material

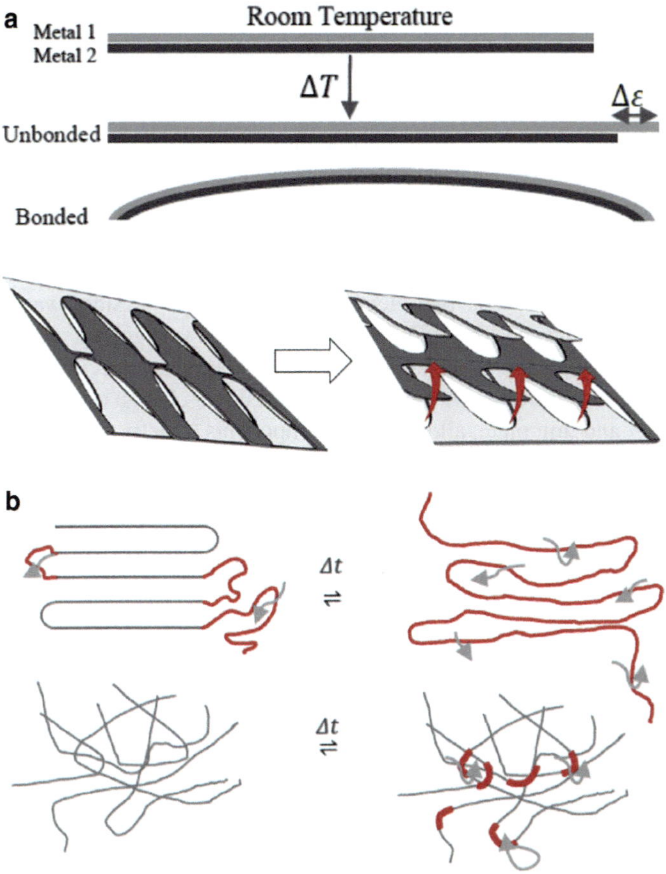

Fig. 6.6 Temperature responsive materials: **a** the concept of thermo-bimetal; and **b** temperature responsive polymers (Meng & Li, 2013)

is subjected to environmental changes. Commercial temperature-responsive SMPs have been used to develop dynamic structures in architectural skins (Beites, 2013).

6.3.3 Electrochromic Materials and Passive Lighting Control

The change of optical properties of materials in response to electric current or temperature change has been utilized for the development of a responsive building envelope. The electrochromic effect, for instance, occurs in partially hydrated transition metal oxides (Svensson & Granqvist, 1985). Reversible electrochemical reactions facilitated by the ion extraction and insertion of metal oxides lead to changes in physical properties such as, conductivity, IR absorption, and color. Electrochromic materials, in the form of thin coatings, have been widely used for building windows and façades. The system shifts from an oxide insulator state to a quasi-metallic one when an external potential is applied. The electroactive layers change their optical properties between their oxidized and reduced form because of electron flow in the system. Tungsten oxide WO_3 in its amorphous state is among the most studied electrochromic coatings (Svensson & Granqvist, 1985). The system can be customized to obtain different time responses as well as absorbing/reflecting rates as demonstrated by many prior researches.

By using a dielectric elastomeric layer derived from an electro-active polymer (EAP) sandwiched between two electrodes, a multilayer system was created by Krietemeyer and Dyson (2011). The dynamic behavior is triggered when an electric current goes through the laminate which rises electrostatic forces that generate a contraction of the elastomer. As a result, a dimensional change occurs going from a thick to a flat and thin plate, allowing the components to deform in predicted directions. This mechanism has been used for daylighting control, where a network made of elastomer coated silver electrodes was developed (Decker, 2013). A bi-directional movement of the laminated components allows the transition between 'on' and 'off' modes for the passage of sunlight into the indoor space.

6.4 Future Outlooks

Responsive building skins represent a promising and evolving paradigm of building envelope technologies which will facilitate the co-evolutionary interaction between the buildings, the inhabitants, and the environment. The research and development of environmentally responsive building envelopes are still in their early stage where numerous technological, economic, and implementation challenges are yet to be overcome. The potential of responsive building envelope technologies to improve building energy efficiency and improve thermal comfort is well demonstrated. Meanwhile, recent decades have witnessed dramatic advances in the precision and reliability of sensors and actuators with falling prices. These technological developments are providing unprecedented opportunities to embed programmable components into building envelopes to further enhance their performance. However, despite these promising trends, many adaptive façade technologies remain designed for, and used in, single architectural projects. The widespread commercialization of RBEs for either new or existing buildings is still limited.

With the majority of our current building stock being obsolete and energy inefficient, the development of adaptive facades for building retrofitting may have a faster and more significant reduction potential on the energy use and greenhouse gas emissions than strategies that solely focus on new buildings.

Future development of responsive building envelope should focus on: (1) technologies or products that can be easily integrated into a wide range of existing building types, (2) intelligent and scalable control techniques that holistically integrate façade technologies/systems performances with the overall building performance, and (3) designs or mechanisms that arise from human-centered design. Materials development, especially the ones that would enable passive and autonomous responses of building envelopes and those that focus on obtaining a higher 'switching ratio' while meeting the requirements for conventional building insulation are still much desired.

Acknowledgments The authors would like to acknowledge the financial support from U.S. National Science Foundation under grant CMMI-1954517.

References

Al-Masrani, S. M., & Al-Obaidi, K. M. (2019). Dynamic shading systems: A review of design parameters, platforms and evaluation strategies. *Automation in Construction, 102*(January), 195–216. https://doi.org/10.1016/j.autcon.2019.01.014.

Angayarkanni, S. A., & Philip, J. (2015). Review on thermal properties of nanofluids: Recent developments. *Advances in Colloid and Interface Science, 225*, 146–176. https://doi.org/10.1016/j.cis.2015.08.014.

Augustin, N. (2018). *Motion with moisture creating passive dynamic envelope systems using the hygroscopic properties of wood veneer*. University of Waterloo.

Ayala Castro, I., Manosong, M., & Chang, Y. C. (2017). *Water-driven breathing skin*. Srchitecture of Catalonia.

Beites, S. (2013). Morphological behavior of shape memory polymers toward a deployable, adaptive architecture. In *ACADIA 2013: Adaptive Architecture—Proceedings of the 33rd Annual Conference of the Association for Computer Aided Design in Architecture* (pp. 121–128).

Benson, D. K., Potter, T. F., & Tracy, C. E. (1994). Design of a variable-conductance vacuum insulation. *SAE Technical Papers, 103*(1994), 176–181. https://doi.org/10.4271/940315.

Berge, A., et al. (2015). Effect from a Variable U-Value in Adaptive Building Components with Controlled Internal Air Pressure. *Energy Procedia, 78*, pp. 376–381. https://doi.org/10.1016/j.egypro.2015.11.677.

Biloria, N., & Sumini, V. (2007). Performative building skin systems: A morphogenomic approach towards developing real-time adaptive building skin systems. *International Journal of Architectural Computing, 07*(04), 643–675. https://doi.org/10.1260/1478-0771.7.4.643.

Biloria, N., & Sumini, V. (2009). Performative building skin systems: A morphogenomic approach towards developing real-time adaptive building skin systems. *International Journal of Architectural Computing, 7*(4), 643–675. https://doi.org/10.1260/1478-0771.7.4.643.

Brigham, J. (2015). *Collaborative Research : Adaptive and Reconfigurable Tiles for Building Surfaces*. NSF.

Cui, H., & Overend, M. (2019). A review of heat transfer characteristics of switchable insulation technologies for thermally adaptive building envelopes. *Energy and Buildings, 199*, 427–444. https://doi.org/10.1016/j.enbuild.2019.07.004.

Das, Sarit K., Stephen U. Choi, Wenhua Yu., & Pradeep, T. (2007). *Nanofluids: Science and technology*. Wiley & Sons, Inc.

Decker, M. (2013) Emergent Futures: Nanotechology and Emergent Materials in Architecture. In *Conference of Tectonics of Teaching: Building Technology Educators Society (BTES).*

Dewidar, K .M, Mohamed, N. M.,& Ashour, Y., et al. (2013). Living skins: A new concept of self active building envelope regulating systems. *SB13 Dubai* (pp. 1–8). https://www.irbnet.de/daten/iconda/CIB_DC26849.pdf.

Eapen, J., et al. (2010). The classical nature of thermal conduction in nanofluids. *Journal of Heat Transfer, 132*(10), 1–14. https://doi.org/10.1115/1.4001304.

Faghri, A. (1995). *Heat pipe science and technology*. Global Digital Press.

Feng, Y. (2005). A user's guide to vacuum technology [book review]. *IEEE Circuits and Devices Magazine*. https://doi.org/10.1109/mcd.2005.1438817.
Fraternali, F., De Chiara, E., & Skelton, R. E. (2015). On the use of tensegrity structures for kinetic solar facades of smart buildings. *Smart materials and structures* 24(10). IOP Publishing. https://doi.org/10.1088/0964-1726/24/10/105032.
Jang, S. P., & Choi, S. U. S. (2004). Role of Brownian motion in the enhanced thermal conductivity of nanofluids. *Applied Physics Letters, 84*(21), 4316–4318. https://doi.org/10.1063/1.1756684.
Joe, J., et al. (2013). Load characteristics and operation strategies of building integrated with multi-story double skin façade. *Energy and Buildings, 60*, 185–198. https://doi.org/10.1016/j.enbuild.2013.01.015.
Kimber, M., Clark, W. W., & Schaefer, L. (2014) Conceptual analysis and design of a partitioned multifunctional smart insulation. *Applied Energy, 114*, 310–319. https://doi.org/10.1016/j.apenergy.2013.09.067.
Krielaart, M. A. R., Vermeer, C. H., & Vanapalli, S. (2015). Compact flat-panel gas-gap heat switch operating at 295 K. *Review of Scientific Instruments, 86*(11), 1–7. https://doi.org/10.1063/1.4936356
Krietemeyer, E. A., & Dyson, A. H. (2011). Electropolymeric Technology for Dynamic Building Envelopes. pp. 75–83.
Liu, Y., et al. (1999). High strain rate deformation of martensitic NiTi shape memory alloy. *Scripta Materialia, 41*(1), 89–95. https://doi.org/10.1016/S1359-6462(99)00058-5.
Loonen, R. C. G. M. (2015). Bio-inspired Adaptive Building Skins. In F. Pacheco Torgal, et al. (Eds.). *Biotechnologies and biomimetics for civil engineering* (pp. 115–134). Springer.
Loonen, R. C. G. M., et al. (2013). Climate adaptive building shells: State-of-the-art and future challenges. *Renewable and Sustainable Energy Reviews, 25*, 483–493. https://doi.org/10.1016/j.rser.2013.04.016.
Loonen, R. C. G. M., Hoes, P., & Hensen, J. L. (2014). Performance prediction of buildings with responsive building elements: Challenges and solutions. In *Proceedings of Building Simulation and Optimization* (pp. 1–8). https://pure.tue.nl/ws/portalfiles/portal/3992582/392669136570421.pdf.
Marshall, M. T. (2015). Bi-directional thermo-hygroscopic facades: Feasibility for liquid desiccant thermal walls to provide cooling in a small-office building. In *ARCC 2015 Conference. FUTURE of Architectural Research* (pp. 45–56).
Meng, H., & Li, G. (2013). Reversible switching transitions of stimuli-responsive shape changing polymers. *Journal of Materials Chemistry A, 1*(27), 7838–7865. https://doi.org/10.1039/c3ta10716g.
Ministerial, G., & Forum, E. (2011). Hygroscopic climatic modulated boundaries: a strategy for differenti-ated performance using a natural circulative and energy captive build-ing envelope in hot and moisture rich laden air environments. 2(1), 41–53.
Mitrofanova, E., Rathee, A., & Santayanon, P. (2014). Hydroceramic digital matter—Intelligent contructions.
Ogwezi, B., et al. (2013). Development of a passive and adaptable façade element for humidity control. *Technologies for Sustainable Built Environments (TSBE)* (p. 7). https://www.reading.ac.uk/web/files/tsbe/Ogwezi_TSBE_Conference_Poster_2013.pdf.
Park, B., Srubar, W. V., & Krarti, M. (2015). Energy performance analysis of variable thermal resistance envelopes in residential buildings. *Energy and Buildings, 103*, 317–325. https://doi.org/10.1016/j.enbuild.2015.06.061.
Pesenti, M., et al. (2015). Kinetic Solar Skin: A responsive folding technique. *Energy Procedia, 70*, 661–672. https://doi.org/10.1016/j.egypro.2015.02.174.
Pflug, T., et al. (2015) Closed translucent façade elements with switchable U-value—A novel option for energy management via the façade. *Energy and Buildings, 86*, 66–73. https://doi.org/10.1016/j.enbuild.2014.09.082.
Pflug, T., et al. (2017). Potential analysis of a new removable insulation system. *Energy and Buildings, 154*, 391–403. https://doi.org/10.1016/j.enbuild.2017.08.033.

Pflug, T., et al. (2018). Modeling of facade elements with switchable U-value. *Energy and Buildings, 164*, 1–13. https://doi.org/10.1016/j.enbuild.2017.12.044.
Philip, J., Shima, P. D., & Raj, B. (2008). Nanofluid with tunable thermal properties. *Applied Physics Letters, 92*(4), 1–4. https://doi.org/10.1063/1.2838304.
Pujadas-Gispert, E., et al. (2020). Design, construction, and thermal performance evaluation of an innovative bio-based ventilated façade. *Frontiers of Architectural Research, 9*(3), 681–696. https://doi.org/10.1016/j.foar.2020.02.003.
Reichert, S., Menges, A., & Correa, D. (2015). Meteorosensitive architecture: Biomimetic building skins based on materially embedded and hygroscopically enabled responsiveness. *CAD Computer Aided Design, 60*, 50–69. https://doi.org/10.1016/j.cad.2014.02.010.
Roth, L. (2015). Hydromembrane. Institute for advanced architecture of Catalonia.
Rybkowski, Z., et al. (2015). *EAGER : Interaction of Smart Materials for Transparent, Self–Regulating Building Skins*. NSF.
Stec, W. J., & Paassen, A. H. C. V. (2005). Symbiosis of the double skin facade with the HVAC system. *Energy and Buildings, 37*(5), 461–469. https://doi.org/10.1016/j.enbuild.2004.08.007.
Sun, L., et al. (2012). Stimulus-responsive shape memory materials: A review. *Materials and Design, 33*(1), 577–640. https://doi.org/10.1016/j.matdes.2011.04.065.
Sung, D. K. (2010). Skin deep: Making building skins breathe with smart thermo bimetals, where do you stand. In *Proceedings of the 2011 ACSA National Conference* (pp. 145–152). Washington, DC: ACSA Press.
Sung, D. (2016). A New Look at Building Facades as Infrastructure. *Engineering, 2*(1), 63–68. https://doi.org/10.1016/J.ENG.2016.01.008.
Svensson, J. S. E. M., & Granqvist, C. G. (1985). Electrochromic coatings for smart windows. *Solar Energy Materials, 12*(6), 391–402. http://www.cylaw.org/nomoi/enop/non-ind/1985_1_111/full.html.
Tang, Y., et al. (2017). Programmable kiri-kirigami metamaterials. *Advanced Materials, 29*(10), 1–9. https://doi.org/10.1002/adma.201604262.
Tomko, J. A., et al. (2018). Tunable thermal transport and reversible thermal conductivity switching in topologically networked bio-inspired materials. *Nature Nanotechnology, 13*(10), 959–964. https://doi.org/10.1038/s41565-018-0227-7.
Varga, S., Oliveira, A. C., & Afonso, C. F. (2002). Characterisation of thermal diode panels for use in the cooling season in buildings. *Energy and Buildings, 34*(3), 227–235. https://doi.org/10.1016/S0378-7788(01)00090-1.
Velikov, K. T. G. (2013). Responsive building envelopes: Characteristics and evolving paradigms. In *Design and Construction of High-Performance Homes* (pp. 75–92).
Villegas, J. E., Gutiérrez, J., & Colorado, H. (2020). (2020) 'Active materials for adaptive building envelopes: A review.' *Journal of Materials and Environmental Science, 6*, 988–1009.
Wigginton, M., & Harris, J. (2002). *Intelligent Skins*. Elsevier.
Wu, Z., et al. (2014). A comparative study on thermal conductivity and rheology properties of alumina and multi-walled carbon nanotube nanofluids. *Frontiers in Heat and Mass Transfer, 5*. https://doi.org/10.5098/hmt.5.18.
Zawidzki, M. (2015). Dynamic shading of a building envelope based on rotating polarized film system controlled by one-dimensional cellular automata in regular tessellations (triangular, square and hexagonal). *Advanced Engineering Informatics, 29*(1), 87–100. https://doi.org/10.1016/j.aei.2014.09.008.
Zohuri, B. (2011). Heat Pipe Design and Technology. *Heat Pipe Design and Technology*. https://doi.org/10.1201/b10806.

Chapter 7
Energy Performance Analysis of Kinetic Façades by Climate Zones

Chengde Wu

Abstract Kinetic façades have been often used as a solution to reduce building energy consumption. Many researchers have been actively studying on energy reduction aspect of kinetic façades, but the energy performance of kinetic façades in a spectrum of climate conditions has not been systematically studied. This research investigated the energy saving aspect of kinetic façades in different climate conditions. Two types of kinetic façades, folding and sliding, were compared to optimized fixed-shading façades and a no-shading façade. The simulation results showed that the kinetic façades, compared to the no-shading façade, reduced 32–56% of energy consumption in ASHRAE zone 1–3, with the folding façade slightly outperforming the sliding façade. In zone 4, the sliding façade outperformed the folding façade. As an exception, the amount of energy reduction in marine zones was negligible. In zone 5–8, the kinetic façades did not show a meaningful energy reduction. Compared to the optimized static façades, the kinetic façades generally showed noticeable energy reduction in hot and warm climate conditions. These simulation results are expected to provide basic rules of thumb to designers when considering kinetic façades as a means of energy reduction strategy.

Keywords Adaptive Façade · Kinetic Façade · Energy simulation · Climate zones · Responsive architecture

7.1 Introduction

Our society has been heavily dependent on fossil fuels, the primary cause for increased greenhouse gases in the atmosphere (EPA, 2017). As a chain reaction, increased greenhouse gas emissions led to global warming and followed by glacier meltdown, sea level rise, and many other negative environmental impacts (Pörtner et al., 2022). Minimizing energy consumption is one of the most effective ways to reduce greenhouse gas emissions and mitigate global warming. Given the fact that

C. Wu (✉)
Iowa State University, Ames, USA
e-mail: cdwu@iastate.edu

J. Wang et al. (eds.), *Advanced Materials in Smart Building Skins for Sustainability*,
https://doi.org/10.1007/978-3-031-09695-2_7

the building sector is responsible for roughly 36% of the global energy use and 40% of energy-related greenhouse gas emissions (IEA, 2018), it is not surprising that researchers and designers have been constantly seeking more energy efficient building solutions. Adaptive façades have been implemented as a solution to reduce building energy consumption. Common adaptive façades can be categorized into different typologies, such as kinetic façades, electrochromic façades (Granqvist et al., 2018), biological façades (Arup, 2013), etc. Kinetic façades regulate the amount of solar radiation (visible light and infrared) entering buildings through windows by adjusting the shape and/or position of the kinetic shading components. During hot summer days, kinetic façades automatically shut down or partially shut down to reduce the incoming solar radiation and to lower building cooling load. During the cold winter days, kinetic façades open up to maximize passive solar heat gain to reduce building heating load. Kinetic façades also operate throughout the day based on the sensors that measure the real-time outdoor weather conditions. Typically, kinetic façades operate at a set interval instead of in continuous motion to increase the longevity of the actuators. For example, Al Bahr towers in Abu Dhabi adjust the façade position every 15 min (Attia, 2018). In case of extreme weather events, the operation schedule will be overridden. Under this operation schedule, the actuators are expected to last 15 years. The kinetic mechanism on the Arab World Institute building in Paris is set to perform a maximum of 18 movements per day (IMA, 2022).

There are numerous studies focused on the energy performance of kinetic façades in hot climates. Bacha and Bourbia (2016) simulated the energy savings of the kinetic façade of a building in the city of Biskra, Algeria, and concluded that the kinetic façade saves 43% cooling energy. Al Bahr towers in Abu Dhabi can save 50% of cooling energy through the kinetic façades (Alotaibi, 2015). Many other studies showed the benefits of kinetic façades in terms of energy performance (Kolarevic & Parlac, 2015). These studies have demonstrated the energy saving benefits of kinetic façades in hot climates.

Intuitively, kinetic façades can reduce cooling energy load in hot climates while the benefits diminish in cold climates and even potentially hurt energy performance in a very cold climate. There are currently many buildings with kinetic façades in operation all around the world. Some of the examples are Al Bahar Towers in Abu Dhabi, Arab World Institute in Paris, Kiefer Technic Showroom in Austria, The Gardens by the Bay in Singapore, and many others. We expect that the kinetic façade in a hot climate such as in Abu Dhabi reduces more cooling load compared to the kinetic façade in a cooler climate such as in Paris. On a spectrum of climate conditions, from very hot to very cold, the energy saving benefits of kinetic façades may have a breakeven line, a climate condition where the benefits of kinetic façades are completely offset by the negative factors of kinetic façades, including the initial cost, more frequent maintenance, and higher embodied carbon footprint. In other words, in any climate that is colder than the breakeven climate, the embodied carbon footprint and the cost of kinetic façades outweigh their benefits in energy saving. This breakeven climate condition is unknown, so one of the goals of this research is to provide insight into where to draw the breakeven line.

The energy savings of the kinetic façades in the studies mentioned above are compared to the no-shading condition as the benchmark. Well-designed static shading façades can block a fair amount of solar radiation during hot summer days while admitting passive solar heat during cold winter days. Static façades have no moving parts (thus far fewer parts) compared to kinetic façades. Therefore, static façades have a lower initial cost, less maintenance, and less embodied carbon. To objectively assess the energy savings of kinetic façades, it is important also compare kinetic façades with static ones.

Architects and designers often rely on rules of thumb when making initial design decisions, especially in the schematic design stage. However, there are no rules of thumb for how much energy kinetic façades can save given a specific project location. It is also unknown if a kinetic façade can outperform a static façade well enough to offset the higher costs and the embodied carbon. Although energy simulations can provide answers, it takes time and resources for architects to build digital models and simulate, assuming someone in the team that has the skills and knowledge in parametric modeling and energy simulation. The primary goal of this research is to set a baseline for architects to understand the impact of kinetic façades so to help them make data-driven design decisions on whether a kinetic façade would be appropriate for their projects in terms of energy saving.

7.1.1 Research Questions

There are three primary research questions:

Q1: How do kinetic façades perform in different climate conditions? Answering this question could provide insight into where the breakeven line is for implementing kinetic façades.

Q2: How much energy do kinetic façades save compared to optimized fixed-shading facades and a no-shading façade?

Q3: How much does the geometry of a kinetic façade affect energy saving efficiency?

7.1.2 Research Scope

The geographical scope of this research is limited to the United States. Ideally, studying different climate conditions across the globe would result in more representative and generalizable results. However, as a pilot study, this research will focus on the cities in the US. Future studies will include cities located globally with a more granular distribution.

Façade orientation in this study is limited to the south façade in the northern hemisphere because the south façade typically receives the most solar radiation.

The energy saving aspect of kinetic façades can be greatly affected by their orientation. Therefore, evaluating the energy performance of kinetic façades at different orientations could provide a more holistic understanding.

Only two different types of kinetic façades, folding and sliding, are used to represent different kinetic façade typologies. A folding façade can create a partial shading device when fully open while a sliding façade can fully reveal the windows to solar exposure.

7.2 Research Method

To answer these research questions, energy simulation was used as the primary research method. A generic shoebox model was used to better control the energy simulation results. Parametric modeling was utilized to facilitate the generation and the simulation of the façade operations. To understand how kinetic façades perform under different climate conditions, 16 US cities were selected, one for each ASHARE zone as shown in Table 7.1. The energy performance of the shoebox model with optimized fixed-shading façades and a no-shading facade was also simulated using the same climate data of the 16 cities, and the results were compared to the results of the kinetic façades.

Table 7.1 ASHRAE climate zones and the US cities selected for each zone

Zone	City	Climate type
1A	Miami, FL	Very hot, humid
2A	Houston, TX	Hot, humid
2B	Phoenix, AZ	Hot, dry
3A	Atlanta, GA	Warm, humid
3B	Las Vegas, NV	Warm, dry
3B-coast	Los Angeles, CA	Warm, marine
3C	San Francisco, CA	Warm, marine
4A	Baltimore, MD	Mild, humid
4B	Albuquerque, NM	Mild, dry
4C	Seattle, WA	Mild, marine
5A	Chicago, IL	Cold, humid
5B	Boulder, CO	Cold, dry
6A	Minneapolis, MN	Cold, humid
6B	Helena, MT	Cold, dry
7	Duluth, MN	Very cold
8	Fairbank, AK	Sub-artic

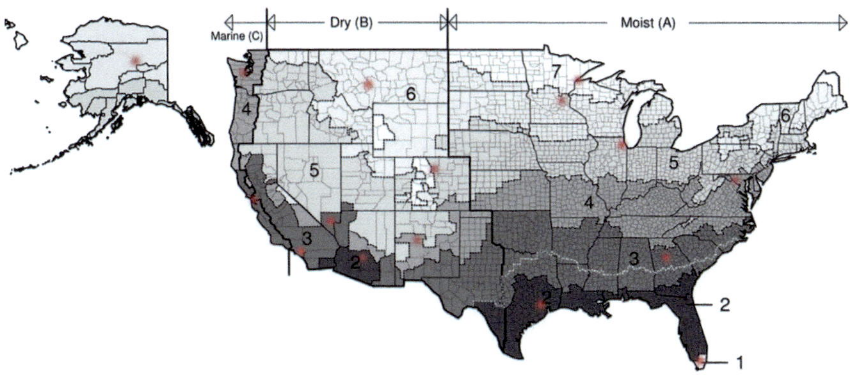

Fig. 7.1 The 16 selected cities on the ASHRAE climate zone map

The geographic locations of the 16 cities are shown in red dots on the ASHRAE climate zone map (Fig. 7.1). The cities show a rather wide spread from east to west and south to north across the US.

7.3 Energy Modeling and Simulation

7.3.1 Energy Modeling

For running energy simulations, a 6 × 10 × 4 m (width × depth × height) shoebox energy model was defined as shown in Fig. 7.2 The front side of the model was set to face the south orientation with a 90% window to wall ratio. The rest of the five surfaces of the shoebox were set to be adiabatic. This is a simplification for simulating a section of a building to better control simulation results.

Fig. 7.2 Shoebox for energy simulations

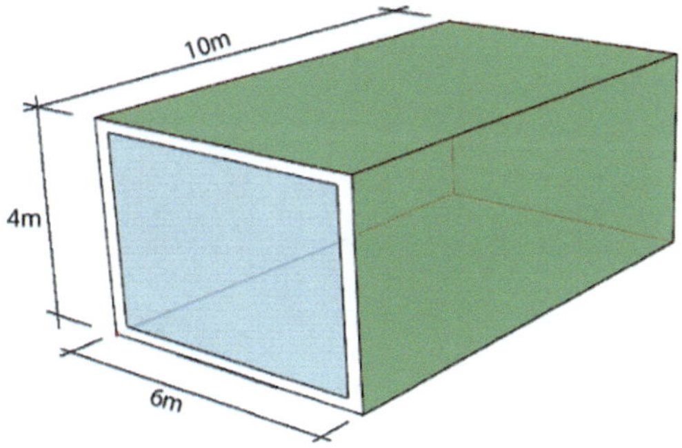

Table 7.2 Simulation parameters in Climate Studio

Program	Office
People density	0.07 p/m^2
Metabolic rate	1.2
Equipment power density	7 W/m^2
Lighting power density	10 W/m^2
Heating setpoint	20 C
Cooling setpoint	26 C
Wall U-value	0.2 W/m^2 K
Window U-value (Double-pane clear glass)	2.69 W/m^2 K
Window SHGC	0.703
Window visible transmittance	0.774
Infiltration	0.5 ACH

Climate Studio V1.5 was used as the simulation software. Climate Studio uses EnergyPlus as the simulation engine. Simulation parameters were set up as shown in Table 7.2.

7.3.2 *Kinetic Façade Modeling*

There are various ways to categorize kinetic façades typologies (Tabadkani et al., 2021). In this research, for the purpose of energy simulation, kinetic façades are categorized into two different typologies: one that creates a new form of shading geometry, such as overhangs or fins, when fully open, and the other type that fully reveals the windows when fully open. The kinetic façade of the Kiefer Technic Showroom in Austria, for example, forms overhangs when fully open and thus partially shade the building (Fig. 7.3). On the other hand, some kinetic façades, sliding louvers as an example, do not shade the windows when fully open.

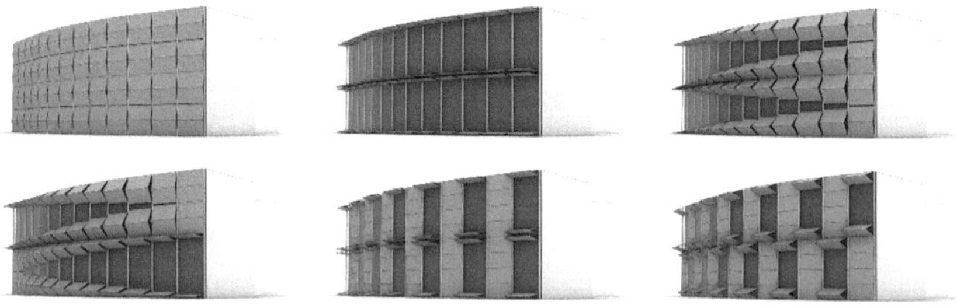

Fig. 7.3 Different modes of kinetic façade operations of the Kiefer Technic Showroom in Austria

In this research, the two types of façades, a folding façade and a sliding façade, were modeled in Rhino/Grasshopper to represent the two different kinetic façade typologies. In reality, the opening motion of kinetic façades is continuous. For the purpose of simulation, the range of motion is divided into 21 discrete steps from fully closed to fully open. Each step represents different opening ratio, i.e., 0%, 5%, 10%, ..., 95%, and 100% open. Figures 7.4 and 7.5 show every other opening step of the two types of façades respectively.

The two types of façades, folding and sliding, were parametrically modeled with different numbers of kinetic panels to investigate the relationship between the number of panels and the energy performance of the façades. Each of the two façades was modeled with the variation of 1, 2, ..., and 10 panels. Figures 7.6 and 7.7 show the variations with a different number of panels.

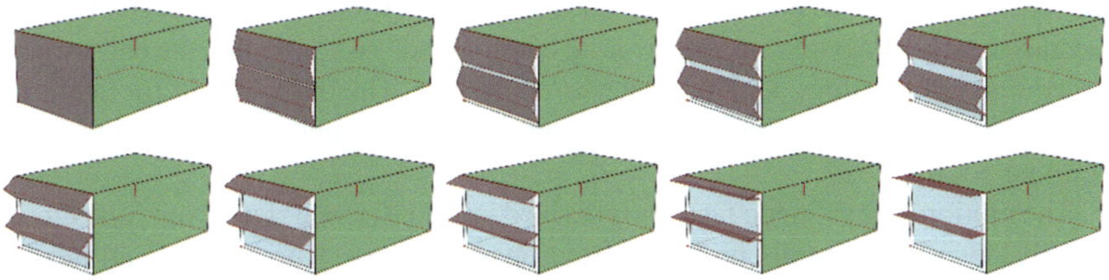

Fig. 7.4 Every other step of the folding façade from fully closed to fully open

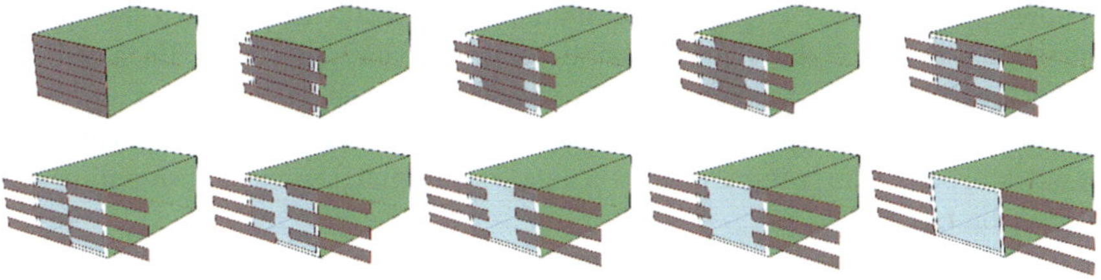

Fig. 7.5 Every other step of the sliding façade from fully closed to fully open

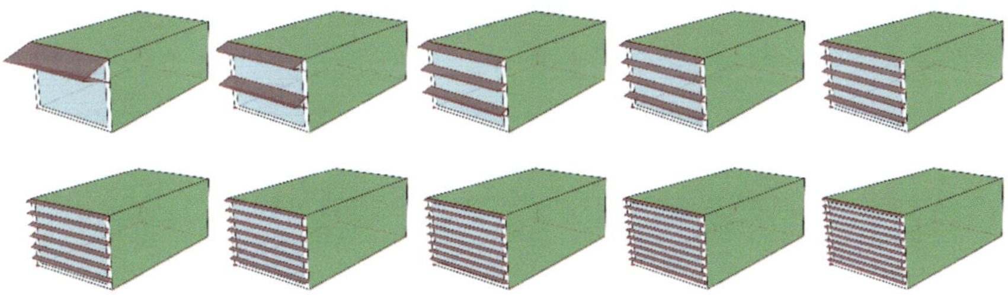

Fig. 7.6 Folding façade with different numbers of foldable panels

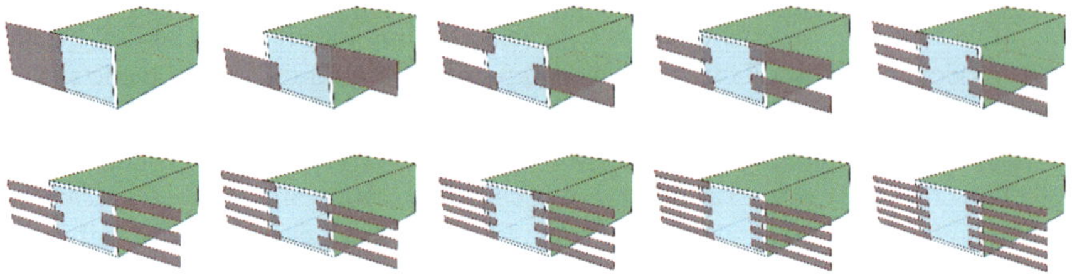

Fig. 7.7 Sliding façade with different numbers of slidable panels

7.3.3 Kinetic Façade Simulation

The term "energy consumption" for the simulation results in this study refers to the sum of the energy used for cooling and heating. This is to focus on the fact that kinetic façades mostly affect cooling and heating energy load, assuming that the energy used for equipment, lighting, and hot water is consistent. The energy used for actuating the kinetic façades is disregarded given that it is often negligible compared to the building cooling load.

The motion of the kinetic façades is divided into 21 discrete steps from fully closed to fully open. This results in 21 different opening ratios. The cooling and heating energy consumption of the kinetic façades was simulated for each of the opening ratios on each hour of the year. The opening ratio with the least energy consumption on each hour was selected. The sum of the least energy consumption of each hour was then used as the yearly energy consumption of the kinetic façade under the optimal operation schedule. Table 7.3 demonstrates how the data is processed. Each cell with a checkmark denotes the opening ratio with the least energy consumption for the hour of the year. The "Sum" column then adds all numbers in the checked cells. The grand total cell at the bottom right corner shows the annual energy consumption with the kinetic façade.

The opening ratio of each hour can be represented as an operation schedule. Figure 7.8 shows the operation schedule of the folding kinetic façade in Phoenix.

Table 7.3 Kinetic façade energy consumption data processing

Step	Hour						
	1	2	3	4	…	8760	Sum
1			√			√	
2	√						
3							
…		√		√			
21					√		
Sum							Annual Cons.

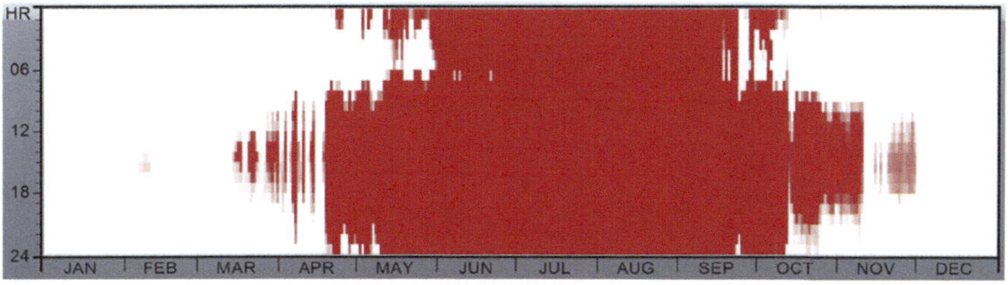

Fig. 7.8 Opening schedule of the folding façade in Phoenix

The X-axis shows the 365 days of a year. Y-axis shows 24 hours of a day. There are a total of 8760 pixels in the graph with each pixel representing the opening ratio of the kinetic façade on each hour of the year. A white pixel denotes that the façade is fully open, and a red pixel denotes that it is fully closed. Any color between red and white means that it is partially open with the shade of the pixel proportional to the opening ratio.

The total number of permutations for each type of kinetic façade is: 21 steps × 10 variations × 16 cities = 3,360 simulation runs. Since there are two types of kinetic façades, folding and sliding, a total number of 3,360 × 2 = 6,720 simulation runs were conducted in this research.

7.3.4 Optimized Static Façade

The simulation results of the kinetic façade were processed with a different method to calculate the energy consumption of an optimized static façade. In Table 3.2, assuming that each cell is filled with the energy consumption value of each opening ratio on each hour of the year, the "Sum" column simply sums up 8760 cells in each row. The row with the least "Sum" denotes the energy consumption of the optimized static façade at that specific opening ratio. The energy consumption of the shoebox model without any shading device was used as the baseline for the comparison with the kinetic facades.

7.4 Simulation Results

7.4.1 Folding Façade Performance

Figure 7.9 shows the energy consumption (heating and cooling) of the shoebox model across 16 cities under three types of shading conditions, i.e., folding façade, optimized fixed façade, and the no-shading model. The cities are arranged in hot

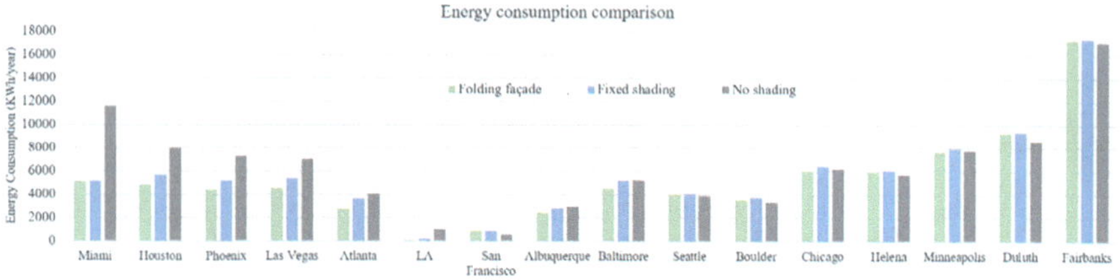

Fig. 7.9 Energy consumption comparison between the folding façade, fixed-shading façade, and no-shading façade

to cold order from left to right in the graph. In general, the folding façade and the fixed shading façade reduce energy consumption in the hot climates (from Miami to Atlanta) and slightly increase energy use in cold climates (from Seattle to Fairbanks). This is due to the fact that the façades block passive solar heat gain in winter.

Table 7.4 shows the amount and the percentage of energy savings of the folding façade compared to the fixed-shading façade and the no-shading model. Compared to the no-shading model, the folding façade reduces energy consumption by 32–56% in hot climates (Miami, Houston, Phoenix, Las Vegas, and Atlanta). Although the kinetic façade saves 91% in LA and −55% in San Francisco, the amount of energy saving (in kWh) is not as substantial as the percentage suggests. The percentages are skewed due to their minimal cooling and heating loads in warm climates. Compared to the optimized fixed-shading façade, the folding façade reduced 15–25% of the energy use in hot climates, with Miami being an exception (1% reduction). The speculation for the reason why the folding façade had very limited energy reduction compared to the fixed-shading facade is the low latitude. More simulation results for low-latitude cities are needed to confirm this speculation in the future. The folding façade always performed better than the fixed façade, but it does not make a meaningful difference in cold climates because both kinetic and fixed façades are underperforming compared to the no-shading model.

Figure 7.10 shows the relation between the energy reduction of the folding façade and the cooling/heating loads of the no-shading model. As seen in the graph, there is a strong positive correlation, a coefficient of 0.97, between the amount of energy saving of the folding façade (red line) and the cooling load. There is a weak negative correlation, a coefficient of −0.48, between the energy saving of the folding façade and heating load. The percentage of saving (green line) generally shows similar results while the data points for LA and San Francisco being the outliers.

Figure 7.11 shows the relation between the energy reduction of the folding façade and the cooling/heating loads of the fixed-shading façade. It shows a weak positive correlation (a coefficient of 0.5) with the cooling load and a weak negative correlation (a coefficient of −0.37) with the heating load.

Given that the folding façades showed the best performance in hot climates conditions, four of the hottest cities were selected to conduct a cross comparison between the folding façades with different numbers of foldable panels (as shown in Fig. 7.11).

Table 7.4 Energy savings of folding kinetic façade

Energy savings	Compared to no-shading		Compared to fixed-shading	
	Amount (kWh/year)	Percentage (%)	Amount (kWh/year)	Percentage (%)
Miami	6443	56	55	1
Houston	3117	39	846	15
Phoenix	2860	39	829	16
Las Vegas	2434	35	802	15
Atlanta	1272	32	924	25
LA	919	91	150	61
San Francisco	−321	−55	29	3
Albuquerque	510	17	428	15
Baltimore	728	14	735	14
Seattle	−128	−3	99	2
Boulder	−235	−7	216	6
Chicago	133	2	373	6
Helena	−264	−5	142	2
Minneapolis	127	2	313	4
Duluth	−738	−9	92	1
Fairbanks	−194	−1	144	1

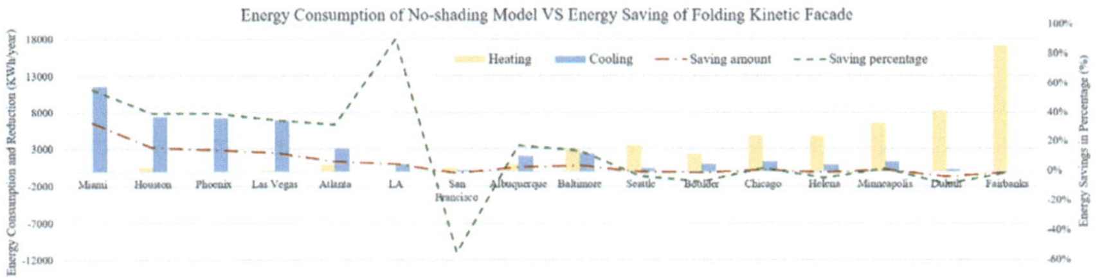

Fig. 7.10 Energy reduction of the folding façade compared to the energy consumption of the no-shading model

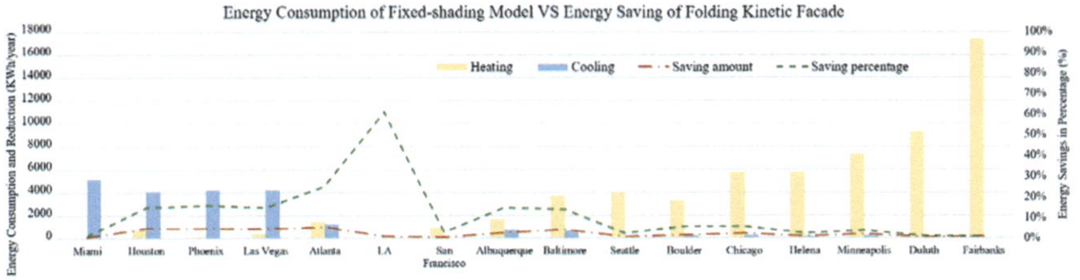

Fig. 7.11 Energy reduction of the folding kinetic façade compared to the energy consumption of the fixed-shading façade

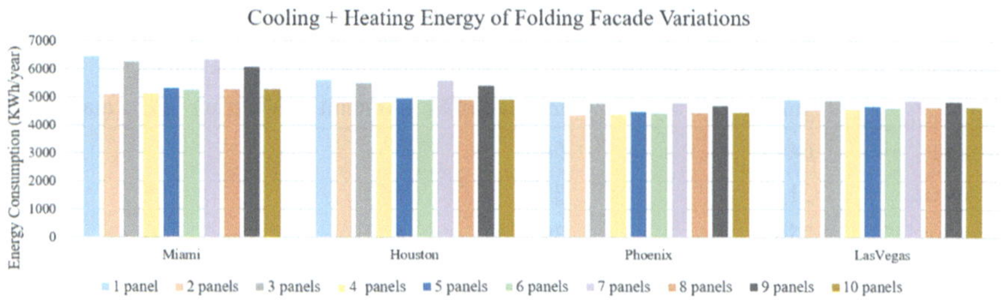

Fig. 7.12 Energy consumption of the folding façade with different numbers of panels

Figure 7.12 shows that there is no meaningful trend between the different numbers of foldable panels when compared to the no-shading model.

7.4.2 *Sliding Façade Performance*

Figure 7.13 shows the energy consumption (heating and cooling) of the shoebox model across 16 cities under three types of shading conditions, i.e., sliding kinetic façade, optimized fixed façade, and the no-shading façade. Contrary to the folding façade, the sliding façade outperformed the fixed shading façade and the no-shading model regardless of the climate condition. In general, the amount of energy savings of the sliding façade reduces as the climate gets colder.

Table 7.5 shows the amount and the percentage of energy savings of the sliding façade compared to the fixed-shading façade and the no-shading model. Compared to the no-shading model, the sliding façade reduces energy consumption in all cities, with 20% or more savings from Miami to Baltimore, on the spectrum of hot to cold climates. The sliding façade saves a substantial percentage of energy, 93% in LA and 29% in San Francisco, but the amount of energy saving (in KWh) is not as impactful as the percentage suggests. The percentages are skewed due to their limited cooling and heating loads in warm climates. Compared to the fixed-shading façade, the sliding façade reduced 15–25% of the energy in hot climates, with Miami being an exception (only 4% reduction).

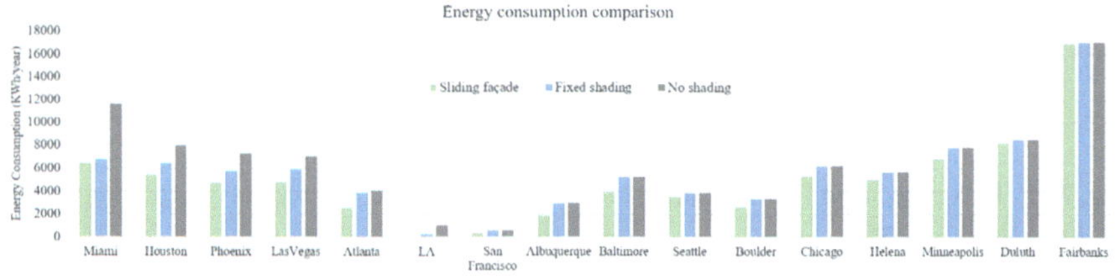

Fig. 7.13 Energy consumption comparison between the sliding façade, fixed-shading façade, and the no-shading façade

Table 7.5 Energy savings of sliding kinetic façade

Energy savings	Compared to no-shading		Compared to fixed shading	
	Amount (kWh/year)	Percentage (%)	Amount (kWh/year)	Percentage (%)
Miami	5148	44	301	4
Houston	2529	32	972	15
Phoenix	2508	35	960	17
Las Vegas	2211	32	1077	18
Atlanta	1448	36	1261	33
LA	942	93	226	75
San Francisco	167	29	167	29
Albuquerque	1071	36	1059	36
Baltimore	1316	25	1315	25
Seattle	355	9	355	9
Boulder	665	20	665	20
Chicago	908	15	908	15
Helena	630	11	630	11
Minneapolis	927	12	927	12
Duluth	264	3	264	3
Fairbanks	133	1	133	1

Figure 7.14 shows the relationship between the energy reduction of the sliding façade and the cooling/heating loads of the no-shading model. As seen in the graph, there is a strong positive correlation, a coefficient of 0.97, between the amount of energy saving of the sliding façade (red line) and the cooling load. There is a weak negative correlation, a coefficient of −0.49, between the energy saving of the folding façade and heating load. The percentage of saving (green line) generally shows similar results while the data point for LA being an outlier.

Figure 7.15 shows the relationship between the energy reduction of the sliding façade and the cooling/heating loads of the fixed-shading façade. It shows a weak

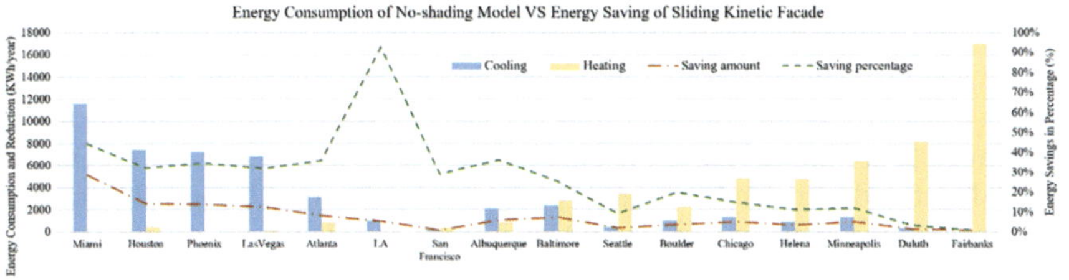

Fig. 7.14 Energy reduction of the sliding façade compared to the energy consumption of the no-shading model

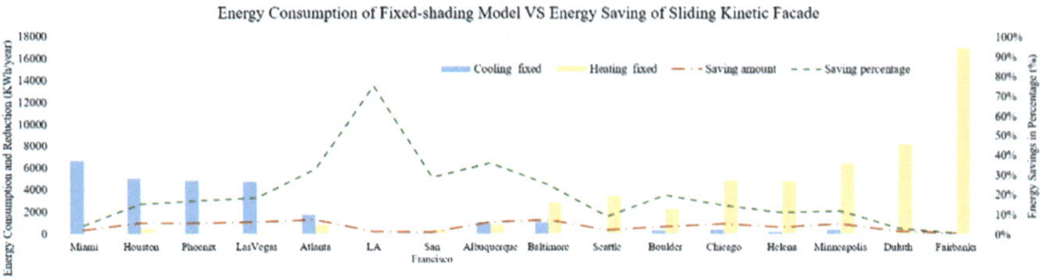

Fig. 7.15 Energy reduction of the sliding façade compared to the energy consumption of the fixed-shading façade

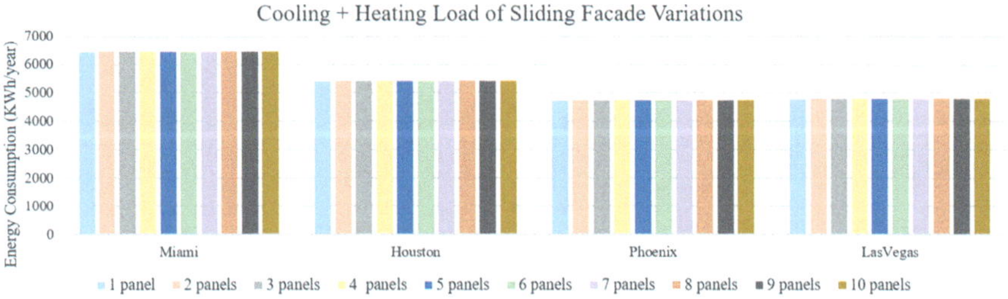

Fig. 7.16 Energy consumption of the sliding façade with different numbers of panels

positive correlation (a coefficient of 0.28) with the cooling load and a weak negative correlation (a coefficient of −0.36) with the heating load.

The top four cities with the hottest climates were used to conduct a cross comparison between the sliding façades with different numbers of slidable panels. Figure 7.16 shows that the façades with different numbers of slidable panels performed virtually the same.

7.4.3 Performance Comparison Between Folding and Sliding Façades

Figure 7.17 shows the energy consumption (cooling and heating) of the folding façade and the sliding façade. The folding façade performs better in the top four hottest cities, i.e., Miami, Houston, Phoenix, and Las Vegas, while the sliding façade performed better in the rest of the cities. Considering that neither the folding nor the sliding façade provided substantial energy savings in cold climates, the practical advantage of the sliding façade is in Albuquerque and Baltimore, saving 36% and 25% respectively. Comparatively, the folding façade saves only about 15% in the two cities.

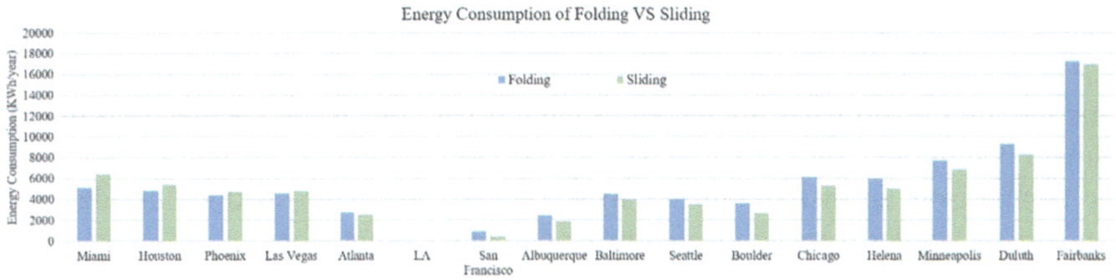

Fig. 7.17 Energy consumption of the folding façade and the sliding façade

7.5 Discussion

This study showed that kinetic façades can reduce cooling energy substantially if applied properly in appropriate climates. Although energy consumption is a very important aspect in architectural design, there are other important aspects of kinetic façades that need to be considered.

Kinetic façades have more components and thus require more resources, maintenance, and embodied carbon. In addition, kinetic façades do consume energy to operate. Although the amount used for operation is typically negligible compared to the reduction in cooling loads, for a more equitable comparison, the operational energy use and the embodied energy need to be considered to fully evaluate the environmental impacts of kinetic façades.

Kinetic façades also have a great impact on daylighting aspect of the built environment (Mahmoud & Elghazi, 2016). A good daylighting environment can increase human performance (Heschong, 2002) and improve human physical health (Aries et al., 2015). Therefore the comprehensive benefits of kinetic façades require a holistic assessment, including daylighting.

Another important factor of kinetic façades is view-out. Figure 3.7 shows the opening schedule of the folding façade in Phoenix. The schedule shows that the façade is mostly closed during work hours for more than half of the year if it is set to the optimal energy performance. This means that in many commercial buildings, occupants have to sacrifice view-out for energy reduction. To alleviate the problem of blocking view-out, many methods have been used as a compromise, such as using perforated panels instead of solid panels or adjusting the panel position to gain partial view out in exchange for minor performance loss.

Besides environmental impacts, kinetic façades can provide a unique character and identity to a building. This could be an undeniable motivation for architects and stakeholders of a building to pursue the unique esthetics of kinetic façades.

This study used two types of kinetic façades, folding and sliding, to represent two of the typologies, one that forms a shading device when fully open, and the other can fully reveal the glazing. Although the performance of different sliding façades does not show a significant difference, the performance of folding façades differs depending on the way that the façades fold or move. To fully evaluate the impact of folding façades, more folding mechanisms need to be studied.

7.6 Conclusion

This research investigated the energy saving aspect of kinetic façades in different climate conditions. Two types of kinetic façades, folding and sliding, were used for comparison against optimized fixed-façades and a no-shading façade. The comparison studies were conducted using computer simulations.

The simulation results showed that the kinetic façades, compared to the no-shading model, reduced 32–56% energy use in ASHRAE zone 1, zone 2, and zone 3, with the folding façade slightly outperforming the sliding façade. In zone 4, the sliding façade outperformed the folding façade. The exception is that the amount of energy reduced by the kinetic façades in marine zones (San Francisco and Seattle) were not as noticeable. In zone 5 to zone 8, the kinetic façades did not show a meaningful energy reduction and even showed an increase in energy consumption in some cases of the folding façade simulations.

This study also provided an insight into how much energy kinetic façades can save compared to optimized fixed-shading facades. While showing noticeable energy reduction compared to the optimized fixed-shading façades in general, the folding and sliding façades in Miami showed only 1% and 4% energy reduction respectively. This result suggests that applying kinetic façades to reduce energy consumption in a climate similar to Miami might not be the best solution. Considering the operational energy use, embodied carbon, and more frequent maintenance of kinetic façades, an optimized fixed-shading façade might be a better solution for reducing environmental impacts.

The performance difference between the kinetic façades with different numbers of panels was investigated. The simulation results showed that there was no clear pattern in energy consumption between the variations of kinetic façades. Knowing that the number of panels does not make a meaningful difference in energy performance can provide designers freedom to decide the number of panels based on other factors such as structural stability, cost, maintenance, etc.

A comparison between the folding façade and the sliding façade showed that the folding façade perform better in hot and warm climates while the sliding façade performs better in mild climate. Neither of the two types of kinetic façades provided a meaningful energy reduction in cold climates.

The simulation results of this research are expected to provide basic rules of thumb to designers when considering kinetic façades as a means of energy reduction strategy. By checking the ASHRAE zone of the project site, a designer can get a sense of whether a kinetic façade would be a potential solution.

In this research only 16 cities were selected, one for each ASHRAE climate zone. To better generalize the results, more cities with a wider geographical distribution need to be investigated in the future. In addition, to validate the speculation for the reason that kinetic façades showed negligible energy reduction in Miami is due to its low latitude, more simulations with low-latitude cities need to be conducted.

Lastly, it is important to recognize that energy consumption is not the only factor that determines whether a kinetic façade is suitable for a project or not. There are many other factors, such as daylighting, view, etc., that need to be considered with a holistic approach.

References

Alotaibi, F. (2015). The role of kinetic envelopes to improve energy performance in buildings. *Journal of Architectural Engineering Technology, 4*(3), 149–153.

Aries, M. B., Aarts, M. P., & van Hoof, J. (2015). Daylight and health: A review of the evidence and consequences for the built environment. *Lighting Research & Technology, 47*(1), 6–27.

Arup. (2013). *Solar Leaf*. Retrieved from https://www.arup.com/projects/solar-leaf

Attia, S. (2018). Evaluation of adaptive façades: The case study of Al Bahr Towers in the UAE. In *QScience Connect, 2017 (2, Special Issue on Shaping Qatar's Sustainable Built Environment-Part I)* (p. 6).

Bacha, C. B., & Bourbia, F. (2016, October). Effect of kinetic façades on energy efficiency in office buildings-hot dry climates. In *11th Conference on Advanced Building Skins* (Vol. 1, pp. 458–468).

EPA. (2017). *Overview of Greenhouse Gases*. Retrieved from http://learning-cleanairasia.org/lms/library/ga3/99-Overview-of-Greenhouse-Gases.pdf

Granqvist, C. G., Arvizu, M. A., Pehlivan, İB., Qu, H. Y., Wen, R. T., & Niklasson, G. A. (2018). Electrochromic materials and devices for energy efficiency and human comfort in buildings: A critical review. *Electrochimica Acta, 259*, 1170–1182.

Heschong, L. (2002). Daylighting and human performance. *ASHRAE Journal, 44*(6), 65–67.

IEA. (2018). *Global Status Report*. Retrieved from https://www.worldgbc.org/sites/default/files/2018%20GlobalABC%20Global%20Status%20Report.pdf

IMA. (2022). Retrieved from https://www.imarabe.org/en/architecture

Kolarevic, B., & Parlac, V. (2015). *Adaptive, responsive building skins, in building dynamics: Exploring architecture of change* (Ch. 6, p. 75). Routledge Taylor & Francis Group

Mahmoud, A. H. A., & Elghazi, Y. (2016). Parametric-based designs for kinetic facades to optimize daylight performance: Comparing rotation and translation kinetic motion for hexagonal facade patterns. *Solar Energy, 126*, 111–127.

Pörtner, H. O., Roberts, D. C., Adams, H., Adler, C., Aldunce, P., Ali, E., Begum, R. A., Betts, R., Kerr, R. B., Biesbroek, R., & Birkmann, J. (2022). *Climate change 2022: Impacts, adaptation and vulnerability.*

Tabadkani, A., Roetzel, A., Li, H. X., & Tsangrassoulis, A. (2021). Design approaches and typologies of adaptive façades: A review. *Automation in Construction, 121*, 103450.

Chapter 8
Integration of Solar Technologies in Facades: Performances and Applications for Curtain Walling

Paolo Rigone and Paolo Giussani

Abstract The building envelope has a dominant impact on a building's energy balance and it plays an essential role towards the nearly Zero Energy Buildings (nZEB) target (Commission Recommendation (EU), (2016); International Energy Agency, (2013)). In this scenario, adaptive façades are becoming increasingly popular because they should provide controllable insulation and thermal mass, daylighting, solar shading, ventilation and humidity control, etc. Moreover, they should collect and convert available renewable energy (mainly solar) in an adaptive way, to follow, as far as possible, the building energy needs. Facing the challenges of decarbonisation for the building sector in the EU (target of 80% for 2050), the building envelope should integrate active functions in terms of energy production (collect, convert, store, distribute). In this regard, building façades are often the largest potential surface for integration of renewable energy generation components (photovoltaic, solar thermal, etc.) in urban areas. According to the definition given in IEC 63092-1, the BIPV module is defined as a "Photovoltaic module that provides one or more of the functions of the building envelope". PV is progressively becoming one of the distinctive and characterizing sign of contemporary architecture, similarly to any other building material or component. The building sector plays a central role in an environmental sustainability perspective, contributing for 36% of global energy consumption and 39% of carbon dioxide emissions when upstream power generation is included (UN Environment and International Energy Agency, (2017)). The BIPV production of solar energy is generally near the utility's peak loads of the building. The renewable solar production reduces energy costs for the building during the

P. Rigone (✉)
Associated Professor of Design of the Building Envelope at Faculty of Building Engineering, Department of Architecture Built Environment and Construction Engineering, Politecnico di Milano, Milan, Italy
e-mail: paolo.rigone@polimi.it

P. Giussani
Facade Engineer Consultant Working in the Field of Advanced Building Envelope Technologies and Components in Partnership With Studio di Ingegneria Rigone, Teaching Assistant—Politecnico di Milano, Milano, Italy
e-mail: p.giussani@studiorigone.it

J. Wang et al. (eds.), *Advanced Materials in Smart Building Skins for Sustainability*,
https://doi.org/10.1007/978-3-031-09695-2_8

time of its greatest demand. The concept of BIPV refers to the capability of photovoltaic systems to be multifunctional and interact with the building, producing free energy from a renewable source. The use of PV in the building sector rises many questions, for example re-imagining the building envelope both in aesthetics and technology, where the photovoltaic element has an additional building functionality, namely replacing an element of the building skin. PV modules should effectively play the role of traditional building components and their performance is required for the integrity of the building's functionality.

Keywords Building envelope · BIPV · Building integration · Building integrated photovoltaics · Curtain walls · Façade · PV glass · Windows

8.1 BIPV Technology

Since the beginning (mid of last 70's), solar energy harvesting has been considered highly expensive, relatively inefficient and accompanied by a general poor design. In the past existing building-integrated photovoltaics (BIPV) have proven to be less practical and economically unfeasible for large-scale adoption due to design limitations and poor aesthetics. New tools and technologies, both for building design and construction, have come to assist designers and clients in the development of buildings able to generate their own energy and to be self-sustaining.

A Building Integrated Photovoltaics (BIPV) system consists of integrating photovoltaics cells into the building skin, such as the horizontal roof or the vertical/inclined facades. At the same time, these components serving as building envelope materials and power generator. BIPV systems can provide savings costs of electricity, reduce use of primary energy (i.e. fossil) and emission while adding architectural goal to the building.

The PV glass panels consist of layers of glass (usually heat-treated safety i.e. laminated with polymeric interlayer foils), which include in the middle a certain number of PV cells (monocrystalline, polycrystalline or amorphous)—(Figs. 8.1, 8.2 and 8.3). The characterisation of BIPV modules must be multifunctional, addressing both electrotechnical and building requirements.

The cells are linked together following electric schemes based on technology of various bus bars connection and plugs (current state of art with 3 or 5 bus bars).

BIPV glass provides the same performance (thermal and sound insulation) as a conventional glass and it can be assembled in Double Glazing Unit (DGU) or Triple Glazing Unit (TGU)—Fig. 8.4. Furthermore, PV systems can also be used as small stand-alone power units. Thus, the BIPV could be inserted in tailored solutions of new glass façades (Fig. 8.5) or replacing old existing glazing into retrofitting of curtain walls of buildings, generating free clean electricity and reducing the carbon footprint.

"Conventional" PV modules are designed to maximise the photovoltaically active area per module, such that non-active areas (opaque or transparent) between and

Fig. 8.1 Amorphous silicon vertical fins—Life Science Building, University of Washington, Seattle, USA (courtesy of Onyx Solar)

Fig. 8.2 Amorphous silicon vertical fins—double skin, DEWA Research center, Dubai, UAE (courtesy of Onyx Solar)

around the solar cells are reduced to the minimum dimensions needed to ensure electrical insulation. By contrast, those BIPV modules that are intended to transmit light and/or solar radiation into a building also include non-negligible transparent or translucent areas. (International Energy Agency, 2020).

The two main photovoltaics technologies available for these types of applications are made of thick crystal products or thin-film products. In the first family, the solar cells are made from crystalline silicon either as monocrystalline or polycrystalline wafers (Figs. 8.6 and 8.7).

Fig. 8.3 Construction phases of GIOIA 22 tower (North/West and North/East elevation)—Crystalline Silicon glass BIPV—Gioia 22 Tower, Milan, Italy

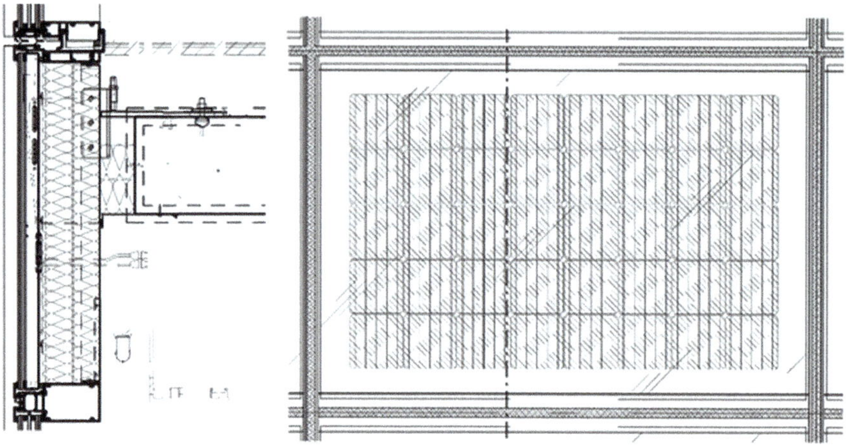

Fig. 8.4 Typical vertical section and frontal view of BIPV spandrel unit

In case of the second family the very thin layer of photovoltaically active material is placed on a glass or metal substrate using a vacuum-deposition manufacturing process similar to the one performed for coating in architectural glass. The main semiconductor used in thin-film products is silicium amorphous (A–Si) or based on

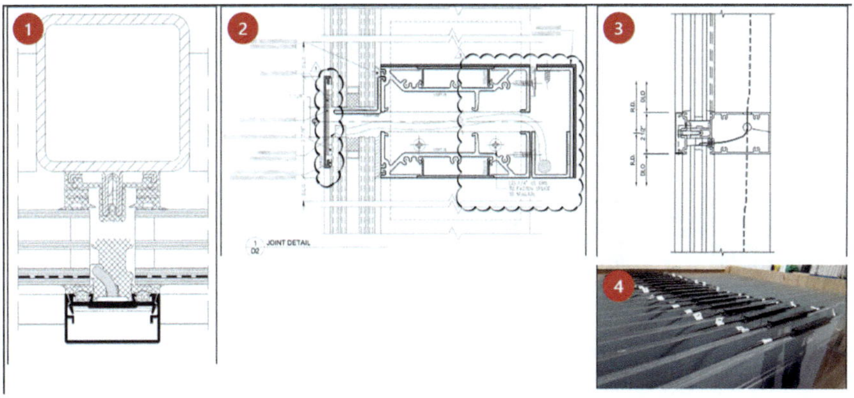

Fig. 8.5 (1) BIPV triple glazing unit applied into slimline all-glass façade on steel substructure. (2) Example of splice joint detail. (3) Example of stick-system construction with BIPV (4) Edge mounted junction box (courtesy of Onyx Solar)

Fig. 8.6 Left: Polycrystalline panel—Right: Monocrystalline panel (*Source* www.solarinnova.net)

cadmium telluride (CdTe), or made of indium and copper diselenide (CIS–CIGS) or Copper–Indium–Diselenide (CIS).

Building retrofit offers the opportunity to reduce energy consumption, improve energy efficiency and increase the use of renewable energy sources. The photovoltaic (PV) technology can be integrated into the building envelope (i.e. ventilated façade), where conventional construction materials can be easily substituted by PV modules (Fig. 8.8). All this makes BIPV an attractive solution for effectively and sustainably retrofitting building envelopes, providing savings in materials and in conventional electricity consumption and, at the same time, improving the energy efficiency of the buildings (Martín-Chivelet et al., 2018).

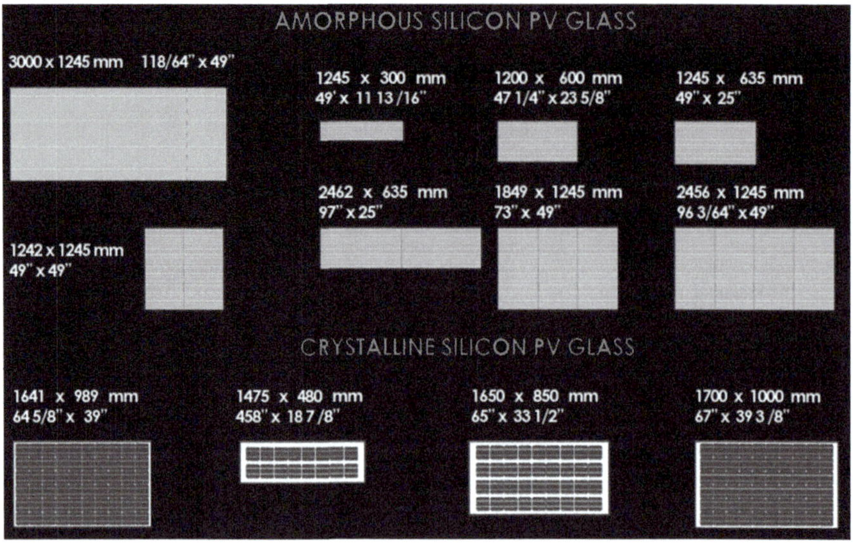

Fig. 8.7 Typical standard size for PV panel (courtesy of Onyx Solar)

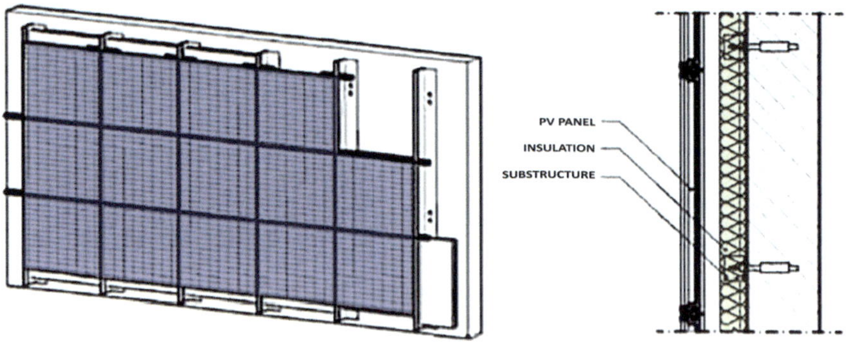

Fig. 8.8 PV technology integrated into ventilated facade

Crystalline silicon PV glass

Its power capacity is given by the number of solar cells used per glass unit. Crystalline Silicon glass (Fig. 8.9) shows a nominal power that usually ranges from 80 up to 160 Wp/m^2, therefore is commonly used in projects seeking maximum power output (Onyx Solar, 2019). The nominal power rate depends on the solar cell density required by design. The average efficiency is up to 16%.

PHOTOVOLTAIC GLASS	1.475 x 480	
044A0-14750480-43-M	5" Mono 158	Crystalline
Electrical data test conditions (STC)		
Nominal peak power	79	P_{mpp} (Wp)
Open-circuit voltage	11	V_{oc} (V)
Short-circuit current	9,41	I_{sc} (A)
Voltage at nominal power	9	V_{mpp} (V)
Current at nominal power	9,10	I_{mpp} (A)
Power tolerance not to exceed	±10	%
STC: 1000 w/m², AM 1.5 and a cell temperature of 25°C, stabilized module state.		
Mechanical description		
Length	1475	mm
Width	480	mm
Thickness	9,8	mm
Surface area	0,71	sqm
Weight	14	Kgs
Cell type	6" Mono 158	Crystalline
No PV cells / Transparency degree	16	43%
Front Glass	4 mm	Tempered Glass Low-Iron
Rear Glass	4 mm	Tempered Glass
Thickness encapsulation	1,80 mm	EVA Foils
Category / Color code		
Junction Box		
Protection	IP65	
Wiring Section	2,5 mm² or 4,0 mm²	
Limits		
Maximum system voltage	1000	Vsys (V)
Operating module temperature	-40...+85	°C
Temperature Coefficients		
Temperature Coefficient of Pmpp	-0,32	%/°C
Temperature Coefficient of Voc	-0,28	%/°C
Temperature Coefficient of Isc	0,07	%/°C

* All technical specifications are subject to change without notice by Onyx Solar

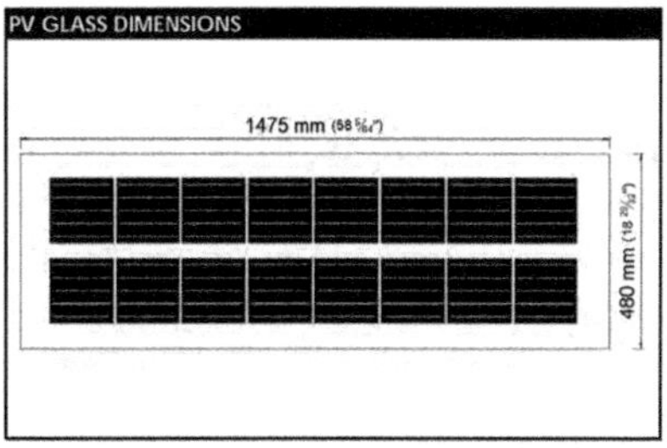

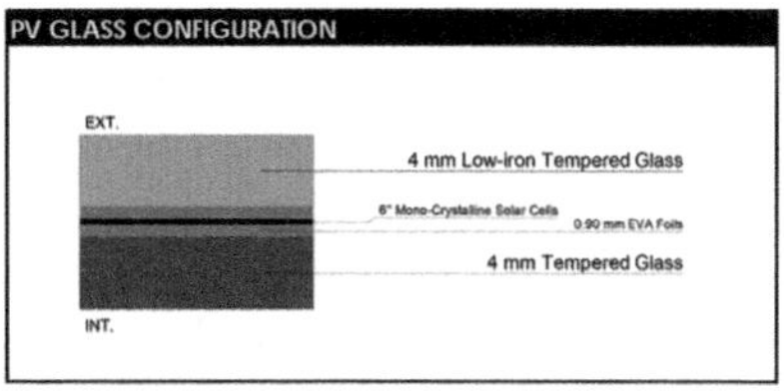

GLASS PROPERTIES	
Light Transmission	43%
U-value (W/sqm.K)	5.2
Peak Power [Wp/sqm]	111,1

Fig. 8.9 Sample data sheet of crystalline silicon PV glass (courtesy of Onyx Solar)

Amorphous silicon PV glass

This PV Glass can be fully opaque/dark (higher nominal power), or present different light transmittance levels, which enables for the natural light to pass through exterior, while maintaining unobstructed views. Amorphous Silicon glass offers a better performance under diffuse light conditions and high temperature (Onyx Solar, 2019). The aesthetic integration of the glass panel with this type of PV is uniform (Fig. 8.10).

PHOTOVOLTAIC GLASS					034_N-12450300-_ _-_
1245 x 300 mm		ref. 00	ref. 10	ref. 20	ref. 30
Electrical data test conditions (STC)		DARK (0%)	M VISION (10%)	L VISION (20%)	XL VISION (30%)
Nominal peak power	P_{mpp}(Wp)	21	15	13	10
Open-circuit voltage	V_{oc}(V)	23	23	23	23
Short-circuit current	I_{sc}(A)	1,50	1,15	0,97	0,77
Voltage at nominal power	V_{mpp}(V)	16	16	16	16
Current at nominal power	I_{mpp}(A)	1,34	0,93	0,79	0,65
Power tolerance not to exceed	%	±5	±5	±5	±5

STC: 1000 w/m², AM 1.5 and a cell temperature of 25°C, stabilized module state.

Mechanical description		
Length	mm	1245
Width	mm	300
Thickness	mm	8,10 (EVA) 8,72 (PVB)
Surface area	sqm	0,37
Weight	Kg	6,00
Cell type		a-Si Thin Film
PV Glass		3,2 mm Float Glass
Rear Glass		4,0 mm Float Glass
Thickness encapsulation	ref. A	0,90 mm EVA Foils
	ref. B	1,52 mm PVB Foils

Junction Box		
Protection		IP65
Wiring Section		2,5 mm² / 4,0 mm²
Limits		
Maximum system voltage	V_{sys} (V)	1.000
Operating module temperature	°C	-40...+85
Temperature Coefficients		
Temperature Coefficient of P_{mpp}	%/°C	-0,19
Temperature Coefficient of V_{oc}	%/°C	-0,28
Temperature Coefficient of I_{sc}	%/°C	+0,09

* All technical specifications are subject to change without notice by Onyx Solar

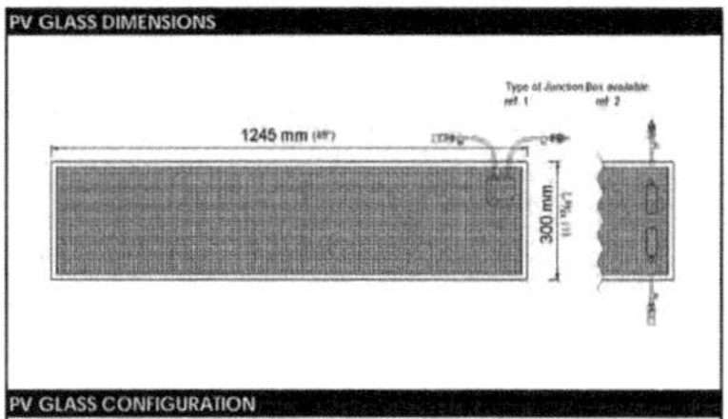

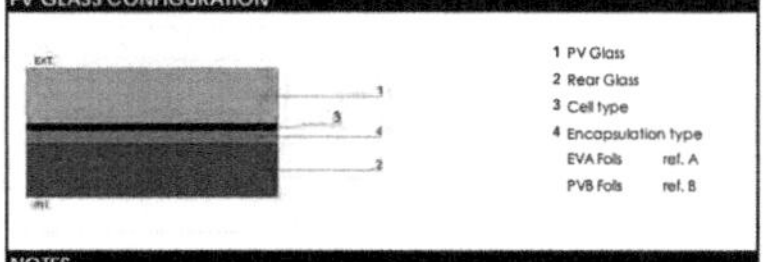

NOTES
* For optical and further mechanical properties, please go to: **Technical Guide. 7.-Other Properties.**
* Optional: Insulating Glass Unit. U value (W/sqm.K), please go to: **Technical Guide. 8.-Insulating Glass Unit.**
* Junction box type and location should be approved by the customer.

Fig. 8.10 Sample data sheet of amorphous silicon PV glass (courtesy of Onyx Solar)

8.2 Innovation and New Frontiers of BIPV Technology

Current research in BIPV technology is now oriented to develop the so called luminescent solar concentrators (LSCs).

LSCs (Fig. 8.11) are transparent sheets of plastics doped with chromophores which, upon absorption of sunlight, re-emit photons at longer wavelengths. These photons are guided by total internal reflection to the edges of the device where they are converted into electricity by conventional photovoltaic cells.

Due to their high integrability in continuous glazing and photovoltaic windows, LSCs are considered one of the most promising tools to realise near-zero energy buildings in highly urbanised areas where roof surfaces are not sufficient to produce all the required electricity.

Furthermore, in conclusion, it is now possible to have large-area LSCs that are stable, non-polluting and with a good efficiency, ready for insertion into photovoltaic double glazing units, providing also thermal and acoustic insulation at the same time.

Aesthetics technology allows the development of photovoltaic insulating glass units with a low aesthetic impact that can be integrated architecturally into passive buildings, being almost invisible both from the outside and from the inside. The degree of transparency of the LSCs can also be determined during production according to the customer's needs, in order to obtain the best compromise between absorbed energy and amount of light for indoor natural day lighting.

Basically, two main technologies have been developed up to now with appreciable potential use into integrated applications to the building skin: Inorganic and Organic Integrated Photovoltaic.

The inorganic technology involves the use of nanoparticles as chromophores in which, thanks to appropriate engineering, the complete decoupling of the process of absorption and emission of light is achieved. In this way, prototypes with good power generation efficiencies have been realised, enabling the production of customised modules. The nanoparticles are made from inorganic materials such as silicon, which are intrinsically stable to solar radiation without danger of degradation, guaranteeing continuity and durability of electricity production.

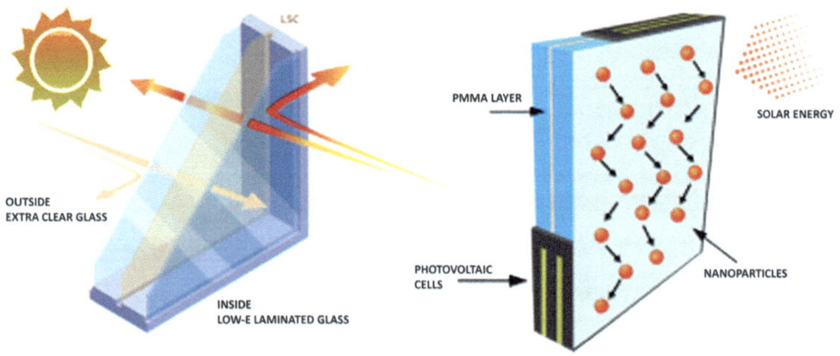

Fig. 8.11 Luminescent solar concentrators (LSCs)

Currently It has been possible to achieve light-to-power conversion efficiencies of up to 3.0% with a degree of transparency in the visible region of around 80% (i.e., only 20% of the light is used to generate electricity while the remaining 80% passes through the panel to illuminate interior spaces). The optical conversion efficiency of the blue/ultraviolet fraction of the solar spectrum reaches values of around 10%.

On the other hand, the Integrated Organic Photovoltaics (BIOPV) is made of an OPV solar cell type where the absorbing layer is based on organic semiconductors (OSC)—typically either polymers or small molecules. For organic materials to become conducting or semiconducting, a high level of conjugation (alternating single and double bonds) is required. Conjugation of the organic molecule results in the electrons associated with the double bonds becoming delocalised across the entire length of conjugation. These electrons have higher energies than other electrons in the molecule, and are equivalent to valence electrons in inorganic semiconductor materials (Spooner, 2022). In the last ten years, the highest efficiency obtained from organic photovoltaics (OPVs), has risen from 2.5 to 11% (Su et al., 2012).

OPV cells are categorized into two classes:

- Small-molecule OPV cells
- Polymer-based OPV cells.

Current research focuses on increasing device efficiency and lifetime. Substantial efficiency gains have been achieved already by improving the absorber material, and research is being done to further optimize the absorbers and develop an organic multijunction architectures (Office of Energy Efficiency Renewable Energy, 2022). Improved encapsulation of OPV (Fig. 8.12) and alternative contact materials should be adopted to reduce cell degradation and push cell lifetimes to industry-relevant values.

8.3 Architectural Integration of Photovoltaics in Façade: The Need of Requirements and Performances as Building Products

International research and best practice have already shown many interesting examples of appreciable and aesthetical integration of PV in contemporary architecture as well as in existing built heritage. The building integration of photovoltaics, where the PV elements actually become an integral part of the building, often serving as the exterior skin with thermal, acoustic and also air permeability and water tightness performances, is growing worldwide. The aesthetic integration is also commonly discussed, as it is a critical point.

The PV panel may be incorporated into many different assemblies and part of the building envelope. PV can be incorporated into façade completing, or replacing, traditional vision areas or spandrel glass.

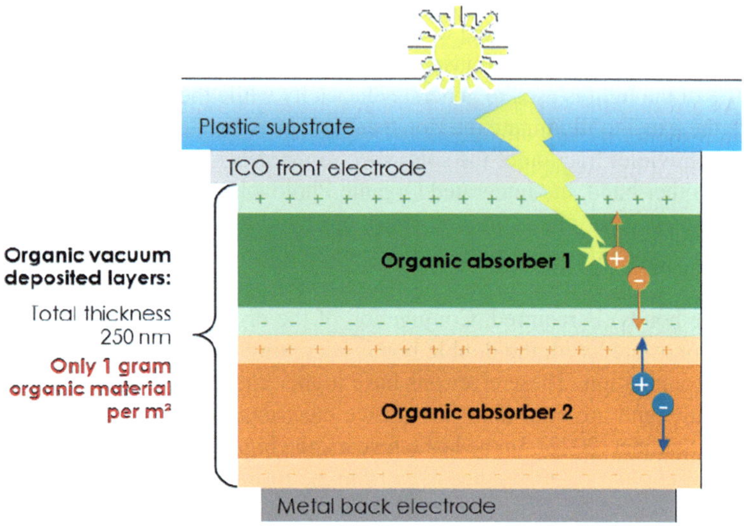

Fig. 8.12 Example of OPV stratigraphy with two layers of organic absorber (*Source* www.heliatek.com)

A photovoltaic module, not only produces electricity using sun power, but it has to behave as all the other curtain walling components, so it must provide one or more of the following performances:

Safety in use

- Resistance to wind load
- Resistance to dead load
- Resistance to horizontal loads
- Resistance to snow load
- Seismic resistance
- Resistance against impact
- Thermal shock resistance
- Mechanical rigidity or structural integrity.

Environment

- Air permeability
- Watertightness.

Fire protection

- Fire resistance
- Reaction to fire of components
- Fire propagation (i.e. to upper levels).

Energy saving

- Thermal transmittance
- Water vapour permeability
- Radiation properties
- Shading & daylighting.

Acoustic

- Airborne sound insulation (Noise protection).

Durability and Maintenance

8.4 Performances and Requirements

As said BIPV module is a PV module and a construction product together, designed to be a component of the building. A BIPV module is the smallest (electrically and mechanically) non-divisible PV unit in a BIPV system which retains building-related functionality. If the BIPV module is dismounted, it would have to be replaced by an appropriate construction product (International Energy Agency, 2018).

Requirements can be listed according to their kind and need for standardization, as described below. Building requirements for modules containing glass panes are related also to their position in the contest of the building envelope, according to the following main topics:

- General requirements
- Electrical requirements
- Building-related requirements
- Requirements for products with glass panes
- Requirements for products without glass panes
- Labelling requirements
- System documentation, commissioning tests and inspection requirements.

As first, safety in use shall be provided by assessing the right choice of profile material and section in relation to the foreseen actions. BIPV modules must be characterised by the property of post-breakage integrity (the broken panel remain safe without fall apart or slip out of its own frame) under a load and for a predefined period of time (mainly it depends on the maintenance strategy and time necessary to replace the broken part). The mechanical behaviour of a laminated glass depends on the glass material as well as the bonding properties of the interlayer. If we take into account the higher operating temperature of the BIPV module, the mechanical performance changes compared to conventional laminated glass. In terms of mechanical resistance of the glass, these changes are not significant. On the other hand, the bonding properties of the interlayer depend on the operating temperature of the BIPV panel. The load duration is also the other parameter effecting the bonding properties and

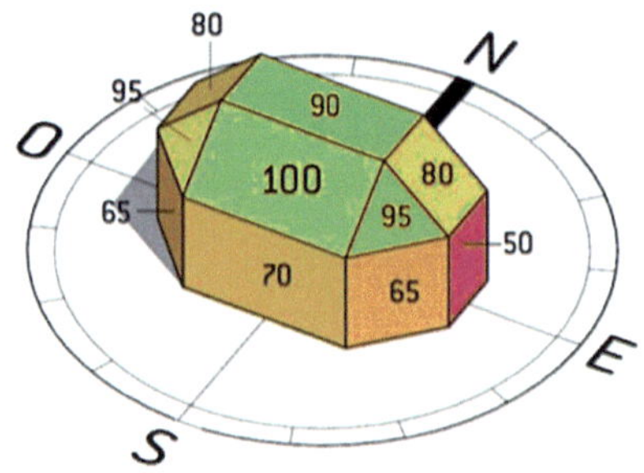

Solar Declination	0°	30°	60°	90°
East	0,93	0,90	0,78	0,55
South-East	0,93	0,96	0,88	0,66
South	0,93	1,00	0,91	0,68
South-West	0,93	0,96	0,88	0,66
West	0,93	0,90	0,78	0,55

Fig. 8.13 Efficiency of BIPV systems for different positions

its mechanical resistance. The interlayer material plays a major role in the mechanical performance of BIPV-laminated glass. Polyvinyl Butyral (PVB) is widely used in glass lamination, alternative interlayer materials, such as ethylene–vinyl acetate (EVA) or polyethylene (PE) with improved temperature stability, are used.

The curtain walling and its own components, shall be safe in use by ensuring the adequate mechanical stability in relation to the foresee actions, such as wind load, dead load, live load and seismic action. The operation level of the curtain walling shall be evaluated in relation to the foresee actions, such as wind load. Fundamental construction requirements for PV modules in order to provide safe electrical and mechanical operation, address the prevention of electrical shock, fire hazards, and personal injury due to mechanical and environmental stresses (AInternational Energy Agency, 2019).

Energy-conscious design has to be carefully considered during design stage (WBDG, 2016). The use of PV technology should be viewed in terms of life-cycle cost and not only initial costs. Often, the installation of BIPV is vertical, reducing access to available solar resources (Fig. 8.13) but the large surface area of the building can compensate the reduced power.

The PV efficiencies are largely reduced by high operating temperature. It means that design of BIPV should provide an adequate ventilation behind the modules to dissipate heat, or use amorphous silicon thin-film (efficiency –0.1%/°C up to –0.2%/°C) that perform better with elevated temperature (Aronica & di Torino, 2018) compare to crystalline cells (efficiency –0.4%/°C up to a –0.5%/°C). The type of connection (i.e., in series or in parallel) could change the level of current generate from PV Cells. The series connection allows increasing the voltage levels of the cells, while the values of generated current can be increased by means of the parallel connection. In order to achieve the voltage and the current values required for a given application, the solar cells are interconnected and encapsulated which constitutes a module or a PV panel (Perea & Alcocer, 2018).

Fire protection shall be ensured by designing horizontal and vertical fire barrier to limit fire propagation to upper floors and by selecting components and materials with a limited reaction to fire. The fire resistance class depends on the type of the building and intended use, the building height, curtain walling type, presence of alternatively controlling fires system such as water fire suppression, sprinkler, etc. generally speaking the curtain wall where BIPV are installed, shall guarantee the adequate level of fire resistance and reaction to fire in relation to the project specifications. Environmental requirements shall be ensured by choosing a facade system in relation to the foreseen actions (weather, wind load, etc.).

Acoustic performances shall be ensured by proper design of the infilling components (glazing), by controlling air permeability through frames and infilling parts and by the use of proper gaskets and sealing for joints of openable parts (windows and doors) and movement and expansion interfaces (butt joints, pressure equalization slots, surround, lap and sleeve joints). To ensure adequately the durability of facade characteristics and performances, the components shall be maintained resistant to UV actions, environment and ageing.

In order to fulfil the energy saving goals, the curtain walling shall be designed by choosing proper components (framing and infills) with adequate thermal transmittance, also known as "Uvalue". This parameter is the classical parameter that measures the quantity of heat that flows through unit area of the facade per unit of difference in temperature outer and inner environment. For IGUs, the thermal transmittance is calculated according to EN 673. Typical thermal transmittance values for common building fenestration products are as it follows: Single glazing: 5.7 W/m^2 K; DGU (double glazing) 1.0 W/m^2 K, TGU (triple glazing) 0.6–0.5 W/m^2 K, complete assembled windows 1.3–1.4 W/m^2 K and 1.0–1.1 W/m^2 K for single skin curtain walling.

In terms of daylight and transparency, the PV panel could perform a certain rate of transparency depending on the coating applied on glass or in case of different manufacturing process. To increase the transparency of the photovoltaic panel, the glass is laser-etched to remove thin lines of active solid cells. Generally, The PV active material is black by nature (faces the sun) while the interior of the glass displays an aluminium-like colour. However, coloured panel could be manufacturing using coloured interlayer (PVB) during the lamination process or other manufacturing process such as screen-printing or ceramic-frits too. It is important to underline that the color could vary on both sides of the panel depending on the intensity of the light (in reflection or transmission). The curtain walling shall provide the adequate level of acoustic insulation in order to guarantee the internal comfort.

Power demands, energy efficiency of the PV system combined with architectural requirements claim to develop a consistent design strategy to take into account the effective available glazed area. To reach an architecturally pleasing composition, the PV modules should be in harmony with the total image of the building according to colour, texture, size, and position. Today PV integration is no more typically limited to windows and glass facades (curtain walls); solar roofs are designed to look essentially indistinguishable from traditional roofing materials such as asphalt and slate shingles. PV integration is applied also to other components of the building shell, replacing conventional building materials such as skylights, external claddings, and glass balustrades too. From this perspective, renewable energy systems can be integrated as functional, aesthetic, and cost-effective elements within buildings.

Fig. 8.14 Monocrystalline horizontal Louvers, Avenida Diagonal 525, Barcelona, Spain (courtesy of Onyx Solar)

Considering mutual obstructions given by buildings, presence of shading devices, architectural constraints such as vertical or horizontal fins (Fig. 8.14), the partial shading occurs more frequently in PV modules integrated in buildings than in ground-mounted plants (International Energy Agency, 2020).

The potential risk of the BIPV technology is the influence of shadings on a BIPV System's Performance. The loss factor in BIPV systems due to the shadings is critical, and this should be well investigated and since its preliminary design phase (Fig. 8.15). Design considerations for BIPV take in consideration the location and orientation of the building itself and the characteristics of the surroundings, any type of utility issues, costs, and electrical loads need and its architecture. Different orientation can have a significant effect on the annual power output of the BIPV system, therefore in the early design phase, it has to be evaluated that solar panel receive maximum exposure to the sun and will not be shaded by obstructions (in particular completely unshaded during the peak of solar grade).

This type of analysis, that are typically performed in the multidisciplinary approach in facade design (Rigone & Giussani, 2019) and energy conscious design techniques, are also necessary to investigate the effects of the indirect radiation coming from different angles of incidence that create inhomogeneous power generation from the different solar cells.

Partial shading occurs when one or more PV cells in a PV module receive less irradiance than the rest of the panel. The impact of shading on a PV has much greater performance effect on the electrical harvest than the footprint of the shadow. Partial shading can lead significant reduction of the energy but also temperature increase of shaded PV cells (hot spot) that could irreversible get damaged.

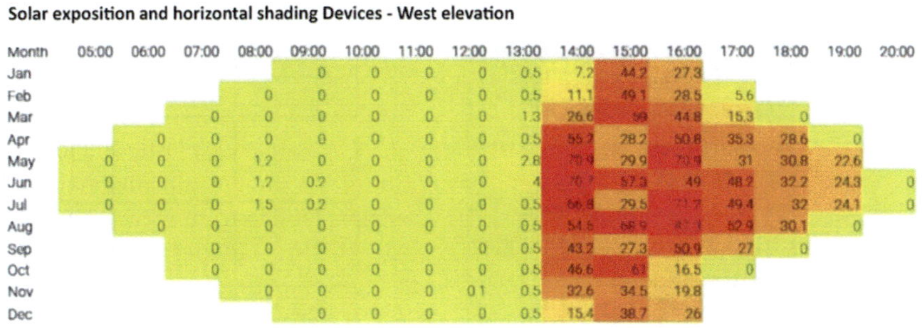

Fig. 8.15 Shading analysis for BIPV

The BIPV System's Performance could also be influenced by the reflectance of a flat-surface PV module under light coming from different angles. This data is generally greater than the reflectance at normal incidence at which the electrical characteristics of a PV module are commonly supplied (and these are the information that we could easily get from the technical sheet of the panel given by the suppliers). Reflection losses on an annual basis can become especially significant in those BIPV applications in which PV modules are positioned far from the optimum orientation regarding sun path or better say sun's position in the sky.

The PV panel showed in Fig. 8.16 is fully integrated in the spandrel part of the curtain wall. The stratigraphy of the panel (Figs. 8.17 and 8.18) is composed by two layers of float glass 6 mm thickness with interlayer foil made in EVA (Ethylene Vinyl Acetate) composes the glass thickness of the BIPV. The glass stratigraphy has to follow the national regulation in term of safety in use in building for façade applications. To increase the light transmission value, the use of specific coating is strategically adopted as for example the external pane in low-Iron version with higher 90% light transmission coefficient. The BIPV panel are then connected to the power line with BOS (inverters, cabling for strings, combiner boxes, etc.). Amorphous Silicon panel is places into a laminated panel separated by the glass pane with polymeric interlayer foils on both sides.

8.5 Quality Control of BIPV Technologies and Components

The PV cells laid in the interlayer foils are manufactured following a specific quality control plan and by setting in place a specific factory production control (FPC) to assess components and their performances. All the controls done for the product comply with the reference standards, but they also include additional quality levels required by the project. The list of controls (Fig. 8.19) performed during the manufacturing process, covers all the scheduled production steps:

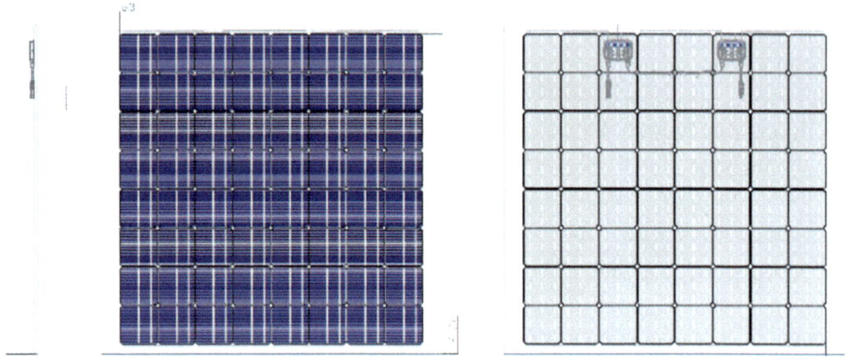

Fig. 8.16 BIPV glass—General view

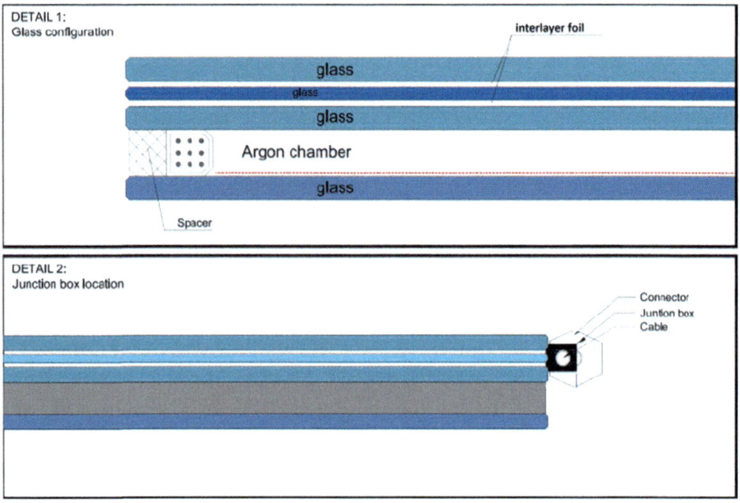

Fig. 8.17 Amorphous Silicon PV Curtain Wall (courtesy of Onyx Solar)

- From the incoming materials and components (i.e. raw glass, EVA expiring day, gel concentration, visual inspection and electrical tests of PV cells, etc.)
- To final product checks (inspection to detect visual defects such as bubbles, dimensional check like planarity-misalignments, flash test, junction box connection polarity test, etc.).

The quality plan includes also a list of laboratory checks, like durability of laminated glass under high temperature and humidity (according to EN 12543); EL micro cracks measurements, electroluminescence testing and IR imaging and flash test, etc.

For particular installation where the BIPV are curved, it is recommended to test the panel under natural sunlight to achieve a realistic variation of incidence angles. In case

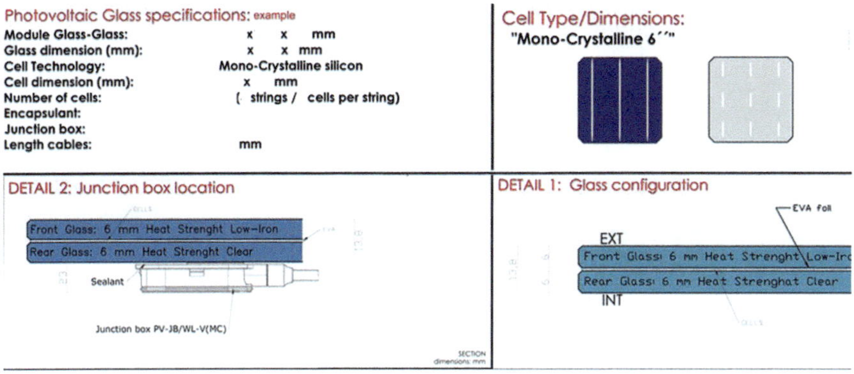

Fig. 8.18 Photovoltaic glass, example of data sheet specifications

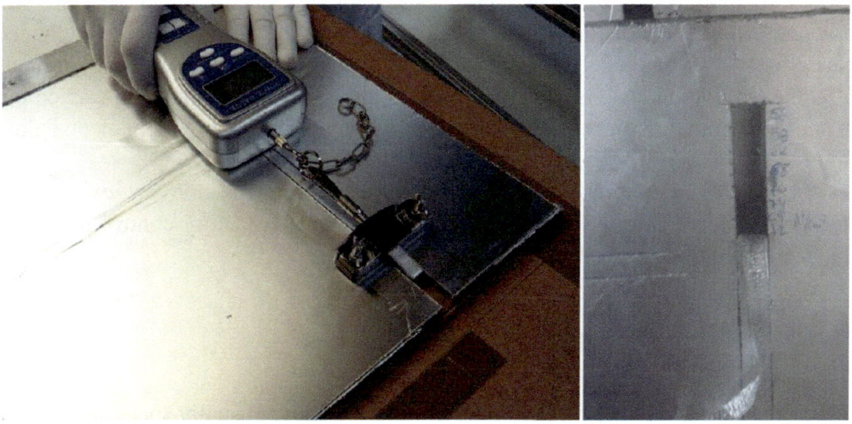

Fig. 8.19 Laboratory adhesion test: determining the peel strength of the interlayer film

of large dimensions, it will be necessary to identify a test "representative-size" of the modules and then apply extrapolation procedures for the rest of the "product family". To assess the mechanical stability and deformation of the panels it is recommended to consider the real mounting conditions and different dimensions affect deformation under load, and perform few tests under load with higher temperature (i.e. test on laminated—safety—glass to determine dependence of deformation and stress).

All glass shall be of the quality specified in accordance with standards such as EN 572-1, EN 572-4, EN 572-9, EN 1096 (Parts 1, 2, 3, 4) while the glazing shall be carried out in accordance with the manufacturer's written recommendations (several quality voluntary protocols and guidelines for glass tolerances are available by glass associations). All glass types shall be cut to accurate sizes with clean cut, arisen edges. Damage such as shark teeth, serration hackle, sharp flare, flake chips, rough chips,

Fig. 8.20 Laboratory test on glass specimen (from left to right: fragmentation tests, edge offset measurement, local bow and roller wave measurement)

feathered edges, shells or other imperfections shall not be acceptable if detrimental to the visual and performance criteria of the glass (Fig. 8.20).

Ensure that thermally treated process (toughened or strengthened) does not produce iridescence, distortion, roll marks or ripples in the glasses. All toughened glass shall be 'heat soaked' for a minimum hours in order to minimize the effects of nickel sulfide inclusions (this process shall be strictly controlled and carried out to standard guideline in regards to minimizing any incidence of spontaneous glass failure).

Prior to the production of PV panels, it is important to demonstrate with documentary evidence that the glass has been submitted to a factory control including traceability of glass, as following:

- Source of supply and evidence of batching
- Dates and records of toughening/heat soaking of all glass
- Certification and laboratory tests report.

For the type of testing regarding Optical performance (usually spectrally dependent and thermal), Solar heat gain coefficient (g_{value}), Thermal transmission (U_{value}). The optical properties, especially regarding reflection and spectrum losses, are also one of the critical parameters for evaluating the power input on a solar cell. These optical properties mainly depend on the angle of incident and zenith angle, which affects the reflection loss and spectrum distribution of the solar irradiation, respectively.

To check product compatibility issues between BIPV and its own components (i.e. silicon Fig. 8.21, EVA foil, glass, polymeric material of JB, etc.) a certain number of laboratory tests should be performed. Thermal cycling and Thermal stress, performed with specific boundary condition (i.e. temperature, humidity, adapted IEC hot-spot test, increase or decrease the number of cycles compared to Standard) are well simulating the effects of a frequent partial shading (by close and distant objects). To assess any degradation of the components properties, the aging tests in climatic chamber should be conducted under specific condition, i.e. a cycle of 85 °C with 85% RH (Relative Humidity) for 3,000 h (with continuous visual inspection while processing the testing campaign).

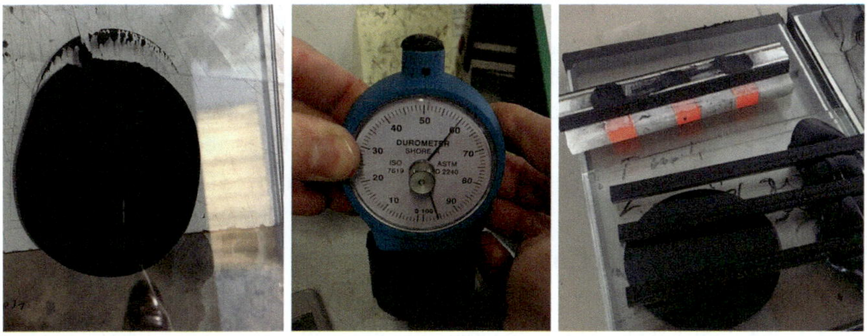

Fig. 8.21 Laboratory test on silicon specimen [from left to right: testing the mixing ratio with marble test for homogeneity (2-component products only), Shore A hardness measurement, tensile adhesion tests]

Concerning the life span of the power output, there is no given test by standards directly related to the minimum specified power output guaranteed during the term of the warranty. Regarding maintenance of the panels is carried out on daily basis taking into account the data acquisition software that the electrical installer or owner will implement. A well performing acquisition data tool could minimize corrective maintenance. Generally speaking, the ordinary maintenance activities should take place every six months or at least once a year (it really depends on the surrounding areas where the panels are installed i.e. traffic areas, countryside, etc.). The summary of the main activities and actions to be considered are:

- Checking cable connections and integrity of electrical junction box and its water tightness condition to avoid any risk to produce corrosion
- Checking the integrity of any gasket or sealant applied on the components
- Checking cracks that may appear on the glass (any glass fracture) or in the covering panels.
- Checking the condition of the substructure that supports the BIPV
- The cleaning procedure of the PV glazing is similar to equivalent any other glazing systems. It is recommended to clean the surfaces with a mixture of neutral detergent and water, in some case it is also suggested to use a mix of 3% in the water of ammonia and a surfactant (no pressurized water nor abrasives). To avoid scratch on the glass surface the use of typical cleaning tool as per instance rubber soft brush could be used. On the other hand, tools must be clean and not contaminated by dust, sand, or any other particles that could scratch the outer face of BIPV.

The expected life span and minimum power output warranties are ensured by complying with the given certifications (IEC 61215, IEC 61730–1&2, UL 1703 and others) where after passing all the testing flow, and specially aging tests, the estimated life span is proven. The aging tests included with these standards are i.e. UV radiation degradation test, Damp Heat test, TC200 and HF10 test within others. Theoretically, the expectation of durability for BIPV modules may be greater than for conventional modules. If we take into account, the procedure to replace a broken or damaged BIPV panel is often mechanically more difficult than a conventional freestanding PV module. Following this criteria, longer usage times of BIPV should be taken into account during the design phase (benchmark criteria), but also for test methods or procedure for both functions (electricity generator and facade component).

8.6 Discussion and Conclusion

BIPV technology help buildings become self-sufficient from an energy point of view, which is key in the struggle against climate change. In fact, buildings are responsible for the consumption of most of the electrical energy of the planet. PV technologies have got a significant improvement in the last decades and they have reached a sophisticated level of efficiency and an adequate expected life span. Today the major difficulties are related to PV integration into the façade design to come to a proper real integrated system, covering the electrical power demands of the building, but also the architectural requirements.

Power demands, energy efficiency of the PV system and design phase take into account main critical topics, such as daily and seasonal sun path, shape, inclination and orientation of the different elevations and shading.

Acknowledgements The authors would like to gratefully acknowledge Onyx Solar and Glass to Power for providing excellent research networking.

Appendix

Standard and Building Codes

- IEC 61215—Crystalline silicon terrestrial photovoltaic (PV) modules—Design qualification and type approval
- IEC 61646—Thin-film terrestrial photovoltaic (PV) modules—Design qualification and type approval
- IEC 61701—Salt mist corrosion testing of photovoltaic (PV) modules
- UL 1703—Standard for Flat-Plate Photovoltaic Modules and Panels
- AAMA 501.1.05—Standard Test Method for Water Penetration of Windows, Curtain Walls and Doors Using Dynamic Pressure
- AAMA 501.4.00—Recommended Static Test Method for Evaluating Curtain Wall and Store-Front Systems Subjected to Seismic and Wind Induced Interstory Drifts
- AAMA 501.5.07—Test Method for Thermal Cycling of Exterior Walls
- EN 410—Glass in building—Determination of luminous and solar characteristics of glazing

- EN 572-1—Glass in building—Basic soda lime silicate glass products—Part 1: Definitions and general physical and mechanical properties
- EN 572-9—Glass in building—Basic soda lime silicate glass products—Part 9: Product standard
- EN 673—Glass in building—Determination of thermal transmittance (U value)—Calculation method
- EN 674—Glass in building—determination of thermal transmittance (U value)—Guarded hot plate method
- EN 1096-1&2&3&4—Glass in building—Coated glass—Part 1: Definitions and classification & Part 2: Requirements and test methods for A, B and S coatings & Part 3: Requirements and test methods for C and D coatings & Part 4: Evaluation of conformity/Product standard
- ISO 6946—Building components and building elements—thermal resistance and thermal transmittance—calculation method
- EN 12543—Glass in building—Laminated glass and laminated safety glass—Part 1: Vocabulary and description of component parts
- EN 12488—Glass in building—Glazing requirements—Assembly rule
- EN 50530—Overall efficiency of grid-connected photovoltaic inverters
- EN 61730-1—Photovoltaic (PV) module safety qualification—Part 1: Requirements for construction.
- EN 61730-2—Photovoltaic (PV) module safety qualification—Part 2: Requirements for testing
- EN 61853—Photovoltaic (PV) modules performance testing and energy rating, Part 1: Irradiance and temperature performance measurements and power rating, part 2: spectral response, incidence angle and module operating temperature measurements.
- EN 13119—Curtain walling—Terminology
- EN 13830—Curtain walling—Product standard
- EN 12152—Curtain walling—Air permeability—Performance requirements and classification
- EN 12154—Curtain walling—Watertightness—Performance requirements and classification
- EN 13116—Curtain walling—Resistance to wind load—Performance requirements
- EN 14019—Curtain Walling—Impact resistance—Performance requirements
- EN 14351-1—Windows and doors—Product standard, performance characteristics

References

Aronica, P., & di Torino, P. (2018). Building integrated photovoltaics (BIPV): Soluzioni innovative e confronto con tecnologie convenzionali.

Commission Recommendation (EU). (2016). Commission Recommendation (EU) 2016/1318 of 29 July 2016 on guidelines for the promotion of nearly zero-energy buildings and best practices to ensure that, by 2020, all new buildings are nearly zero-energy buildings. Retrieved 09 June, 2020, from http://data.europa.eu/eli/reco/2016/1318/oj

IEC 61215/IEC 61730-1&2—"Crystalline Silicon Terrestrial Photovoltaic (PV) Modules—Design Qualification and Type Approval" and "Photovoltaic (PV) module safety qualification—Part 1: Requirements for construction & Part 2: Requirements for testing".

IEC 61701—"Salt mist corrosion testing of photovoltaic (PV) modules".

International Energy Agency. (2013). Technology roadmap—Energy efficient building envelopes. Retrieved 09 June, 2020, from https://www.iea.org/reports/technology-roadmap-energy-efficient-building-envelopes

International Energy Agency. (2019). Analysis of requirements, specifications and regulation of BIPV. Report IEA PVPS: 2019.

International Energy Agency. (2020). Multifunctional Characterisation of BIPV. Report IEA-PVPS T15-11.

International Energy Agency 2018International Energy Agency. (2018). International definitions of "BIPV". Report IEA-PVPS T15-04: 2018.

Martín-Chivelet, N., Carlos Gutiérrez, J., Alonso-Abella, M., Chenlo, F., & Cuenca, J. (2018). CIEMAT-photovoltaic unit. Building Retrofit with Photovoltaics: Construction and Performance of a BIPV Ventilated Façade.

Office of Energy Efficiency & Renewable Energy. (2022). *Organic Photovoltaics Research*. Retrieved 01, 2020, from https://www.energy.gov/eere/solar/organic-photovoltaics-research

Onyx Solar. (2019). Photovoltaic glass for buildings.

Perea, V. E. A., & Alcocer, J. L. B. (2018). Influence of shading on a BIPV system's performance in an urban context: Case study of BIPV systems of the Science Center of Complexity Building of the National and Autonomous University of Mexico in Mexico City.

Rigone, P., & Giussani, P. (2019). Multidisciplinary approach in facade design: the upcoming role of the specialist 'Façade Engineer'. *Costruzioni Metalliche,* 2. ISSN: 0010-9673.

Spooner, E. (2022). University of Sheffiel, Ossila. Retrieved 01, 2020, from https://www.ossila.com/pages/organicphotovoltaicsintroduction#WhatisOPV

Su, Y.-W., Lan, S.-C., & Wei, K.-H. (2012). *Organic photovoltaics*. Department of Materials Science and Engineering, National Chiao Tung University

UN Environment and International Energy Agency. (2017). Towards a zero-emission, efficient, and resilient buildings and construction sector. Global Status Report 2017.

WBDG. (2016). Building integrated photovoltaics (BIPV).

Chapter 9
Interdependencies Between Photovoltaics and Thermal Microclimate

Elisabeth Fassbender and Claudia Hemmerle

Abstract Climate change can only be reduced to a minimum when increasing the share of renewable energies. Thus, the definition of climate protection goals led to an increasing number of photovoltaic installations worldwide in recent years. At the same time, especially urban areas are already suffering from the consequences of climate change: Heat waves during summer months and reduced night-time cooling lead to negative impacts on the human organism and health. Within this context, the present review investigates interdependencies arising between photovoltaic panels and outdoor thermal microclimate, particularly in urban areas. In a first explorative, then systematic literature research approach, relevant publications were identified and selected based on distinct criteria. The studies were classified into the following main categories: (i) impacts of photovoltaics on the thermal microclimate; (ii) impacts of thermal microclimate on photovoltaic performance. The literature is then accordingly analyzed and main interdependencies, parameters and metrics are identified. The results, however, are incongruent, reaching in category (i) from 'solar cooling' of the surrounding ambient air temperature of up to −1.8 K to 'solar heating' of up to 4 K. Hence, further research, both experimental and simulation-based, is required to comprehensively understand the interactions between photovoltaics and its thermal surroundings.

Keywords Urban heat island (UHI) · Outdoor thermal comfort · Building related photovoltaics · Building greenery · Heat balance · Photovoltaic heat island (PVHI) · (Urban) solar cooling · Simulation · Metrological investigations

E. Fassbender (✉) · C. Hemmerle
TUM School of Engineering and Design, Chair of Building Technology and Climate Responsive Design, Technical University of Munich, Arcisstrasse 21, 80333 Munich, Germany
e-mail: elisabeth.fassbender@tum.de

C. Hemmerle
e-mail: claudia.hemmerle@tum.de

J. Wang et al. (eds.), *Advanced Materials in Smart Building Skins for Sustainability*,
https://doi.org/10.1007/978-3-031-09695-2_9

9.1 Introduction

Rising global temperatures enhance the need for climate protection measures and renewable energies to achieve climate protection goals and to reduce global warming to a minimum. In the past few years, photovoltaic (PV) electricity generation faces an exponentially increasing rate, rising from 20 TWh in 2008 to 681 TWh worldwide in 2019 (IEA, 2021). In urban areas, solar energy presents the largest renewable energy source that can flexibly be utilized by applying photovoltaic panels on building envelopes.

At the same time, particularly cities are already facing the consequences of climate change: They are subject to the Urban Heat Island Effect (UHI), that is defined as the difference in ambient air temperature between metropolitan and rural areas. The increase in urban population will further foster the UHI, leading to negative impacts on human organism and health. Thus, heat waves increase mortality rates and morbidity in the elderly population in particular and reduced night-time cooling limits sleep-recovery and leads to a reduction in performance (Kuttler, 2011).

At local scale, the thermal microclimate is particularly affected by surrounding surfaces and their materials and physical characteristics. The replacement of conventional building materials with photovoltaic modules in the interest of climate protection thus raises the question how the new construction material interacts with the thermal microclimate and whether it might contribute to the UHI due to enhanced absorption—thus competing with building greenery as a climate adaptation measure—or even function as 'solar cooling' due to the conversion of the absorbed solar radiation into electricity. The topic has only recently gained importance in urban planning practices and few experimental and simulation-based studies have been conducted. Therefore, the chapter aims at presenting an extensive literature review on the state of research concerning the interactions between photovoltaics and thermal microclimate (eventually altered due to the implementation of greenery) with the goal of providing an overview about existing knowledge, revealing research gaps and incongruences, and pointing towards new research directions.

9.2 Methodology

The present work applies the methodologic framework of a literature review as described in Ritschl et al. (2016). It combines an explorative and a systematic approach according to the 'Preferred Reporting Items for Systematic Reviews and Meta-Analyses' (PRISMA) (Page et al., 2021), thus allowing to identify most existing studies related to the topic.

Phase 1: Development of research questions

In a first approach, an explorative literature research provides a broad overview of existing literature concerning the topic of photovoltaics in the built environment. The explorative approach led to the development of three distinct research questions:

- Which interdependencies arise between photovoltaics and its thermal environment?
- Which metrics are used to describe the interdependencies?
- Which methods are applied to quantify the interdependencies and what are their respective strengths and weaknesses?

In this review, the term 'thermal environment' comprises the thermal microclimate and human outdoor thermal comfort that is relevant for evaluating the city climate. Moreover, impacts of the implementation of vegetation on the thermal microclimate and on photovoltaic performance are considered.

Phase 2: Search strategy and identification of studies

In the second phase, a systematic literature research based on keywords and a criteria-based selection of relevant studies allowed to identify literature suitable to answer the raised research questions. The online database *Web of Science* was screened for studies concerning the interdependencies between photovoltaics and thermal microclimate using different combinations of the keywords '*photovoltaic/photovoltaics/PV/solar panel*', '*microclimate/heat island/air temperature/heat flux/heat balance/thermal comfort*' and '*vegetation/greenery/green*'. The keywords were applied to title and author's keywords of the literature. No temporal limit was set, assuming that fundamental, physical findings will not be outdated regardless of the year of publication. After removing duplicates, 263 articles were identified.

Phase 3: Selection of studies

In a third phase, the titles, abstracts, and eventually full texts of the identified literature were screened, and studies of major interest were selected based on the following inclusion and exclusion criteria:

- Studies should focus on the interdependencies between photovoltaics and the surrounding outdoor thermal microclimate or outdoor thermal comfort. The modification of the outdoor thermal microclimate due to vegetation is considered as well. Research concerning the impact of photovoltaics on the indoor environment (indoor climate, heating/cooling load) however is excluded from the selection.
- Studies should focus on experimental or simulation-based research settings (primary research). Review articles are excluded.
- Studies should focus on building related photovoltaics and free-standing photovoltaics in open spaces featuring similar constructive settings as installations on flat roofs. Research concerning further PV applications such as PV pavements are excluded as well as research on concentrating photovoltaics.

Even though the present work puts emphasis on the interdependencies between photovoltaics and the urban thermal microclimate, studies investigating the impacts of free-standing photovoltaics in rural areas were selected as well to gain a general understanding of the physical impacts. The term 'building integrated photovoltaics' (BIPV) will be used for all sorts of building related photovoltaics. As the number of studies related to the topic is scarce, journal articles as well as conference papers were included in the review. However, the quality of the studies has been checked and studies with indefinite study design were excluded. After the content- and quality-related evaluation, 19 studies were selected. In an iterative search approach, the bibliographies of the previously identified studies (see phase 2) were evaluated based on the titles of the cited literature. Potentially suitable literature was then analyzed according to the inclusion and exclusion criteria and its quality and either excluded or integrated into the review. This iteration resulted in the selection of further seven articles (see Fig. 9.1).

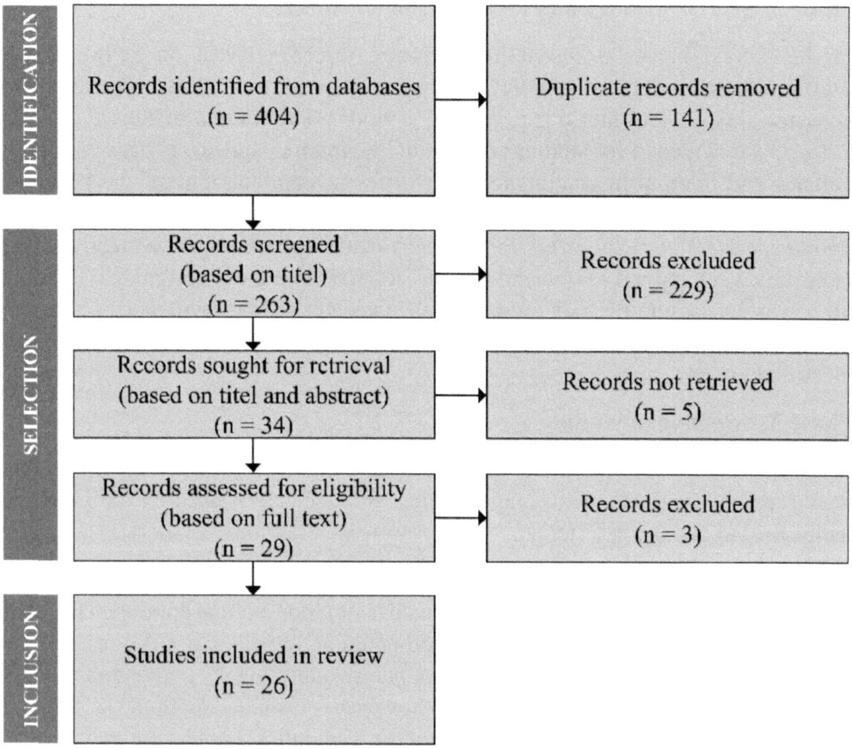

Fig. 9.1 Identification and selection of studies according to the PRISMA flow chart (Page et al., 2021)

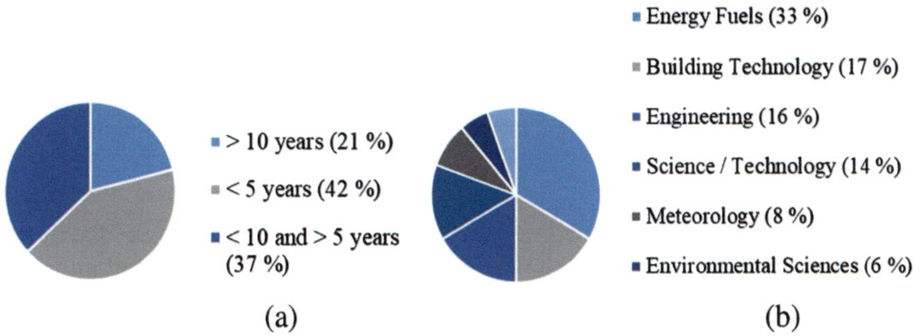

Fig. 9.2 Analysis of selected studies: **a** year of publication; **b** field of publication

Phase 4: Classification and analysis of studies

In a fourth phase, the selected studies were classified based on their focus area: Investigation of the impacts of photovoltaics on the thermal microclimate or vice versa. Articles belonging to category one were further classified according to their study environment: Studies in rural environments, studies within a city context and studies within a district context. This classification allows to compare the impacts of photovoltaics on the surrounding microclimate in different contexts and spatial scales. Amongst the studies investigating the impacts of thermal microclimate on photovoltaics, two main classifications emerged: The impact of ambient air temperature on PV and the interdependencies between greenery and PV.

A brief analysis of the selected publications provides a contextual background for the literature review. The increasing number of published papers in recent years is proof of the rising interest in the interdependencies between photovoltaics and thermal microclimate in the light of climate change (see Fig. 9.2a). Researchers based in the USA and China take particular interest in studying the topic. Moreover, the broad spread of publications in various research areas, inter alia involving energy fuels, building technology and environmental sciences, underlines its interdisciplinarity (see Fig. 9.2b).

Phase 5: Interpretation and presentation of results

The results of the literature research (see Sect. 9.3) are described in two parts according to the developed classifications (see Table 9.1): First, the impacts of the use of solar panels on the thermal microclimate are investigated. Then, conversely the effects of the microclimate on the efficiency of the photovoltaic modules are analyzed.

Table 9.1 Identified studies organized according to category, focus and study type

Category	Focus	Type of study	Sources (alphabetically)
PV → thermal microclimate	Rural environment	Experimental	Barron-Gafford et al. (2016)
			Broadbent et al. (2019)
			Jiang et al. (2021)
			Wu et al. (2020)
	City scale	Simulation-based	Berardi and Graham (2020)
			Brito (2020)
			Genchi et al. (2003)
			Masson et al. (2014)
			Roulet (2001)
			Salamanca et al. (2016)
			Taha (2013)
	District scale	Experimental	Brown et al. (2020)
			Pham et al. (2019)
		Simulation-based	Cortes et al. (2015)
			Scherba (2011)
			Tian et al. (2007)
			Weihs et al. (2018)
Thermal microclimate → PV	Air temperature	Simulation-based	Dubey et al. (2013)
			Evans (1981)
			Skoplaki et al. (2008)
	Greenery & PV	Experimental	Chemisana and Lamnatou (2014)
			Hui and Chan (2011)
			Köhler et al. (2002)
			Ogaili (2015)
			Penaranda Moren and Korjenic (2017a)
			Penaranda Moren and Korjenic (2017b)
		Simulation-based	Berardi and Graham (2020)
			Hui and Chan (2011)

9.3 Results

9.3.1 Impacts of Photovoltaics on the Thermal Microclimate

The use of photovoltaics not only contributes to renewable energy production, but also has an impact on the radiation balance of the surroundings, thus altering the thermal microclimate and outdoor thermal comfort. The alteration of the radiation balance is caused by several effects (Barron-Gafford et al., 2016; Genchi et al., 2003; Wang et al., 2006) (see Fig. 9.3):

- **Albedo**: The albedo of photovoltaic modules is typically lower than the albedo of conventional building envelopes or rural surroundings, thus leading to an increase of absorbed solar radiation. This effect is desirable for gaining a maximum energy yield, but it may also result in higher surface temperatures compared to conventional building envelopes and natural materials, and thus to higher long-wave radiation emitted to the sky and the surroundings.
- **Energy conversion**: Depending on the efficiency η of the photovoltaic modules, part of the short-wave incident solar radiation is converted into electrical energy. The energy conversion reduces the increase in module temperatures (T_{PV}) due to the higher albedo and, in the case of photovoltaics integrated into the building envelope, the surface temperatures of buildings, thus diminishing long-wave radiation. Böer (2001) states that the temperature reduction is typically in the order of 10 °C when using cells with an efficiency $\eta = 20\%$ at AM1 insolation.
- **Shading**: Free-standing and rack-mounted photovoltaic modules shade the soil and rooftops beneath them, thus reducing conductive heat transfer into the building or ground as well as the convective and radiative heat transfer to the environment. Additionally, the photovoltaic cover lowers the sky view factor of the surfaces beneath them, hence reducing night-time cooling.

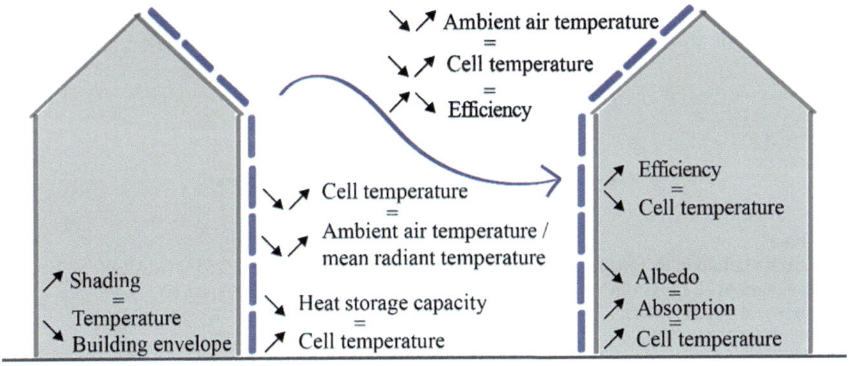

Fig. 9.3 Interdependencies between building integrated photovoltaics and thermal microclimate

- **Heat storage capacity**: Photovoltaic modules have a lower heat storage capacity than soil or conventional building surfaces. Therefore, the modules rapidly heat up, but also faster release the stored heat back into the environment.

These theoretical assumptions concerning the impact of photovoltaics on the microclimate have been investigated in experimental and simulation-based studies in various research environments.

9.3.1.1 Rural Environment

In the rural environment, four studies identified a warming effect of photovoltaic modules on the surrounding thermal microclimate (Barron-Gafford et al., 2016; Broadbent et al., 2019; Jiang et al., 2021; Wu et al., 2020). The authors conduct metrological investigations at photovoltaic power plants in rural parts of Arizona (USA), Xinjiang (China) and Qinghai (China). They conclude that the operating photovoltaic modules increase the day-time air temperatures (T_{air}) in the immediate vicinity of the power plant by 0.2–4 K. Therefore—analogous to the Urban Heat Island (UHI)—the term 'Photovoltaic Heat Island' (PVHI) has been defined (Barron-Gafford et al., 2016). The PVHI represents the difference between the air temperature close to a photovoltaic power plant and the air temperature in a rural region without photovoltaic applications. As the shading of the soil leads to a reduction in night-time heat release, the PVHI is more pronounced during day-time (Barron-Gafford et al., 2016; Broadbent et al., 2019). The studies in China even identified a slight cooling effect on the nocturnal ambient air temperatures: In Xinjiang, the temperature decreased by averagely −0.24 K (−0.08 K) in two meters (ten meters) height above the photovoltaic modules (Jiang et al., 2021). In Qinghai, a maximum decrease of night-time air temperature of 1.82 K was measured (Wu et al., 2020) (see Fig. 9.4).

However, the results of the mentioned studies on PVHI have limited validity when investigating the effects of photovoltaic applications on the thermal microclimate in urban environments: A natural environment has a higher albedo than the anthropogenically built environment and in urban areas already sealed surfaces are

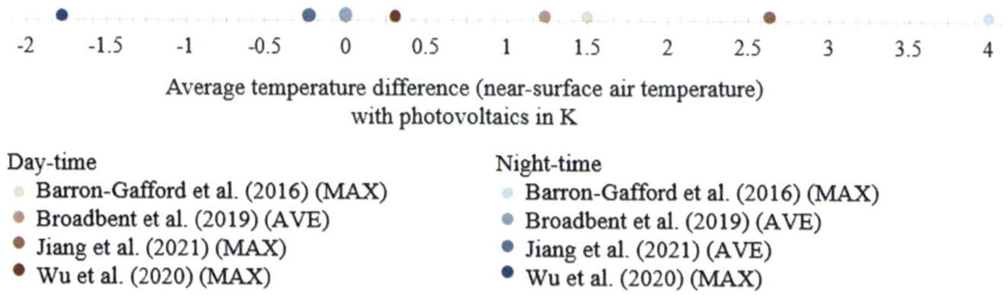

Fig. 9.4 Impact of photovoltaics on near-surface air temperature in rural areas detected by several studies: Solar cooling <0 K, PVHI > 0 K

replaced or complemented by photovoltaics. This suggests a lower to non-existent PVHI in the built environment.

9.3.1.2 City Context

In 2001, Roulet investigated the effects of active solar energy use on roofs and facades on the thermal microclimate at city scale based on simplified steady-state calculations of the urban energy balance. He hypothesized that the heat dissipated into the urban area will decrease as soon as the efficiency η of the photovoltaic modules is higher than the average albedo of the city. This hypothesis is based on the theory that, due to the conversion of solar radiation into electricity, a smaller fraction of the radiation is converted into sensible heat. The reduction of the sensible heat flux leads to reduced temperatures in urban areas, thus resulting in urban solar cooling.

Roulet's hypothesis was tested several times by means of mesoscale simulations, revealing marginal impacts of BIPV on the surrounding ambient air temperature (T_{air}) (Brito, 2020; Masson et al., 2014; Salamanca et al., 2016; Taha, 2013). Unlike photovoltaic installations in rural areas (Barron-Gafford et al., 2016; Broadbent et al., 2019), the use of photovoltaics in urban environments seems leading to a slight decrease in urban air temperatures and thus to a reduction of the UHI.

A simulation-based study relying on the mesoscale climate model *uMM5* identifies a reduction of the ambient air temperature of up to 0.15–0.2 K five meters above roof level for the metropolitan area of Los Angeles. The photovoltaic roofs are modeled in a simplified way by estimating the effective albedo, thus implicitly including the efficiency of the photovoltaic modules (Taha, 2013). In a similar research design, the *Town Energy Balance model (TEB)* is used to conduct studies for the metropolitan area of Paris (Masson et al., 2014). The assumption of a mixed installation of photovoltaic modules and solar thermal energy systems on the rooftops leads to the identification of an urban solar cooling of up to 0.3 K. The photovoltaic modules are modeled based on an energy balance with a simplified calculation of the module temperature T_{PV} according to Eq. (9.1):

$$T_{PV} = T_{air} + 0.05\ K/\left(Wm^{-2}\right) * G_T \tag{9.1}$$

The approach used by Masson et al. for modeling the photovoltaic energy balance and module temperatures is implemented into the *Weather Forecast and Research Model (WRF)*. Thus, the summertime impacts of photovoltaic use on the near-surface ambient air temperatures in Phoenix and Tucson (Arizona) are simulated, identifying a temperature reduction of 0.4–0.8 K (Salamanca et al., 2016). In a further approach, the establishment of steady-state radiation balances leads to the conclusion that the impact of rooftop photovoltaics significantly depends on the modules' efficiency. For typical albedo values of rooftops (0.3) and PV modules (0.1), the installation of PV fosters the UHI as soon as their efficiency is lower than $\eta = 22\%$. For photovoltaic facades, the effect rather depends on the albedo of the surrounding surfaces, especially the ground albedo. Along bright roads (0.3), there is nearly always a heating

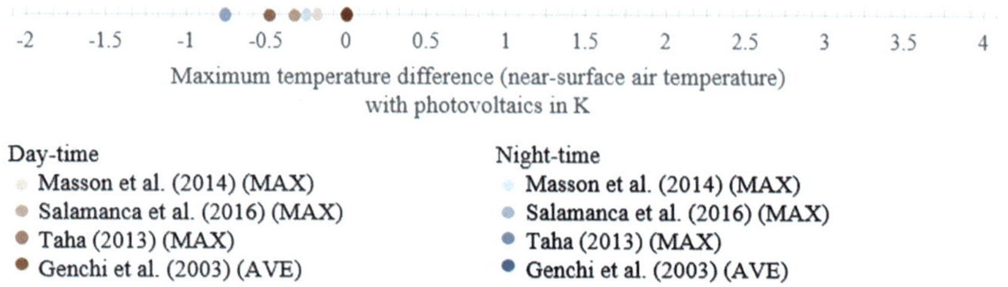

Fig. 9.5 Impact of photovoltaics on near-surface air temperature within city context detected by several studies: Solar cooling <0 K, PVHI >0 K

effect whereas in the extreme case of a dark road (0.0), modules of any efficiency lead to urban cooling (Brito, 2020).

In addition to the direct impacts on urban climate, the shading of the building envelope due to the addition of photovoltaic modules seems to lead to a reduction of the summer cooling energy demand of up to 12%, thus reducing anthropogenic heat emissions (Genchi et al., 2003; Masson et al., 2014; Salamanca et al., 2016) (see Fig. 9.5). However, none of these calculations and simulation-based approaches have been validated by measurements.

9.3.1.3 District Context

Further studies focus on the effects of energy activated building envelopes on the local microclimate within a district and street context. At this spatial scale however, the results are inconsistent, ranging from cooling effects of up to -1 K to significant warming effects. The studies identifying warming effects however expressed their results by comparing sensible heat fluxes without conversion into temperature differences.

The effective albedo approach developed by Taha (2013) (see Sect. 9.3.1.2) is integrated into the microclimate simulation software *ENVImet* to assess the impacts of rooftop photovoltaics on the ambient air temperature inside a street canyon. Compared to the reference scenario with an albedo of the rooftop surfaces of 0.2, the application of photovoltaic modules featuring an efficiency of 18%—hence an effective albedo of 0.3—reduces the near-surface air temperature by up to 0.09 K (Berardi & Graham, 2020).

Further studies focus on photovoltaic facades. They suggest that energy activated facades might have lower surface temperatures than conventional building facades, however that the impact on the surrounding ambient air temperature is only slight (Cortes et al., 2015; Tian et al., 2007; Weihs et al., 2018): Simulation-based investigation for China affirm that the module temperatures are up to 4.6 K lower than the surface temperatures of reference facades with an albedo of 0.1 (Tian et al., 2007). By coupling computational fluid dynamic simulations (CFD) with the *WRF* model, the surface temperatures of the building envelope behind the photovoltaic modules

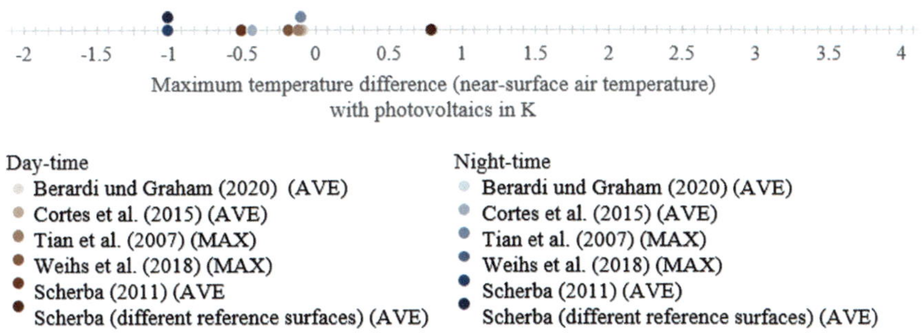

Fig. 9.6 Impact of photovoltaics on near-surface air temperature within district context detected by several studies: Solar cooling <0 K, PVHI >0 K

were investigated. The simulations suggest a maximum reduction of the surface temperature of 0.8 K during day-time and 4.3 K during night-time, thus leading to reduction of ambient air temperature of up to 0.4 K at night (Cortes et al., 2015). A simulation-based study for Vienna (Austria) additionally investigated the effect of photovoltaic modules on human outdoor thermal comfort. Even though the identified influence of photovoltaics on the ambient air temperature was slight (±0.2 K), the Universal Thermal Comfort Index (UTCI) was reduced by up to 1 K. The decrease is caused by a reduction of reflections compared to conventional facades with high albedo. Less reflections lead to a decrease of the mean radiant temperature (T_{mrt}), a relevant parameter for the determination of thermal comfort. Yet, validations show deviations of up to 30% between metrological investigation and simulation (Weihs et al., 2018) (see Fig. 9.6).

9.3.1.4 Reflection of Studies in City and District Context

The aforementioned studies indicating an urban solar cooling however might underestimate the sensible heat fluxes due to the simplified modeling of photovoltaic modules and the neglect of the modules rear side as radiative surfaces (Brown et al., 2020; Pham et al., 2019). Hence, further studies refute Roulet's hypothesis of urban solar cooling and identify an increase in air temperature due to urban photovoltaic use: Metrological investigations on a significantly down-scaled mock-up of a canopy structure with photovoltaic panels in Arizona and the subsequent calculation of sensible heat fluxes based on the measured data taking into account both sides of the PV modules as radiative surfaces suggest an cumulative, daily increase in sensible heat flux by up to 230 W/m^2 per day when compared to a canopy with reflective materials (albedo of 0.82) (Pham et al., 2019). Similar metrological investigations conducted on a down-scaled mock-up of free-standing photovoltaic modules on top of a test cube in Arizona identify a cumulative, daily increase of sensible heat flux of up to 140 W/m^2 per day in comparison to a light-colored roof with an albedo of 0.63. Studies using the thermal building simulation program *EnergyPlus* focus on the sensitive heat fluxes between roof, photovoltaic modules, and the direct surroundings in six different cities in the USA during summer months. The simulations suggest that the addition of photovoltaic modules to light-colored or green roofs increases the sensible heat flux by 95–120 W/m^2. The installation of photovoltaic modules on dark-colored roofs however reduces the sensible heat fluxes by up to 11% (Scherba, 2011).

9.3.2 *Impacts of Thermal Microclimate on Photovoltaic Performance*

9.3.2.1 Air Temperature and Photovoltaic Performance

Inversely, the surrounding thermal microclimate and in particular the ambient air temperature has an impact on the photovoltaic cell temperature (T_c). The cell temperature in turn affects the efficiency of the modules. Equation (9.2) represents the correlation between efficiency and cell temperature (Evans, 1981):

$$\eta = \eta_{\text{T-ref}}[1 - \beta_{\text{ref}}(T_c - T_{\text{ref}})] \tag{9.2}$$

The reference efficiency $\eta_{\text{T-ref}}$ is determined at so-called standard test conditions (STC) with solar irradiation (G_T) of 1.000 W/m^2, a cell temperature of 25 °C and mass of air (AM) of 1.5. The temperature coefficient β_{ref} describes the temperature sensitivity of the photvoltaic modules, that depends on the cell technology. Usually, the temperature coefficient is in the range of 0.2–0.5%/°C (Dubey et al., 2013).

The relationship between cell temperature and ambient air temperature is commonly expressed by means of the NOCT equation defined by IEC 61215 standard (see Eq. (9.3)). The nominal operating cell temperature T_{NOCT} used as a reference temperature is the cell temperature at a solar irradiation (G_T) of 800 W/m^2, ambient air temperature of 20 °C, wind velocity of 1 m/s and zero electrical load for free-standing photovoltaic modules (IEC, 2021). T_{NOCT} is individual for each photovoltaic module and depends on its characteristics of the materials and construction.

$$T_c = T_{\text{air}} + (T_{\text{NOCT}} - 20\,^{\circ}\text{C}) * \left(G_T/800\ \text{W/m}^2\right) \tag{9.3}$$

Building integrated photovoltaics however require different equations for the accurate determination of cell temperatures (Skoplaki et al., 2008). As the urban heat island effect leads to higher air temperature in urban areas, decreased photovoltaic efficiencies and thus lower solar yields are anticipated.

9.3.2.2 Green Infrastructure and Photovoltaic Performance

Vegetation has a cooling effect on the ambient air temperature. The cooling effect primarily relies on the evapotranspiration of greenery and might be beneficial for the efficiency of photovoltaic modules. Hence, the interdependencies between photovoltaics and greenery are investigated.

On rooftops, these interdependencies have been evaluated with several metrological investigations and simulation-based studies (Chemisana & Lamnatou, 2014; Hui & Chan, 2011; Köhler et al., 2002; Ogaili, 2015). The authors identified an increase in solar yield due to plant-photovoltaic interactions ranging from 0.08 to 8.3%, depending on different factors: In general, plant species with a high evapotranspiration rate have the highest cooling impact on the ambient air temperature. Moreover, a high albedo of the plant canopy increases the reflected radiation, thus leading to an increased solar yield. For the combination of a photovoltaic-green roof, extensive green roofs with plant species with low heights are preferable to reduce shading losses. Few studies focused on the impact of vertical greenery systems on photovoltaic modules in front of the vegetation layer. First results suggest a reduction of photovoltaic operating temperatures ranging from 1 to 4 K (Penaranda Moren & Korjenic, 2017a, b). However, the constructional complexity is high and the plants as living matter might overgrow the photovoltaic modules.

Compared to vertical greenery systems and green roofs, street strees have the highest cooling impact on the local microclimate (Brasche et al., 2017). Yet, the tree shading reduces the incident solar radiation on building envelopes depending on their height and location. By integrating the effective albedo approach into the microclimate simulation software *ENVImet* and using a python-based post-processing script, it was found out, that the addition of street trees in a specific district in Brampton (Ontario) reduces the solar yield of rooftop photovoltaics by up to 32% while the air temperature is reduced by up to 0.44 K (Berardi & Graham, 2020). It can thus be concluded that the reduction in solar yield due to tree shading and the positiv cooling impact of street trees need to be negotiated.

9.4 Discussion

9.4.1 Main Findings: Impacts of Photovoltaics on the Thermal Microclimate

The results of studies in rural environments unanimously suggest a day-time heating effect, while studies in city and district context mostly identify a day-time solar cooling effect. Independent of the research context however, one tendency is evident: The low heat storage capacity of photovoltaic modules leads to lower night-time module temperatures (compared to the reference surface temperatures within the same study) and hence to a lower heating or even to a cooling effect of the surrounding ambient air temperature at night.

Moreover, the efficiency of the photovoltaics and the albedo of the surrounding surfaces are relevant parameters influencing the effects of photovoltaic modules on the thermal microclimate. Hence, it was found out that the threshold for rooftop photovoltaics contributing to urban solar cooling typically lies at a module efficiency of $\eta = 22\%$. The energy balance for facades is more complex due to several surrounding surfaces and is less dependent on the module efficiency but significantly depends on the reflectivity of the environment. Generally, installing photovoltaic modules in bright environments with higher albedos of the surrounding rooftops, walls, and ground as well as lower module efficiency result in higher heating effects. Photovoltaic modules with high efficiencies may generate a solar cooling effect.

Throughout the selected studies, the metrics 'ambient air temperature' and 'surface temperature' have consistently been used as relevant indicators for describing the impacts of photovoltaics on the thermal microclimate. Later studies also include the analysis of sensible heat fluxes. Only one study considers the impact of photovoltaics on the outdoor thermal comfort (Weihs et al., 2018). According to Chokhachian et al. (2017) however, increasing the albedo of surfaces leads to a significant reduction of outdoor thermal comfort. Thus, the neglect of outdoor thermal comfort considerations including mean radiant temperature leads to an insufficient description of the thermal microclimate when comparing photovoltaic modules with surfaces with high albedo.

Moreover, the reviewed studies show inconsistent results, paricularly when the impacts of photovoltaics on the thermal microclimate are analyzed on a smaller spatial scale. The discrepancies between different simulation-based studies can be primarily attributed to significant simplifications in the applied methods. For example, Masson et al. (2014) and Salamanca et al. (2016) apply a photovoltaic model that calculates the module temperature using a simple addition of ambient air temperature and irradiance at module level multiplied by a constant factor. Thus, effects of wind speed and direction and of the modules efficiency on the photovoltaic temperatures are neglected. The effective albedo approach used by Taha (2013) and Berardi and Graham (2020) is also subject to significant simplifications and assumptions and has not yet been validated. Brito (2020) excludes longwave radiation as well as diffuse radation from his energy balance calculations, that are only performed with the sun being at its zenith and Taha (2013) and Cortes et al. (2015) assume that the albedo of the surrounding building surfaces is equal to the albedo of the photovoltaic modules. The analyses of Salamanca et al. (2016) are based on the assumption that each building disposes of mechanical air conditioning and the waste heat is released as sensible heat to the environment. Hence, the identified reduction of the ambient air temperature is primarily due to the reduced cooling energy demand and thereby reduced anthropogenic heat emissions. Moreover, the different research designs with varying reference locations, climates, albedo values and diverging photovoltaic efficiencies lead to a reduced comparability of results. For experimental studies, differing measurement concepts and measurement heights further complicate the comparison of results. Additionally, not all studies describe the construction type of the photovoltaic installations and clarify whether the announced temperature differences are the maximally reached impacts or values averaged in time and spatial scale.

It remains thus unclear to what extent and under what boundary conditions large-scale applications of photovoltaics on façades and roofs can affect the local microclimate—in particular ambient air temperature and mean radiant temperature—and thus the thermal comfort at district scale. It even remains ambiguous whether PV applications lead to a warming of the urban climate—in city or district context—or contribute to urban solar cooling and might thus be classified as a climate adaptation measure. To close this research gap, metrological investigations at real scale for validation purposes as well as the inclusion of detailed and validated photovoltaic temperature models into microclimate simulations are indispensable.

9.4.2 *Main Findings: Impacts of Thermal Microclimate on Photovoltaic Performance*

Higher urban air temperatures slightly influence the photovoltaic efficiency. This effect can partly be encountered by increasing the amount of vegetation into urban areas, thus altering the thermal microclimate. However, as the quantification of cooling effects of vegetation remains challenging due to various influencing parameters and the living organisms of plants, the examination of interdependencies between photovoltaics and plants remains vague. Existing studies only identify slight reductions in cell temperatures, hence increasing the solar power yield.

9.5 Conclusion

The present review synthesizes existing work concerning the impacts of photovoltaics on the thermal microclimate and vice versa with a focus on the urban built environment. In total, 26 publications reporting on experimental research and simulation-based studies were selected and analyzed, identifying the interdependencies along with metrics to describe and applied methods to quantify them.

To summarize, the analysis of the impact of photovoltaics on the thermal microclimates reveals major incongruences between the studies and diverging results. Notably, according to experimental studies, photovoltaic panels contribute to heating the surrounding air temperature, while simulation-based studies rather suggest a 'solar cooling', particularly at night-time.

The diverging results are mainly due to four reasons: (i) Simplified modeling of photovoltaic modules (cell temperature and efficiency) within microclimate models and (ii) lack of validation of simulation-based studies; (iii) lack of real-size experimental studies in urban environments; (iv) limited comparabilty of studies due to differing research designs (different locations, reference values, measurement concepts and modelling approaches). All studies identify the parameters 'photovoltaic efficiency' and 'albedo' as relevant. The albedo even needs to be considered in a threefold approach: The albedo of the surrounding city and street, the albedo of the surface on which the photovoltaic modules are installed and in case of BIPV the albedo of the substituted material respectively reference material.

Studies related to the impact of air temperature and vegetation on photovoltaic performance however obtain similar results: Although marginal, the surrounding air temperature affects the efficiency of photovoltaic panels. Thus, the UHI leads to lower solar yields, while the combination with vegetation might—under certain circumstances—negate the negative impacts.

The obtained results reveal research gaps and indicate future research paths to follow: Notably the topic 'impacts of photovoltaics on its thermal environment' gains increasing relevance within the context of climate change, but to date comprises several unanswered questions. Future work needs to assess the effects of photovoltaic panels on its thermal surroundings with experimental measurements on real-size photovoltaic installations in the urban environment and accordingly calculated heat balances. Numerical microclimate simulations with detailed and experimentally validated photovoltaic models can furthermore generalize the results gained by experimental case studies.

Acknowledgements This research is sponsored by the Bavarian Ministry of Science and the Arts in the context of Bavarian Climate Research Network (bayklif).

References

Barron-Gafford, G. A., Minor, R. L., Allen, N. A., Cronin, A. D., Brooks, A. E., & Pavao-Zuckerman, M. A. (2016). The photovoltaic heat island effect: Larger solar power plants increase local temperatures. *Scientific Reports, 6*, 35070.

Berardi, U., & Graham, J. (2020). Investigation of the impacts of microclimate on PV energy efficiency and outdoor thermal comfort. *Sustainable Cities and Society, 62*(8), 102402.

Böer, K. W. (2001). Self-cooling of photovoltaic cells when power is drawn. *Physica Status Solidi, 184*(1), 201–209.

Brasche, J., Hausladen, G., Maderspacher, J., Schelle, R, & Zölch, T. (2017). *Zentrum Stadtnatur und Klimaanpassung: Teilprojekt 1: Klimaschutz und grüne Infrastruktur in der Stadt*. Technische Universität München.

Brito, M. C. (2020). Assessing the impact of photovoltaics on rooftops and facades in the urban micro-climate. *Energies, 13*(11), 2717.

Broadbent, A. M., Krayenhoff, E. S., Georgescu, M., & Sailor, D. J. (2019). The observed effects of utility-scale photovoltaics on near-surface air temperature and energy balance. *Journal of Applied Meteorology and Climatology, 58*(5), 989–1006.

Brown, K. E., Baniassadi, A., Pham, J. V., Sailor, D. J., & Phelan, P. E. (2020). Effects of rooftop photovoltaics on building cooling demand and sensible heat flux into the environment for an installation on a white roof. *Journal of Engineering for Sustainable Buildings and Cities, 1*, 021001-1.

Chemisana, D., & Lamnatou, C. (2014). Photovoltaic-green roofs: An experimental evaluation of system performance. *Applied Energy, 119*, 246–256.

Chokhachian, A., Perini, K., Dong, M. S., & Auer, T. (2017). How material performance of building facade affect urban microclimate. *Proceedings of the PowerSkin Conference, 2017*, 83–95.

Cortes, A., Murashita, Y., Matsuo, T., Kondo, A., Shimadera, H., & Inoue, Y. (2015). Numerical evaluation of the effect of photovoltaic cell installation on urban thermal environment. *Sustainable Cities and Society, 19*, 250–258.

Dubey, S., Sarvaiya, J. N., & Seshadri, B. (2013). Temperature dependent photovoltaic (PV) efficiency and its effect on PV production in the world—A review. *Energy Procedia, 33*, 311–321.

Evans, D. L. (1981). Simplified method for predicting photovoltaic array output. *Solar Energy, 27*(6), 555–560.

Genchi, Y., Ishisaki, M., Ohashi, Y., Takahashi, H., & Inaba, A. (2003). Impacts of large-scale photovoltaic panel installation on the heat island effect in Tokyo. *Fifth Conf. Urban Clim.*

Hui, C. M., & Chan, S. C. (2011). Integration of green roof and solar photovoltaic systems. In *Proceedings of Joint Symposium 2011: Integrated Building Design in the new Era of Sustainability* (pp. 1.1–1.10).

IEA. (2021). *Data and statistics: Data tables: Electricity*. Retrieved December 23, 2021, from https://www.iea.org/data-and-statistics/data-tables/?country=WORLD&energy=Electricity&year=2009

IEC. (2021). *61215-2:2021: Terrestrial photovoltaic (PV) modules—Design qualification and type approval.*

Jiang, J., Gao, X., Lv, Q., Li, Z., & Li, P. (2021). Observed impacts of utility-scale photovoltaic plant on local air temperature and energy partitioning in the barren areas. *Renewable Energy, 174*, 157–169.

Köhler, M., Schmidt, M., Laar, M., Wachsmann, U., & Krauter, S. (2002). Photovoltaic panels on greened roofs: Positive interaction between two elements of sustainable architecture. In *Proceedings of RIO 02—World Climate & Energy Event* (pp. 151–157).

Kuttler, W. (2011). Klimawandel im urbanen Bereich: Teil 1, Wirkungen. *Environmental Sciences Europe SpringerOpen Journal, 23*, 11.

Masson, V., Bonhomme, M., Salagnac, J.-L., Briottet, X., & Lemonsu, A. (2014). Solar panels reduce both global warming and urban heat island. *Frontiers in Environmental Science, 2.*

Ogaili, H. H. K. (2015). Measuring the effect of vegetated roofs on the performance of photovoltaic panels in combined systems. Master thesis, Portland State University, Portland.

Page, M. J., McKenzie, J. E., Bossuyt, P. M., Boutron, I., Hoffmann, T. C., Mulrow, C. D., Shamseer, L., Tetzlaff, J. M., Akl, E. A., Brennan, S. E., Chou, R., Glanville, J., Grimshaw, J. M., Hróbjartsson, A., Lalu, M. M., Li, T., Loder, E. W., Mayo-Wilson, E., McDonald, S., McGuinness, L. A., Stewart, L. A., Thomas, J., Tricco, A. C., Welch, V. A., Whiting, P., & Moher, D. (2021). The PRISMA 2020 statement: An updated guideline for reporting systematic reviews. *BMJ (Clinical Research Ed.), 372.*

Penaranda Moren, M. S., & Korjenic, A. (2017a). Green buffer space influences on the temperature of photovoltaic modules: Multifunctional system: Building greening and photovoltaic. *Energy and Buildings, 146*, 364–382.

Penaranda Moren, M. S., & Korjenic, A. (2017b). Hotter and colder—How do photovoltaics and greening impact exterior facade temperatures: The synergies of a multifunctional System. *Energy and Buildings, 147*, 123–141.

Pham, J. V., Baniassadi, A., Brown, K. E., Heusinger, J., & Sailor, D. J. (2019). Comparing photovoltaic and reflective shade surfaces in the urban environment: Effects on surface sensible heat flux and pedestrian thermal comfort. *Urban Climate, 29*, 100500.

Ritschl, V., Weigl, R., & Stamm, T. (Eds.). (2016). *Wissenschaftliches Arbeiten und Schreiben.* Springer Berlin Heidelberg.

Roulet, C.-A. (2001). Solar energy and global heat balance of a city. *Solar Energy, 70*(3), 255–261.

Salamanca, F., Georgescu, M., Mahalov, A., Moustaoui, M., & Martilli, A. (2016). Citywide impacts of cool roof and rooftop solar photovoltaic deployment on near-surface air temperature and cooling energy demand. *Boundary-Layer Meteorology, 161*(1), 203–221.
Scherba, A. (2011). Modeling the impact of roof reflectivity, integrated photovoltaic panels and green roof systems on the summertime heat island. Master thesis, Portland State University, Portland.
Skoplaki, E., Boudouvis, A. G., & Palyvos, J. A. (2008). A simple correlation for the operating temperature of photovoltaic modules of arbitrary mounting. *Solar Energy Materials and Solar Cells, 92*(11), 1393–1402.
Taha, H. (2013). The potential for air-temperature impact from large-scale deployment of solar photovoltaic arrays in urban areas. *Solar Energy, 91*, 358–367.
Tian, W., Wang, Y., Xie, Y., Wu, D., Zhu, L., & Ren, J. (2007). Effect of building integrated photovoltaics on microclimate of urban canopy layer. *Building and Environment, 42*(5), 1891–1901.
Wang, Y., Tian, W., Zhu, L., Ren, J., Liu, Y., Zhang, J., & Yuan, B. (2006). Interactions between building integrated photovoltaics and microclimate in urban environments. *Energy and Buildings, 128*(2), 168–172.
Weihs, P., Zamini, S., Krispel, S., Oswald, S., Peyerl, M., Revesz, M., Schneider, A., & Trimmel, H. (2018). *Optimierung reflektierender Materialien und Photovoltaik im Stadtraum bezüglich Strahlungsbilanz und Bioklimatik: PVOPTI-Ray*. Universität für Bodenkultur.
Wu, W., Yue, S., Zhou, X., Guo, M., Wang, J., Ren, L., & Yuan, B. (2020). Observational study on the impact of large-scale photovoltaic development in deserts on local air temperature and humidity. *Sustainability, 12*(8), 3403.

Chapter 10
Material Driven Adaptive Design Model for Environmentally-Responsive Envelopes

Maryam Mansoori, Zofia Rybkowski, and Negar Kalantar

Abstract This chapter discusses material-driven adaptive design (MDAD) as an emerging interdisciplinary in material science and design disciplines. The authors define MDAD as a design system with a dynamic interrelationship with its environmental context. They evaluate prospects and challenges associated with MDAD, along with some of the shifts in the design model that it necessitates. Examples as research prototypes are provided to demonstrate the potential of integrated development processes that combine materials with geometry. Drawing from their research on the use of shape-memory polymers authors explore the use of "smart" materials in adaptive design to produce self-responsive and flexible surfaces. The design process includes correlating the micro-scale behavior of materials with macro-scale needs in architecture to create surfaces with the ability to change their geometries in response to temperature. This design process can pave a path to move beyond what is currently used as the primary means of adaptation such as complex mechanical systems (sensors, circuitry, motors, etc.) in architecture. Leading toward a more sustainable and more responsive built environment, this approach also can produce synergistic effects by bringing together expertise in two fields of study—design, and materials science—that are currently divided by differences in terminology, methods, and research practices.

Keywords Material driven adaptive design · Smart materials in architectural design · Environmentally responsive architecture · Shape memory behavior · Shape memory polymer

M. Mansoori (✉)
Assistant Professor, Washington State University, School of Design and Construction, Carpenter Hall 114, Pullman, WA 99164, USA
e-mail: maryam.mansoori@wsu.edu

Z. Rybkowski
Associate Professor, Texas A&M University, College of Architecture, 400 Bizzell St, College Station, TX 77843, USA
e-mail: zrybkowski@tamu.edu

N. Kalantar
California College of Art, 1111 8th St, San Francisco, CA 94107, USA
e-mail: kalantar@cca.edu

J. Wang et al. (eds.), *Advanced Materials in Smart Building Skins for Sustainability*,
https://doi.org/10.1007/978-3-031-09695-2_10

10.1 Introduction

This chapter will address the concept of material-driven adaptive design (MDAD) by defining its principles and qualities. To better understand, MDAD and its prospects and challenges design prototypes have been designed fabricated and tested. The first half of the chapter provides a theoretical framework, while the second half presents specific processes and methods to create self-transformable surfaces. These surfaces have the ability to change their geometries in response to temperature.

Adaptive architecture in this study can be defined as constructed geometries that change in response to environmental stimuli. Such adaptive approaches have primarily been used in architectural skins, for example, façade elements that adjust to the position of the sun (Persiani, 2019; Yilmaz, 2017). Similar to biological skins, these surfaces are well-positioned to interact with a changing environment. While often modeled on natural systems, adaptive architecture has primarily used mechanical means (sensors, circuitry, motors, etc.) to respond to environmental changes. In recent years, however, designers have started to consider alternative ways to implement adaptive designs. The current state of the field can be characterized as being in transition from a purely mechanical approach toward more "passive" environmental response strategies, including material-driven adaption (Kretzer & Hovestadt, 2014; Menges, 2015; Hensel et al., 2006).

Advancements in materials science have been crucial for instigating this transition, particularly with the increasing availability and fundamental research into "smart" materials. These materials are designed at a molecular level to respond to environmental stimuli such as moisture, heat, or light by changing their physical qualities (such as form, transparency, or flexibility). Smart materials can offer significant potential advantages in adaptive architecture due to the ability to reduce mechanical components, lower construction and maintenance costs, decrease weight, and improve reliability (Barozzi et al., 2016; Lopez et al., 2015; Menges, 2015). Essentially, molecular properties in these materials can act as a substitute for the more expensive and less reliable sensors, wiring, and actuators that characterize mechanical-based adaptive architecture. New computational design tools are also making an important contribution to this paradigm shift, as readily-available software now allows designers to analyze material performance, simulate material responses to dynamic environmental conditions, and directly control fabrication tools to implement materials-based design solutions (Kretzer, 2016; Lopez et al., 2015; Sunguroğlu Hensel and Vincent, 2016).

As has historically been the case in architecture, these new innovations are being driven not only by new technologies but also by new concepts, most especially in our understandings of nature as a creative inspiration and model for innovation. The concept of the natural world as being composed of static, conflicting, individual components—an outlook that predominated in earlier centuries—has now been replaced by views that emphasize the flexibility and interconnectedness of biological structures. Many of today's adaptive designs are biomimetic—they mirror the functions, and sometimes the design, of biological systems (Kadri, 2012; Persiani, 2019).

For example, Achim Menges referred to the conifer cone and its repetitive opening and closing cycles in response to humidity changes as an inspiration for the adaptive qualities of his "HygroSkin Pavilion" (Menges, 2015). Similarly, Doris Sung's design of "Bloom" was inspired by leaf stomata (Sung, 2012), Skylar Tibbet's and Nery Oxman's projects (Totems, Silk Pavilion for example) mimics natural processes of organic growth and expansion (Tibbits, 2021; Oxman, 2015).

The HygroSkin and Bloom projects mostly make use of materials that have the ability to deform and then return to their original shape when a particular stimulus is applied. These are generally called shape memory materials (SMMs). Like other types of smart materials, their potential applications in architecture are novel, and we are only beginning to explore their potential uses. The Hygroskin Pavilion relies primarily on the hygroscopic behavior of wood, demonstrating that SMMs do not necessarily have to be manufactured in a lab. By carefully designing bi-layer wooden components, Menges was able to take full advantage of wood's tendency to expand when it is wet. Since the expansion of the inner layers is less severe than the outer layers, the overall bi-layer wood material bends in response to changes in humidity (Menges, 2015; Ugolev et al., 1986; Hensel et al., 2006). The more recent direction through which researchers are exploring wood's responsiveness to humidity. In 3D printed wood project Correa and others use 3D-printing technologies to create specially formulated wood–plastic composite materials and direct the response behavior. The 3D-printed composite materials also require a complicated analysis and technological fabrication process to be effective for intended purpose.

The "Bloom" project takes a similar approach to use material for adaptive design. Sung uses metal materials and takes advantage of their different responses to temperature. Using more than 400 parabolic-shaped panels made of laminated sheets of two metals with different coefficients of thermal expansion, the shell "opens" and "closes" in response to the heat produced by direct sunlight (Sung, 2012).

As these examples demonstrate, architects and designers have been experimenting with SMMs for some time. Experiments to precisely measure shape-memory behaviors for specific combinations of materials and architectural geometries are few, and as a result, the prototype structures that have been created are limited in terms of their response time, strength, range of curvature, and ability to repeatedly change shape without incurring damage.

There are significant challenges in this area mainly due to the current disciplinary split that exists between the design field and the materials science field—pedagogies, practices, and research literature in these disciplines are almost entirely separate from one another, which makes it difficult to establish an integrated method or workflow that considers both material design and structural design. Often designers lack even rudimentary knowledge of current research developments in materials science, let alone more sophisticated understandings of current directions in smart-materials research. On the other side of the split, material developers typically study transformations that occur at a scale of one-billionth of a meter. The application of novel materials to solve design problems and the integration of materials with larger structural geometries is often well beyond the research scope of the scientists who create these materials (Kretzer, 2016; Carolina Ramirez-Figueroa & Dade-Robertson, 2013).

To address these challenges, we propose that we recognize a specific research area called "material-driven adaptive design" (MDAD), which will stake out a unique place in the overall architecture and design firmament. MDAD can be considered as a particular process or system for designing, characterized by an integration of materials science research into design process. Designers who take this approach are likely to focus on inspiration from adaptive features found in natural systems that are difficult to mimic in conventional architectural practice, and they will pursue new materials development that can make such functionalities possible in the built environment. They are also likely to emphasize direct involvement of designers in the fabrication process and its parameters. By combining the two fields in this fashion, we can expect a synergistic outcome in which large-scale design needs contribute more strongly and directly to the direction of materials development, and in which a more robust and precise knowledge of new materials informs the possibilities and horizons of design.

10.2 Material Driven Adaptation as a Design System

Adaptive systems are dynamic. This quality may appear self-evident, but when applied to architectural discourses that have long privileged stability, it can entail significant cognitive shifts to perceive the ever-changing building/environment dynamic as intrinsic to design. This section will discuss some notable aspects of that cognitive shift and how they are relevant to materials-oriented thinking. Most importantly, since environmental features are dynamic and often unpredictable, a truly fluid system has better ability to respond to them in real-time. The building skin may need to open or close at varied times or change its transparency or a surface may need to transform to another shape when exposed to a specific level of heat or humidity. MDAD designers must therefore have a detailed and precise knowledge of material response parameters within specific structural geometries, which can be developed through prototyping and tested through empirical experimentations.

10.2.1 Decentralized Control

MDAD thinking diverges from the centralized perspectives of mechanical adaptive systems. Such systems typically operate through the interaction of their three main parts: (a) the sensor, which transmits input data, (b) the controller, which processes the input and selects a structural activation, and (c) the actuator, which generates a change, for example, a physical change in the built environment. This model replicates cognitive decision-making in the biological realm, in which an input is processed the brain, leading to a decisive response action. An MDAD approach is based on a more decentralized or distributed control. The control is based on materials-intrinsic, and usually continuous function of environmental response,

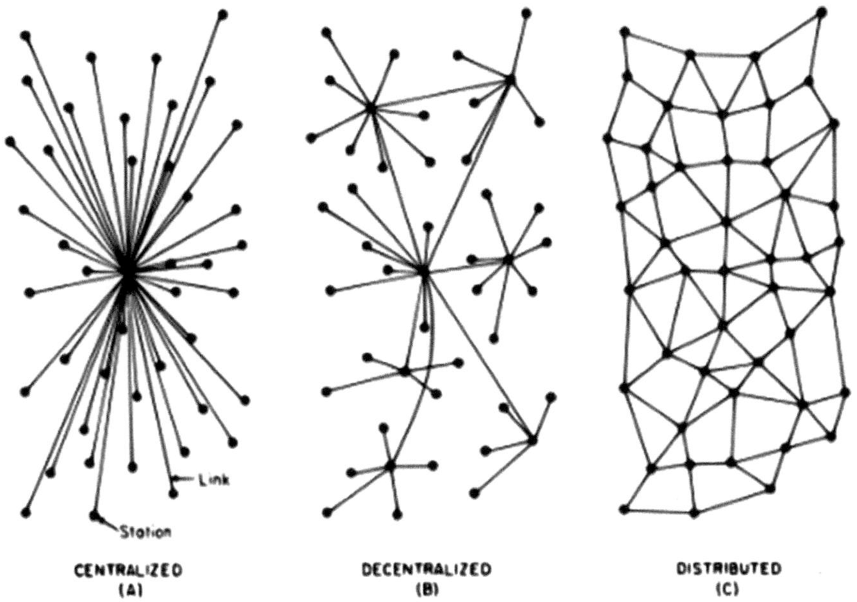

Fig. 10.1 Decentralized control versus Centralized control, illustrated by Paul Baran for network typologies

which can be compared to autonomous biological functions such as sweating in warm temperatures or opening and closing in plant stoma to survive environmental stress. Such intrinsic functions comprise the majority of the ways in which organic materials adapt to environmental changes (Grumezescu, 2016; Rehm, 2013; Tibbits, 2016). Rather than relying on distinct mechanical components and lines of communication between them, the parts of an MDAD system are based on decentralized, self-responsive, and self-sufficient design (Fig. 10.1).

10.2.2 Self-Responsiveness

All biological systems contain self-responsive components, and many parts of the natural world (such as plants) rely exclusively on such responses. A flower may gradually turn to follow the position of the sun; a rise in humidity can trigger pine cones to curl up their scales to prevent ineffective seed dispersal in wet weather. An MDAD approach embraces this model, leading to designs in which materials, structural elements, and/or entire building programs respond to external stimuli in a continuous and dynamic way (Menges, 2015; Hensel et al., 2006). The design process as well as outcomes adjust to accommodate a range of environmental conditions, rather than being bound to the static stand-or-fail dichotomy of unresponsive structures.

10.2.3 Self-Sufficiency

The self-sufficiency of MDAD projects is most notable in terms of energy sources. The adaptations have been defined as "passive" in the sense that they do not or have minimum rely on external energy inputs, beyond what the materials themselves can harness from the immediate environment (Barozzi et al., 2016). Transformations in geometry and other structural characteristics are carried out through internal material reactions (a process of adaptation or self-adjustment), and both the original and altered forms are internally stable.

10.2.4 Micro–macro Effect

The properties of the various materials used in an MDAD project can create a complex, interrelated network that results in overall system-wide behavior. This is often referred to as the micro–macro effect, and it is a vital part of bridging the gap between materials science research and architectural-scale applications. The self-sufficiency of parts and the absence of a top-down governing mechanism in MDAD projects does not mean that the parts operate in isolation from one another. Designers should be aware of this, and they should generally not expect to use smart materials in a monolithic, building-wide fashion. Instead, they should consider the synergies that can be created through the interaction of diverse material components. Like other aspects of the MDAD approach, this can entail significant shifts in perspective as designers learn to think in terms of complex organic development and emergent properties at various scales (Steiner et al., 2020). One strong advantage of this approach is that it can lead to flaw-tolerant designs, in which the system continues to adapt and function even when some individual components become damaged or impaired.

10.2.5 Strength and Flexibility

In architecture, flexibility has often been perceived as disruptive to the strength of a structure and to the designer's control over the behavior of its elements. While architecture often favor impermanence, adjustability, and in some cases transportability in different scales, it is more common to focus only on the rigid structures and define architecture as collections of elements that cannot alter their shape without being destroyed. This continues to be true with mechanical-based adaptive approaches, in which component mobility is obtained through assemblies of rigid parts (hinges, bearings, gears), and in which the failure of a single rigid part can break the entire system. The MDAD approach de-emphasizes rigidity and encourages a productive dialectic between flexible and rigid elements. In other words, while some rigid elements are needed to ensure the stability of the structure, there is less dependency on the survival of such elements. Instead, stability is established through the flexible interconnection of many supporting components, as can be seen in natural phenomena such as wood, bones, shells, and scales.

10.2.6 Free-Form Transformation

Architectural geometries are primarily flat, rectilinear, and orthogonal. There are many reasons for this right-angle-centric design, including the standardization of mass-produced components; ease of shipping, storage, and construction; and simplicity in force calculations. When it comes to adaptive architecture, this has resulted in transformations that are mostly based on the sliding and folding of rectilinear components. However, as Antoni Gaudí famously observed, "there are no straight lines or sharp corners in nature" (Crippa et al., 2003).

Similarly, the transformations that occur in natural structures rarely take place along right-angles but are instead smooth and continuous. Such natural forms and transformations have evolved to be efficient, stable, and resilient, often resulting in various geometries including curved outcomes. MDAD approaches are motivated by similar concerns of efficiency and resilience, and they emphasize materials' capabilities to exhibit smooth and continuous transformation behavior. To better understate the physics and design possibilities of free-form transformation, MDAD designers can benefit from an understanding of curvilinear surfaces and their common techniques in architectural design.

10.3 Experiments

The use of prototyping and scientific experimentation is fundamental in MDAD, as a crucial means of understanding and developing the synergetic relationship between materials and geometry. In our lab we have primarily examined the use of SMMs, focusing on their ability to: (a) respond to environmental changes without using a distinct sensor component, (b) self-adjust into pre-defined forms without the use of an external/artificial energy source, (c) embody both flexibility and strength to produce stable surfaces, and (d) transform geometrically without tearing or cutting. While other adaptive designers have embraced the shape-changing properties of commonly used materials in architecture such as metal or wood, we have focused on exploring the potential of shape-memory polymers (SMPs) that are currently not widely used in architectural research. The research process in our lab begins with tests to evaluate the environmental response properties of a particular SMP when fabricated into specific geometries. We use a variety of fabrication techniques, ranging from conventional SMP mixing and hardening methods to novel 3D-printing approaches. We then evaluate possibilities for combining the SMPs with other materials to create hybrid self-transforming surfaces.

10.3.1 Shape Memory Polymers

There are a variety of shape-memory material options that currently exist, ranging from the aforementioned traditional options of wood and metal to more recently developed shape-memory alloys and polymers. Among these options, SMPs demonstrate the greatest ease of formability and workability, and they are usually lower in cost compared to other shape-changing materials. Another important aspect of SMPs is their ability to readily bond with diverse conventional materials to produce hybrid structures (Wei et al., 1998).

Our investigation into the behavior of SMPs led us to select a temperature-responsive polyurethane epoxy resin-based polymer, which has been widely used as an independent structure, as a coating, and as an adhesive material. Epoxies can effectively bond with other materials as a resin, and the temperature required for stimulating their shape-memory behavior is within the range that we desired. The materials needed to make this type of SMP typically begin their life cycle as liquid mixtures and are solidified through an initial application of heat. This means that they are very easy to shape; the process is pouring the mixture into a mold or applying it with a brush prior to hardening. Once they have gone through the initial hardening process, temperature responsive SMPs remain in a rigid state as long as they are below their "glass transition temperature" (Tg). When the temperature rises above Tg, they become flexible and can be reshaped through the application of force. They then retain the new shape after cooling. However, if the temperature rises above Tg, the material will rapidly self-transform back into its original manufactured shape and no external force is present. Thus, if a flat sheet of SMP is manufactured, it can be heated and pressed into, for example, a double-curved dome shape, and then cooled in order to retain that shape. Reheating the material in the absence of force will lead it to rapidly revert back to its original flat position (Figs. 10.2 and 10.3). These transformations between different shapes can be repeated nearly indefinitely (Wei et al., 1998).

The formula used in the experiments reported here was based on an epoxy called Epon 826, which was mixed with a curing agent (Jeffamine D-230) along with a strengthening agent (neopentyl glycol diglycol ether, NGDE). The general temperature-mediated shape-memory effect of the resulting material has been widely documented in prior materials science literature (Xiao et al., 2012), so we could begin our investigation with confidence that its parameters would be similar to the material effects that we wished to pursue.

The self-transformation properties of this material are a result of a reorientation of the molecular chains in SMP. The Jeffamine D-230 bonds and hardens the epoxy materials to create the shape-memory behavior. The NGDE dilutes the polymer and helps to strengthen it against cracking and brittleness. The relative proportions of these material components in the mixture determine its specific Tg, the extent of flexibility, and other material properties.

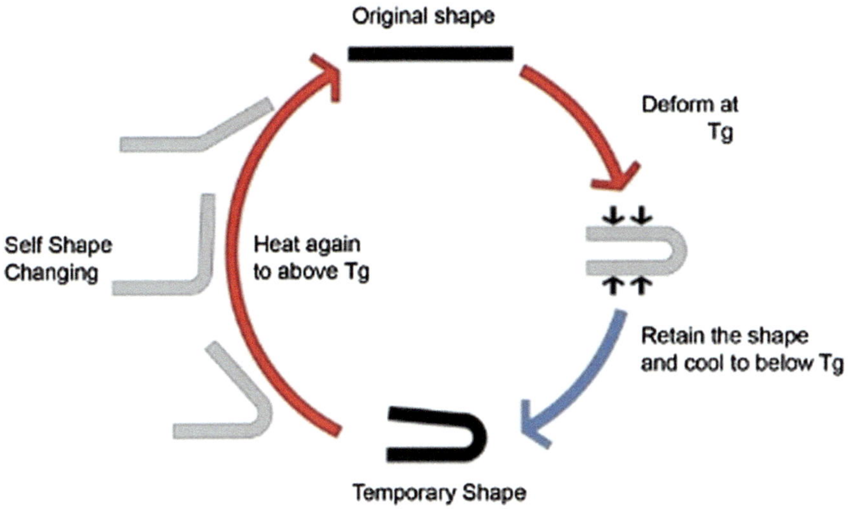

Fig. 10.2 The SMP shape-changing cycle. Illustrated by author

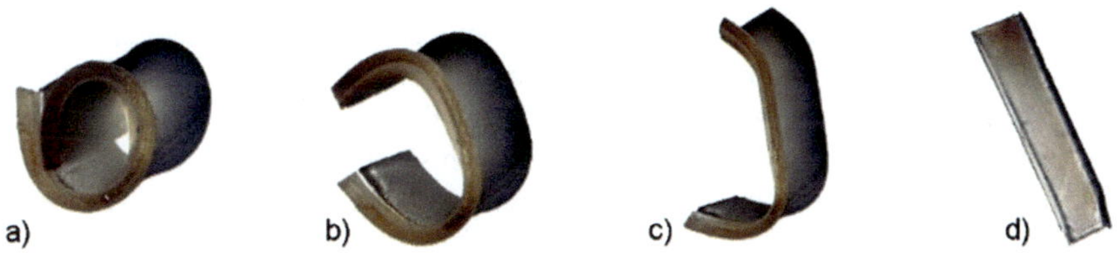

Fig. 10.3 Prototype A:Initial test of SMP responsiveness: **a** the deformed shape of a fabricated sample after applying heat and pressure, and the shape recovery process when re-applying heat for **b** 2 s, **c** 6 s, and **d** 8 s (Mansoori et al., 2019)

10.3.2 Testing SMP Surfaces

Some of our initial experiments and fabrication processes have been discussed at length in our previous publications (Mansoori et al., 2019). In short, to fabricate the SMP surfaces we combined various amounts of the three components in a plastic cup; stirred vigorously with a wooden stick for about 10 s; used a vacuum process during the initial heating to reduce air bubbles; and cured the material in an oven at 100 °C, using a silicone rubber mold to shape the resulting form (Fig. 10.4). For these prototypes, we wanted to achieve a Tg of approximately 40°–60 °C. This range allows the material to remain fully rigid at room temperature, while not requiring overly excessive heating to reach a flexible/shapable state. Tailoring the Tg is vital for design applications, as it directly impacts the range of functional temperatures in accordance with the design goals and the intended operating environment. We tested multiple samples of the polymer to evaluate this behavior, and determined that for

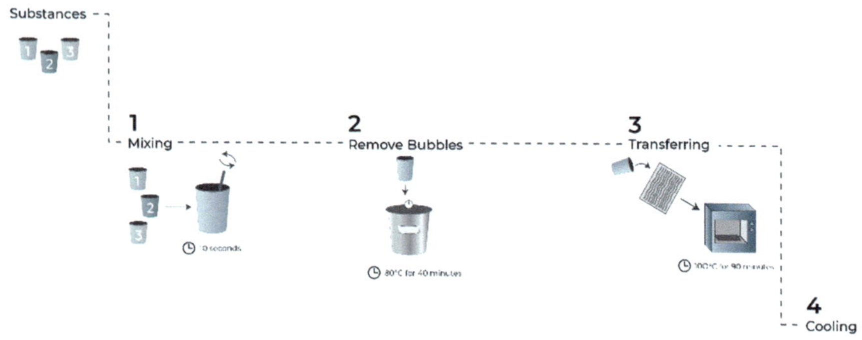

Fig. 10.4 SMP fabrication process in the lab. Illustrated by the author

our purposes a desirable Tg could be obtained with a combination of 1.00 part Epon 826, 0.63 parts Jeffamine D-230, and 0.60 parts NGDE.

The resulting material exhibited full rigidity at 20–25 °C (room temperature), and a significant ability to deform when above 40 °C. To test the shape-memory behavior, a flat sample was heated to 40 °C using a heat-lamp. Pressure was applied to push the sample into a curved shape and hold it in that shape after the heat lamp was removed. After few seconds at room temperature, the composite material cooled sufficiently to become rigid again and retain the new curved form. When once again heated to 40 °C in the absence of force, the material "remembered" its original flat shape and returned to it within 8 s. These initial experiments confirmed the functionality of our material mixture and laid the groundwork for subsequent research into using the SMP as part of hybrid materials and geometric forms.

10.3.3 SMP and EcoFlex Composite

One of our prototyping experiments conducted with the SMP mixture was called "SMP + Flex." In this project we laminated a stretchable silicone material known as "EcoFlex" to a structural framework made of SMP. The goal was to leverage the shape-memory capacities of the composite to produce a surface that could reversibly transform between flat and double-curvature shapes. The stretchable nature of the silicone material allowed it to accommodate these transformations, while the SMP provided the shape-memory effect and room-temperature rigidity.

An Archimedean spiral geometry was used in the SMP to assist the shaping process and encourage a smooth double curvature (Fig. 10.5). This form gives the composite plane surface a desired flexibility and kinematic properties that allow it to transform into a dome-like geometry. The spiral-cut SMP was created first and was placed into a flat mold. The EcoFlex silicone precursors, which also begin in a liquid state, were then mixed together and poured into the mold over the cut SMP. After curing at room temperature for 3 h the composite surface was ready for testing. As we had hoped, the

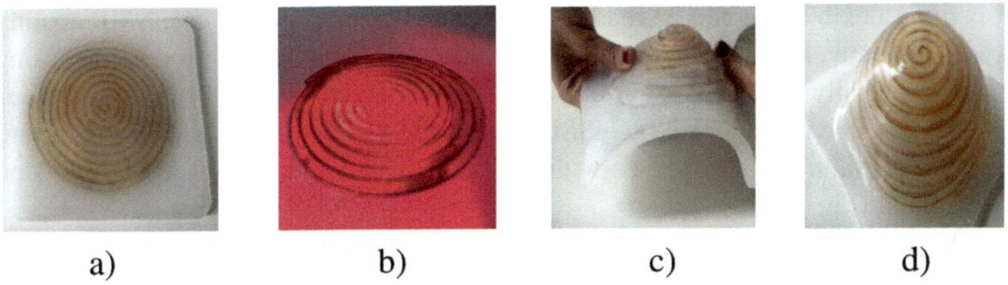

Fig. 10.5. Prototype B. **a** The "SMP + Flex" project began with (**a**) SMP cut into an Archimedean spiral pattern to encourage smooth transitions between flat and dome-like forms. This SMP component was then "cooked into" a surface comprised of EcoFlex silicone. **b** The resulting composite integrated shape-memory behavior into the silicone surface. The composite is in its flexible state when heated above (40 °C). **c** The composite can be deformed to double curvature in its flexible state. **d** The doubled curve surface is cooled and fixed in its new position. The double curvature can self-transform to the flat surface when heated again to (40 °C). We used infra red lamp to heat the surface

inclusion of the SMP base provided greater rigidity to the resulting material at room temperature (compared to EcoFlex alone). The composite regained its full flexibility when the environmental temperature was above the SMP's Tg range (40 °C). In its elastomeric state, force was applied and the material deformed smoothly into a dome geometry. When cooled again to room temperature it retained this shape. It is difficult to fix the composite in its cured position; the material has the tendency to get flat when is heated. After being heated once again to Tg, the composite "remembered" its original flat form and returned to it within 6 s. This cycle could be continuously repeated with no discernable material degradation.

One outcome that we noted in this project was that the stretching of the EcoFlex material in the dome geometry resulted in a latent force that limited the flexibility of the SMP in its glass state. It took additional pressure to form the material into a dome compared to shaping the SMP alone, and the resulting transformation was somewhat less extensive after the material was cooled and the force removed. Correspondingly, however, this latent force resulted in the SMP more quickly returning to its original flat shape when re-heated in the absence of external force. The experiment demonstrated that in pursuing these types of composites, particular attention should be given to the interface regions where the SMP bonds to the silicone material, as this interaction will be very significant in determining the overall geometric properties of the composite as well as being likely points of structural weakness.

10.3.4 Using SMP with Wood Veneers

For this experiment we laminated a veneer made of red oak with a layer of SMP. The veneer, which is often used for patching, repairing, and restoring common items such as doors or shelving, consisted of a very thin (1 mm) sheet of authentic wood.

It is quite flexible and can be readily cut by hand using scissors. The SMP coating was also relatively thin at 0.5 mm for both sides. The combination produced a strong bond, especially in comparison to the SMP–silicone composite from the previous experiment. We heated the wood–SMP composite to Tg (40 °C), and shaped it into a gentle curvature in accordance with the range of flexibility allowed by the wood component (Fig. 10.6).

The hybrid material was quite rigid at room temperature (more so than the wood veneer in isolation, and also more so than the SMP–silicone composite from the previous experiment). Unsurprisingly, the flexibility range of the wood composite at Tg was somewhat less than that of the silicone composite; however the wood was still able to form and hold a significant curvature. No cracking or structural deterioration of the wood veneer was noted after multiple deformation cycles, though this integrity is undoubtedly related to the extent of curvature obtained.

We created two veneer-SMP prototypes. In prototype C-1 the original shape was a flat surface. We used this prototype to test self-transformation from double curved surface to a flat surface. The original shape for in prototype C-2 was double curved surface. The test for this prototype shows how the surface retunes to its curved shape from flat against gravity (Figs. 10.6 and 10.7).

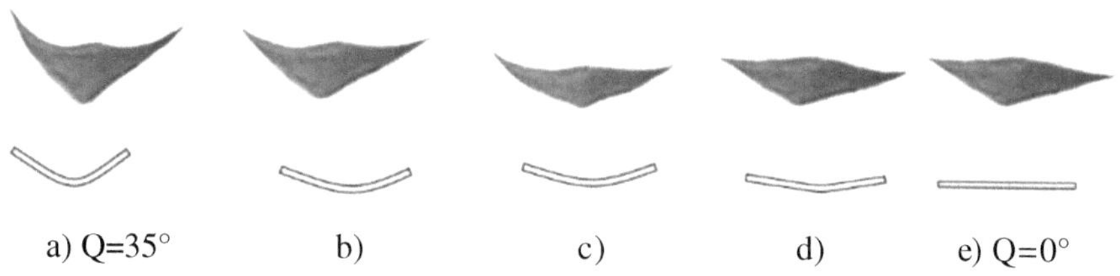

Fig. 10.6 Prototype C-1. Self-transformation from double curved surface (Q = 35°) to flat surface (Q = 0°). **a** the deformed shape of a composite after applying heat and pressure, the other images show self-transformation to flat surfaces when re-applying heat for **b** 2 s, **c** 4 s, **d** 6 s **e** 7 s

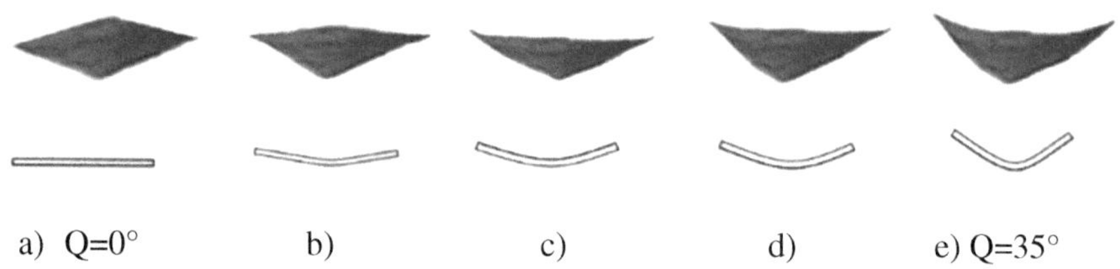

Fig. 10.7 Prototype C-2. Self-transformation flat surface (Q = 0°) to curved surface. **a** the flat shape of a composite after applying heat and pressure, the other images show self-transformation to double curvature surfaces when re-applying heat for **b** 2 s, **c** 5 s, **d** 7 s **e** 11 s

10.4 Conclusion

The experiments explained in this chapter were generative exploratory modes of the design process that demonstrate the potential of combining design (in this case, geometry) with materiality. While the study has not exhausted all possibilities but shows how MDAD makes it possible to (1) achieve predefined self-responsiveness without distorting the material, using electrical or mechanical sources, (2) combine contradictory properties of flexibility and rigidity and adopt the combination as an alternative design solution, (3) employ the dynamism in micro-scale to adaptation needed in macro-scale.

We showed how SMP and hybrid materials based on SMP can morph between a flat geometry—useful for ease of manufacturing, shipping, and storage—and curved surfaces, which may be useful in a variety of architectural applications such as shell surfaces. The use of particular SMP, however, was only one of the potential uses of "smart" materials for future adaptive design. More systematic with precise measurements of transformations are needed to continue the research which should address hybrid material and mechanical properties. Other functionalities are increasingly available for integration into our adaptive design repertoire, ranging from two-way polymers to other types of smart materials. While the transformation to the intended form as a result of environmental stimuli, its reversal to the previous condition required external force in our current experiment. Future studies on two-way polymers for example allow us to make the whole cycle automatic. There are tremendous and diverse research opportunities under the MDAD umbrella, particularly when it comes to scaling-up the applications of material-driven design, rigorously measuring the functionality of specific materials/geometry combinations and integrating environmental changes into design.

The MDAD approach can serve as a means for facilitating knowledge-transfer between the materials science and design disciplines. To embrace this process, architects and designers must recognize that it will entail adjustments to our established research and design practices which might need shifting toward different design processes such as an open-ended paradigm of prototyping and testing: the design process to include innovative ways not only form or the material but their behavior.

References

Barozzi, M., Lienhard, J., Zanelli, A., & Monticelli, C. (2016). The sustainability of adaptive envelopes: developments of kinetic architecture. Retrieved December 6, 2018, from https://www.researchgate.net/publication/307528257_The_Sustainability_of_Adaptive_Envelopes_Developments_of_Kinetic_Architecture.

Carolina Ramirez-Figueroa, L. H., & Dade-Robertson, M. (2013). Adaptive morphologies: Toward a morphogenesis of material construction. Retrieved December 6, 2018, from http://papers.cumincad.org/data/works/att/acadia13_051.content.pdf.

Crippa, M. A., Gössel, P., Klinkhamels, S. & Blass, C. (2003). *Antoni Gaudí, 1852–1926 : From nature to architecture.* Köln; London, Taschen.

Grumezescu, A. M. (2016). *Fabrication and self assembly of nanobiomaterials*. Amsterdam Netherlands, Boston MA, William Andrew.

Hensel, M., Menges, A., & Weinstock, M. (2006). Towards self-organisational and multiple-performance capacity in architecture. *Architectural Design, 76*, 5–11.

Kadri, L. B. (2012). *Towards the LIVING envelope: Biomimetics for building envelope adaptation*. Doctorate, Wöhrmann Print Service B.V.

Kretzer, M. (2016). Information materials: Smart materials for adaptive architecture. In: *Smart materials for adaptive architecture*. Illustrated ed.: Springer.

Kretzer, M., & Hovestadt, L. (2014). Alive: Advancements in adaptive architecture. In: Manuel Kretzer, L. H. (ed.) *Spaces of adaptivity: Thoughts on the relationship of life and architecture*.

Lopez, M., Croxford, B., Rubio, R., Martín, S., & Jackson, R. (2015). Active materials for adaptive architectural envelopes based on plant adaptation principles. *Journal of Facade Design and Engineering*, 3.

Mansoori, M., Kalantar, N., Creasy, T., & Rybkowski, Z. (2019). Adaptive wooden architecture: Designing a wood composite with shape-memory behavior.

Menges, A. (2015). Fusing the computational and the physical: Towards a novel material culture. *Architecture Design, 85*, 8–15.

Oxman, N. (2015) Templating design for biology and biology for design archit. *Design 85*, 100–108

Persiani, S. (2019). *Biomimetics of motion: Nature-inspired parameters and schemes for kinetic design*. Springer.

Rehm, B. (2013). *Bionanotechnology : Biological self-assembly and its applications*. UK, Caister Academic Press.

Steiner, H. A., Antonelli, P., Burckhardt, A., Boom, I., Oxman, N., & The Mediated Matter Group (Cambridge Massachusetts). (2020). *Neri Oxman/material ecology*. The Museum of Modern Art.

Sung, D. K. (2012). *Skin deep: Breathing life into the layer between man and nature*. Building Research Information Knowledgebase, American Institute of Architects and the National Institute of Building Sciences, Los Angeles.

Sunguroğlu Hensel, D., & Vincent, J. F. V. (2016) Evolutionary inventive problem-solving in biology and architecture: ArchiTRIZ and Material-Ontology. *Intelligent Buildings International 8*, 118–137.

Tibbits, S. (2016). *Self-Assembly Lab : Experiments in programming matter*. NY, Routledgevolume.

Tibbits, S. (2021). *Things fall together : A guide to the new materials revolution*. Princeton University Press.

Ugolev, A., Zzaripov, B., & Niezuitova, N. (1986). A revision of current data and views on membrane hydrolysis and transport in the mammalian small intestine based on a comparison of techniques of chronic and acute experiments: Experimental re-investigation and critical reviews. *Comparative Biochemistry and Physiology*, 593–612.

Wei, Z. G., Sandstroröm, R., & Miyazaki, S. (1998). Shape-memory materials and hybrid composites for smart systems: Part I Shape-memory materials. *Journal of Materials Science, 33*, 3743–3762.

Xiao, X., Xie, T., & Cheng, Y. T. (2012). *Patent of Self-healing and scratch resistant shape memory polymer system*. US patent application.

Yilmaz, I. (2017). Move-tecture: A conceptual framework for designing movement in architecture. *IOP Conference Series: Materials Science and Engineering*.

Chapter 11
Design Principles, Strategies, and Environmental Interaction of Dynamic Envelopes

Pengfei Wang, Junjie Li, and Zehui Peng

Abstract In the era of sustainable development, the building envelope system has gradually become the focus of attention, because of its important function of serving as the separation interface between indoor and outdoor. The dynamic adjustment mechanism of environmental interaction is an effective strategy for buildings to deal with complex external environments. As the dynamic envelope system has technical and aesthetic advantages that form unique space experience and functional adjustment, it provides a new way for building development. With the progress of the times, it will show more abundant prospects. This chapter takes the practical development of the dynamic envelope system as the research object and summarizes the current climate-adaptive design strategies, realization forms, problems, and future trends by sorting out the categories. Finally, combined with the specific practice, the operability application method is proposed, which can provide a reference for the design method of the dynamic envelope system combined with buildings.

Keywords Dynamic envelope · Climate response · Sustainable design · Material construction · Design strategy

11.1 Appearance and Space, Static to Dynamic

In recent years, the issue of building energy consumption has become increasingly prominent. How to adapt the interior of the building to changes in the external environment and reduce energy consumption has become one of the issues that architects focus on. As the climate boundary, today's dynamic envelope system can perceive external information and respond to changes in the surrounding environment through its own adjustment, and gain design autonomy and independence due to technological upgrades in architecture, giving full play to the interface effect.

P. Wang
School of Architecture, Southeast University, Nanjing, China

P. Wang · J. Li (✉) · Z. Peng
School of Architecture and Design, Beijing Jiaotong University, Beijing 100044, China
e-mail: lijunjie@bjtu.edu.cn

J. Wang et al. (eds.), *Advanced Materials in Smart Building Skins for Sustainability*,
https://doi.org/10.1007/978-3-031-09695-2_11

The dynamic envelope system of a building, including walls, doors and windows, and roof, refers to an interactive change system that can implement measures such as opening and closing the boundary according to people's needs for adjusting the indoor climate (Moloney, 2011). In 1966, American architect Robert Venturi decomposed architectural issues into two levels: space and skin in Complexity and Contradiction in Architecture, and pointed out that in addition to the limited creation of space, there are infinite and rich possibilities for the creation of skins in architecture (Venturi, 1977). The architect regards the envelope system which is independent of the space more and more importantly, trying to create a two-layer interface that exists at the same time. The inner layer solves the function, while the outer layer solves the form. At the same time, the dynamic boundary of the outer layer makes the interior space of the building no longer isolated from the natural environment.

On the other hand, the rapid development of technical means and communication media has given new meaning to the dynamic envelope system. Due to the dynamic characteristics of the envelope system itself, it gets rid of the fixed image of the original facade, and has unique advantages in terms of space requirements and user experience. With the close integration of mechanical dynamics, parametric design and other fields with architecture, the dynamic envelope system integrates various technical design methods to become an interface for improving building performance. The dynamic environment interaction can become an important fulcrum for architectural mass change and image communication. Buildings are no longer satisfied with the limited space framed by walls, and rely on facades for information dissemination. These make people feel about the building from the inside to the outside, from the whole to the fragments, and from the space to the appearance (Chen & Mo, 2008). The design of a dynamic envelope system that interacts with the environment has gradually become an important focus of architectural design.

11.2 The Value Pursuit of the Dynamic Envelope System

11.2.1 Ecological Value Pursuit: Light, Heat, and Wind Environment

Under the dominant trend of sustainable development needs, building envelope systems are closely related to energy consumption. Like the cell membrane that controls the entry and exit of substances, it controls the entry and exit of natural elements such as indoor and outdoor light, heat, and wind. It is the interface between the building and the environment (Fig. 11.1).

The envelope interface is changed according to the comfort of building users, so that the building integrates into the natural environment, reduces the dependence on the artificial environment, and avoids the energy consumption of the facade during the four seasons. It constitutes a sufficient condition for building energy efficiency, and therefore has become an important architectural design goal. Among the many

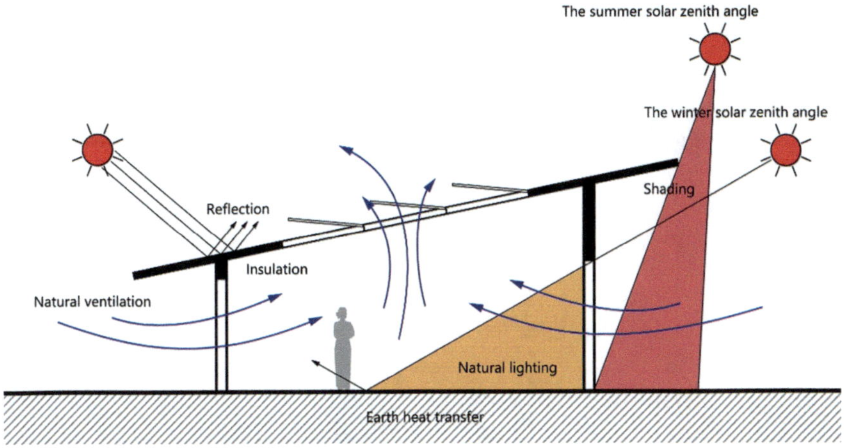

Fig. 11.1 Interaction between the envelope system and the environment

factors that affect the climate, the most decisive factor is the sun's light and heat radiation (Li et al., 2008). In response to the four seasons of the external environment, the interaction requirements inside the building are not fixed, such as the need to introduce sunlight to enhance heat preservation in winter, the need to block sunlight to enhance ventilation in summer (Feng, 2004), and other conflicting conditions. A single design strategy cannot fully satisfy the use or avoidance of natural conditions in different seasons or weather conditions. While the dynamic envelope system that interacts with the environment gets rid of the fixed passive adjustment style and can better exert the climate response advantages to enhance the comfort and energy-saving effect of the building.

For example, in the new office building of the Dutch Charity Lotteries company (Goede Doelen Loterijen) (Fig. 11.2), the green roof composed of many triangular blades can use diffuse reflection to avoid direct glare, guide natural ventilation, and also have the effect of collecting rainwater. The roof spans the original building volume, turning the old courtyard into a sunny Mediterranean square.

11.2.2 Diversified Spatial Adaptability

The dynamic interface will have the opportunity to take advantage of some temporary architectural functions and rely on enclosures to meet diverse spatial needs. For a long time, buildings are often only satisfied with certain specific functional requirements, but with the development of multi-functional space requirements throughout the life cycle, there are higher requirements for buildings in the new era. Dynamic

Fig. 11.2 The green roof of the Dutch Charity Lotteries company

buildings can provide conditions for different activities and behaviors by controlling the opening and closing of the shape according to the climate or use conditions (Zuk & Clark, 1970). At this time, the dynamic envelope is generally not an isolated component unit, but a certain volume that constitutes the building. Due to the changing opening and closing of the envelope system, the boundary between indoor and outdoor is blurred, forming different spatial effects.

The typical method of using dynamic envelopes to form a variable space is to form a semi-outdoor space by moving the skin. For example, "The Shed" (Fig. 11.3) located in Hudson Yards in the United States has the same dynamic principles as the sliding house in Suffolk, England many years ago. The telescoping outer shell of "The Shed" is separated from the base building and glides along the rail to form an iconic space for large-scale performances and exhibitions. The unfolded building volume increases the interior area by 1,600 square meters compared to the original. The designed kinetic system comprises a sled drive and bogie wheels guided by the rails to realize the changing process of retraction and extension. Besides, the energy-saving design scheme adopted by "The Shed", through the radiant heating system, variable forced air heating and cooling system, provides the most efficient environmental regulation for the extended part of the outer skin (Eric Baldwin 2019).

Fig. 11.3 The Shed

11.2.3 Improved Aesthetic Feeling

One of the goals of the dynamic enclosure system is to enhance aesthetics. The design puts beauty above the adjustment of environmental factors such as light and heat. The form of the envelope structure is relatively obvious, and it is mostly two independent logics with the inner space, which breaks away from the modern rule of "form follows function".

The dynamic envelope system itself has unique aesthetic characteristics (Wang, 2011), which conforms to the dynamic aesthetic characteristics of the crowd (people tend to notice moving objects more easily). Compared with the single form of the general building volume, the envelope forms can be rich and diverse, showing the complexity of change and the beauty of order. Some pay attention to dynamic changes and positive interactions with human senses. The opening and closing of the interface satisfy people's different psychology of being close to or isolated from nature. Just as the American ecological psychologist Theodore Roszak put forward the concept of ecological subconsciousness in his book "The Voice of the Earth", he believes that people have an instinct to love life (Biophilia) (Roszak, 1992). In the architectural practice of recent years, the closeness and the interaction between man and nature have become more and more concerned.

In addition, the use of lights, colors, textures, etc. in the envelope brings users a unique sensory experience due to the rich and variable effects they form. For example, in the Shanghai Bund Finance Center, the dancing facade refers to the traditional Chinese bridal crown decoration. Decorative pillars of different lengths can move

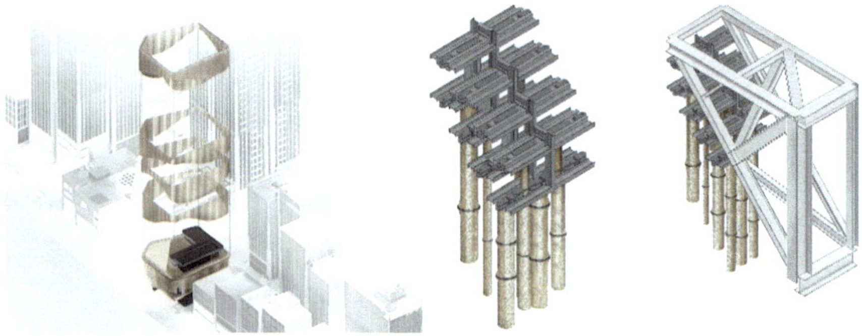

Fig. 11.4 Facade of Shanghai Bund Finance Center

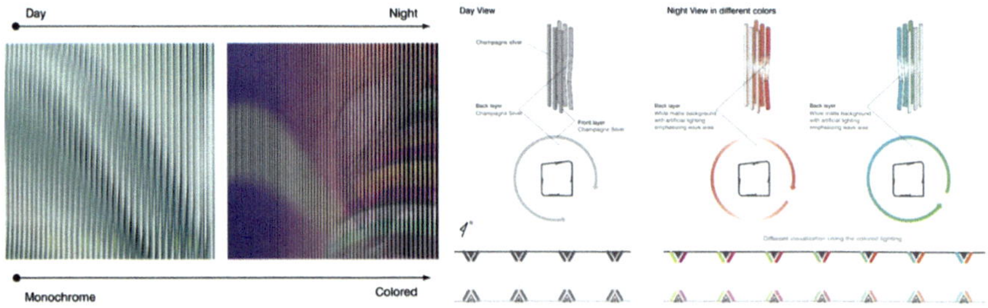

Fig. 11.5 Optical illusion appearance of Galleria Centercity department store

independently. The appearance of the building presents different transparency and visual effects according to the rotation and overlap of the interfaces at various levels (Fig. 11.4). South Korea's Galleria Centercity Department Store adopts a double-layer multimedia facade on the exterior of the building, that is, the vertical sides on the glass shell and the inclined sides of the inner skin (UNStudio, 2019). This combination forms a wave-like optical illusion effect, which will change according to the difference in viewing position (Fig. 11.5).

11.3 The Changing Principle of the Dynamic Envelope System

The famous 20th-century scholar Gilles Deleuze reflected in his book "The Fold: Leibniz and the Baroque" that when the surface of a building has the dual characteristics of construction and material, the surface can replace the spatial structure and become the dominant of the building, and the dominant of space and time (Deleuze, 1993). Based on this, the following section divides the dynamic principle of building envelope into three types: variable construction, variable material, and variable construction combined with materials (Payne & Firefly, 2013).

11.3.1 Variable Construction Depending on the Mechanical Device

The variable construction in envelope systems refers to relying on other external energy sources such as motors or manual drives to change the spatial position of the unit through a mechanical system, and respond to the external environment accordingly. There are no special requirements for the material properties of the components, and according to the principle of change (Moloney, 2011), it can be subdivided into basic motion forms of rotation, sliding, and compound motion forms, such as folding, telescoping, etc.

(1) Rotation. As a common dynamic change mode, due to the uniform rules of the variable unit, most of the components are uniformly repeated geometric shapes. Rotation can obtain more varied effects according to different angles without changing the shape of the unit. The way of change can be around the central axis, around a certain unit frame, etc. (Fig. 11.6), to form a kaleidoscope-like facade visual effect, and interactively adjusts environmental factors such as light, heat, and ventilation.

(2) Sliding. The slide rail, which usually contains a two-dimensional plane, is a more direct way of opening and closing the interface. In order to reduce the influence of gravity on components, most of them move in the horizontal direction, and interface units are often overlapped or complemented, so as to achieve the requirements for versatility in lighting, ventilation, and functional space.

(3) Folding. The system is usually composed of simple panel components, which are controlled by telescopic rods, and the shape changes through unfolding

Fig. 11.6 Schematic diagram of rotation principle of Hazza Bin Zayed stadium envelope system

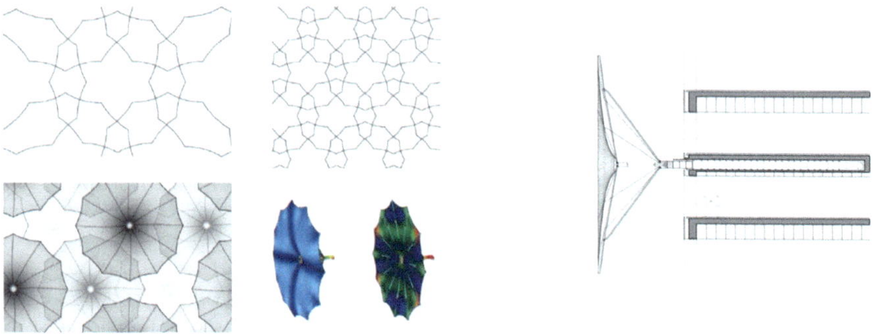

Fig. 11.7 Dynamic envelope unit construction of the Conjoined Media Towers

and stacking to realize the opening and closing for the control of natural light and ventilation systems. Because its change method is similar to the principle of folding doors, it is sometimes used to meet the functional requirements of building internal space expansion and flexible division (Miao & Feng, 2016).

(4) Stretching. It can be understood as a composite form of sliding and folding, which is widely used. This principle usually changes in a three-dimensional space. For example, the Conjoined Media Towers design scheme for the "unending sunlight" in the Middle East. Contrary to the traditional method of retracting umbrellas, the retractable sunshade envelope system passes outwards. The driving force retracts the umbrella surface, and then the whole is hidden in the corresponding structure along with the umbrella handle (Fig. 11.7).

11.3.2 Variable Materials Based on Their Own Characteristics

The material-variable refers to the interaction between the dynamic envelope system and the environment based on the characteristics of the material itself. Building regulation in this case often does not require external energy sources such as motors to perform work, but it is usually not as precise and controllable as the construction method.

The vertical planting of the building is a typical variable material. Plants are in a state of change due to their own growth characteristics. In summer, plants grow luxuriantly, which can bring good shading effect, and the transpiration and photosynthesis of plants can effectively reduce the surrounding environment temperature and improve the environmental quality. In winter, the leaves of plants are fallen, so that the building can be fully lighted, and the different types of plants give the facade a vibrant scene (López et al., 2015).

Another example is that under the action of a lower driving voltage or current applied to electrochromic materials, the optical properties (reflectivity, transmittance, absorptivity, etc.) will undergo stable and reversible color changes (Smart glass [EB &

OL], 2020). The electrochromic smart glass made from it has the adjustability of light absorption and transmission. Selectively absorb or reflect external heat radiation and internal heat diffusion, reduce a large amount of energy that the building must consume in summer cooling and winter heating, and at the same time play an active role in regulating the degree of natural light.

11.3.3 Combination of Variable Construction and Material

It not only uses the characteristics of the material itself but also adopts a structural method to form a dynamic envelope system that is driven by energy to generate climate-responsive adjustments. In the construction method, the appropriate change method should be selected according to the value pursuit. In the component materials selection, it should be durable and energy-saving or use renewable energy. Because it combines the advantages of the above two response methods, it has a wide range of application prospects.

As mentioned above in "The Shed", the ETFE material used can realize the adjustment of the building environment by means of lamination and so on (Cui & Miao, 2014). The combination with the sliding strategy enhances the climate response of the extended space and makes the indoor environment more selectively controlled. Meanwhile, the energy consumption of equipment is reduced, and the effect of seeking benefits and avoiding harm to the natural environment is achieved.

11.4 Organizational Mode of a Dynamic Envelope Unit

The unit organization mode is related to the aesthetic feeling of the building facade and the way of control. When the unit modules are densely laid on the building surface, it is required that the building shape should not be too complicated, and the units can be changed in groups vertically or horizontally along the floor. In the design process, it is necessary to avoid friction loss, enhance the replaceability of components, and consider the influence of natural factors such as wind.

11.4.1 Unit Form

In the selection of unit form, regular geometric shapes are often adopted due to factors such as cost and construction. The tessellations (Wikipedia, 2020), which have been studied as early as the era of Pythagoras, can interpret this well.

For example, using a single regular polygon to fill the plane:

The sum of the interior angles of a regular n-sided polygon is $(n-2)180°$, and the degree of each interior angle is $(n-2)180°/n$. When it meets the dense paving

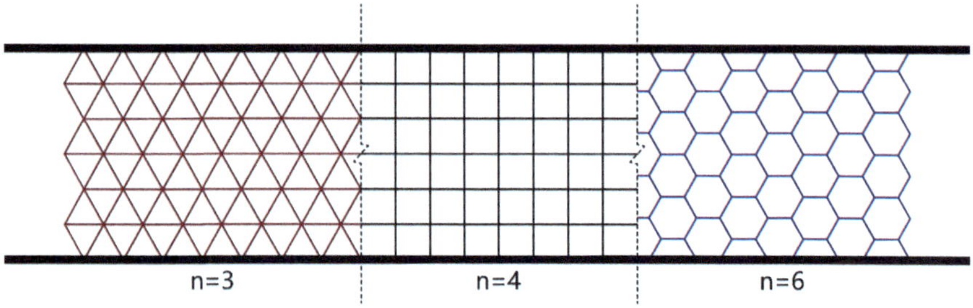

Fig. 11.8 Different unit form with the tessellation principle

condition, there is a positive integer x, which satisfies the equation x(n−2) 180°/n = 360°.

$$\text{Get } n = 2 + 4/(x - 2)$$

Since n is a positive integer, x = 3, 4, 6, corresponding to n = 6, 4, 3.

Therefore, except for irregular unit geometries, these methods can achieve complete tiling on the plane. In reality, single triangles, rectangles and hexagons have also become the choice of most dynamic building envelope systems (Fig. 11.8). The regular modular unit method can reduce the cost, is easy to construct, and is conducive to the aesthetics of the facade.

11.4.2 Scale Division

Judging from the current practice of the dynamic envelope system, the unit size is usually selected according to the principle of change and the size of the building facade. In addition, the needs of production, transportation and subsequent maintenance and replacement should also be considered. The floor height of the building is mostly equal to or an integral multiple of the height of the facade dynamic unit, so as to facilitate installation and allow users to control independently according to the floor.

11.5 Design Strategy of Dynamic Envelope Systems

In the design process of the dynamic envelope system, the unit can be set to a regular geometric form, and the changing principles such as rotation and sliding can be applied, or new material properties can be actively explored (Table 11.1).

In practice, we divide the building envelope systems into three types according to different locations: facade interface, roof interface and atrium interface. The next part

Table 11.1 Environmental interaction design elements of dynamic envelope systems

Project name	Project picture	Value pursuit	Changing principle	Specific strategy
Galleria Centercity department store		Aesthetic feeling	Material	Optical illusion material
Shanghai Bund Finance Center		Aesthetic feeling	Construction	Sliding
The shed		Spatial adaptability	Construction and material	ETFE material and sliding
Residential in Suffolk, England		Spatial adaptability	Construction	Sliding
Hazza Bin Zayed stadium		Shading	Construction	Rotation
Kiefer technic showroom		Lighting, shading	Construction	Folding
Conjoined Media Towers		Ventilation, shading	Construction	Stretching

of the article will discuss the three types of dynamic envelope patterns, each of which is detailed with the engineering practice that the author participated in as an example, and each of which is different in the value pursuits and changing strategies, in order to conduct targeted research on the design strategy of environmental interaction.

11.5.1 Rotating Roof Interface Based on the Pursuit of Ventilation and Shading

In the project of a green building competition in Changzhou, Jiangsu Province that the author participated in, in view of the climate characteristics of the hot-summer and cold-winter zone, the natural ventilation of the building becomes the design focus. The design strategy applies dynamic envelopes to the fifth facade of the building (Fig. 11.9), using the principle of rotating mechanical construction to achieve interaction with the environment.

The design strategy places the roofs on both sides perpendicular to the solar zenith angle in winter and summer. In terms of the shape and scale of the unit, in accordance with the principles of compactness, reliability, and economy, an isosceles triangle with a minimum side length of 1.2 m is selected as the basic unit module. Taking the vertical line where the unit's vertical center is located as the axis of rotation, it opens and closes in different amplitudes according to the changes in natural light, thereby creating suitable lighting conditions for the interior (Fig. 11.10). The independent closed office space inside the building reduces the indoor heat load in summer and the penetration of cold wind in winter due to the "protective umbrella" function of the roof interface.

The triangular unit cell is combined with the thin-film photovoltaic system (Fig. 11.11), and the angle is rotated according to the light intensity in different seasons. In the high-temperature season, the angle with the highest solar power generation efficiency is selected, taking into account the sunshade (Elghazi & Mahmoud, 2016). In the low-temperature season, the direct glare is reduced while satisfying indoor lighting. In addition, the opening of the roof skylight forms the vertical through space inside the building, which enhances the effect of buoyancy-driven

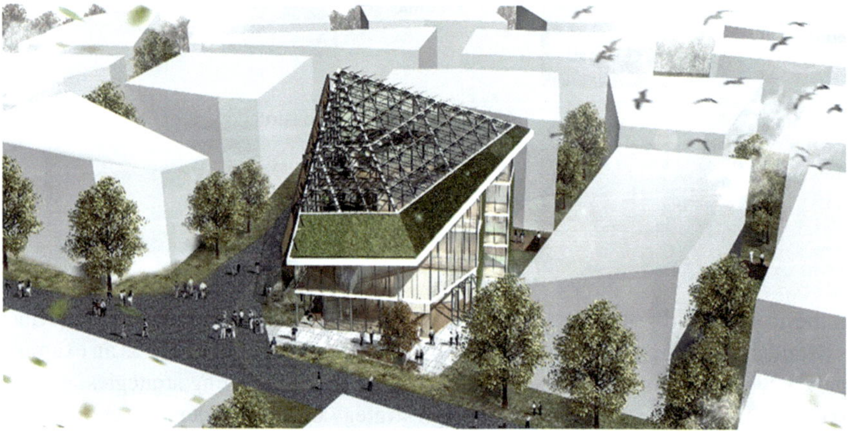

Fig. 11.9 Bird's eye view of the design plan

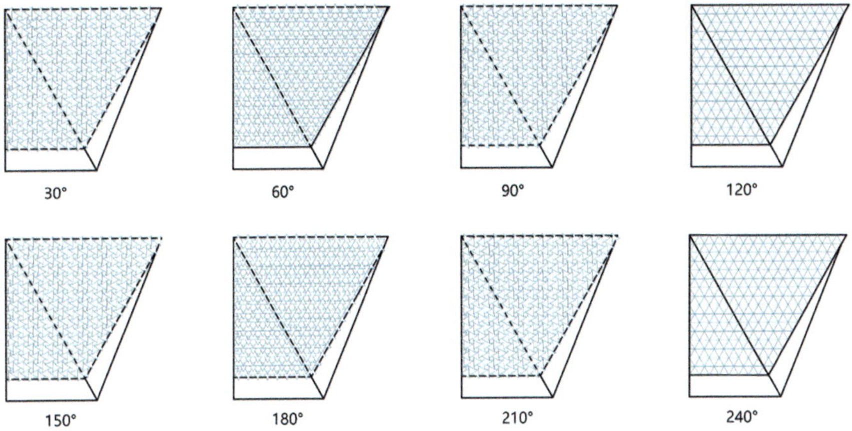

Fig. 11.10 Different rotation angles of dynamic roof units

ventilation. And use software to simulate and verify the effect of different rotation angles on indoor lighting and natural ventilation, and find the optimal set of choices (Figs. 11.12, 11.13 and 11.14).

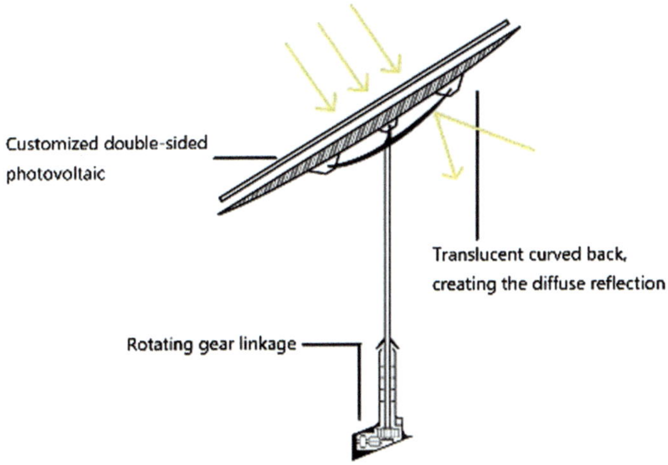

Fig. 11.11 Unit structure design

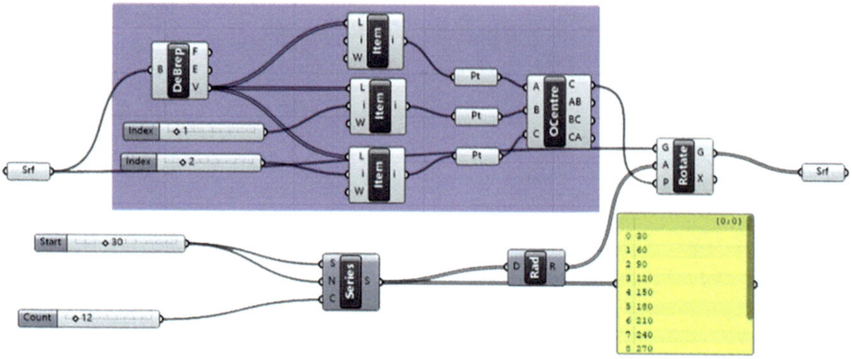

Fig. 11.12 The unit rotates at different angles under the control of Grasshopper

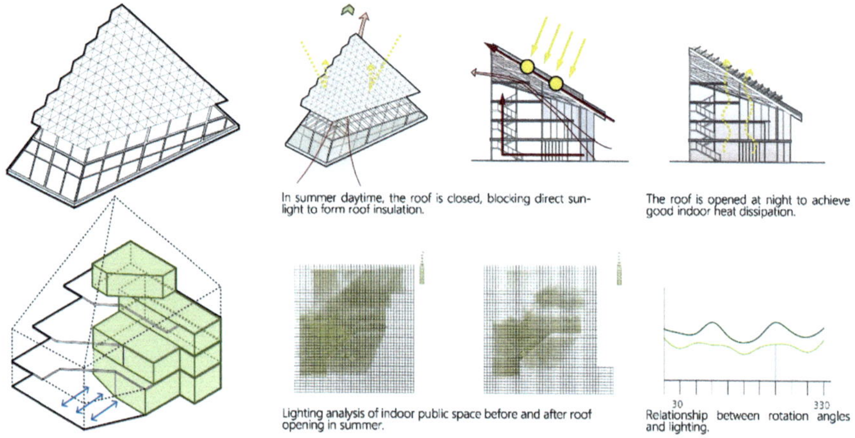

Fig. 11.13 Analysis of summer environmental interaction mode

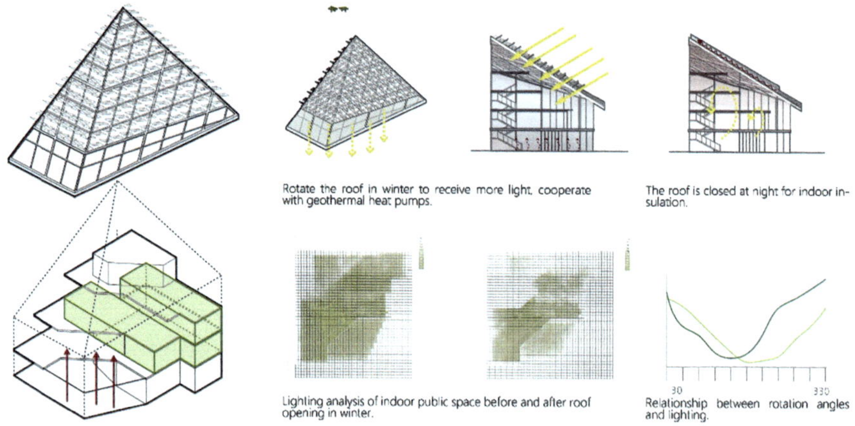

Fig. 11.14 Analysis of winter environmental interaction mode

11.5.2 Folding Facade Interface Based on the Pursuit of Shading

The Miura fold is a folding technology invented by Kōryō Miura, an honorary professor of structural engineering at the University of Tokyo, Japan. This technology is to unfold the items by opening the diagonal ends, and pushing them in the reverse direction when they are contracted, so as to realize the rapid expansion and stacking of materials and the most intensive storage space (Fig. 11.15).

The author takes this as a prototype to design the dynamic building envelope system that interacts with the environment. A high-rise hollow office building is selected, and the proposed column span is 8.0 m and the story height is 3.2 m. Solar photovoltaic panels are selected as the material for the upward part, which collects the electricity generated by the facade and has the effect of shading. Lightweight1 solar panels are selected for the downward part to meet the effects of light transmission, ventilation and heat insulation (Fig. 11.16).

From the formula for the effective use of photovoltaic panels, it can be seen that the larger Rα, the greater the area of the projection surface, and the higher the utilization efficiency of photovoltaic panels. And the utilization efficiency has nothing to do with the edge length. The angle $\alpha = 60°$ of the rhombus unit is selected comprehensively, and β is combined with the formula to select the best choice for each season (Fig. 11.17).

$$f(\beta) = x^2 \sin\alpha * k;\ k = \frac{\cot\beta * \cos\varphi + \sin\varphi}{\sqrt{1 + \cot^2\left(\arccos\frac{\cos\alpha}{\cos\beta}\right) + \cot^2\beta}}$$

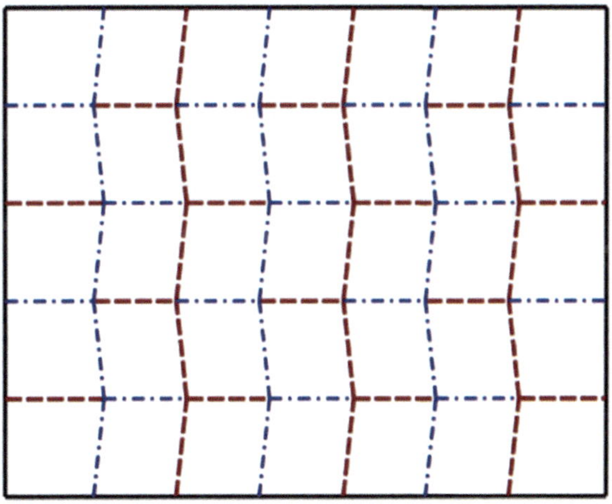

Fig. 11.15 Miura fold principle (Wikipedia, 2020)

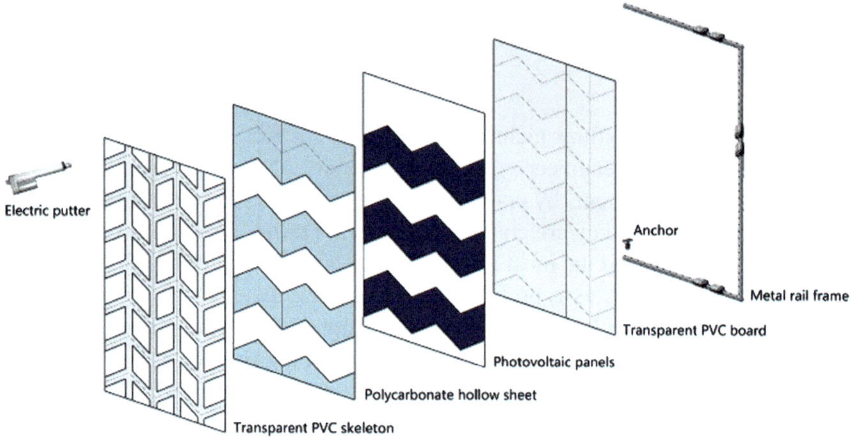

Fig. 11.16 Material selection and connection of double-layer envelope system

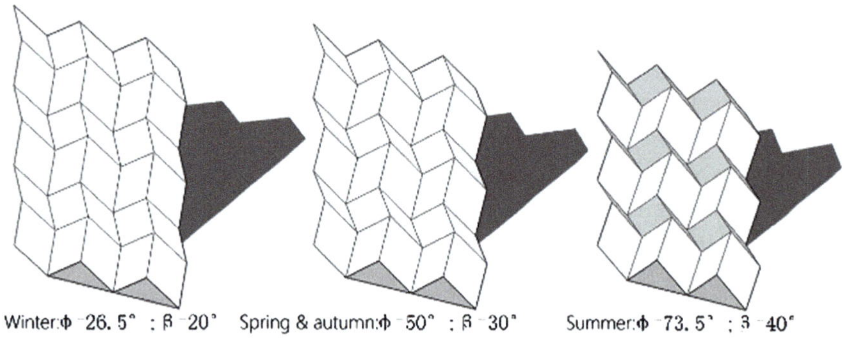

Fig. 11.17 Seasonal changes of a double-layer envelope system

In the above formula: α is the interior angle of the rhombus; β is the angle between the edge length and the vertical direction; x is the side length of the rhombus (Fig. 11.18); φ is the solar zenith angle; f (β) is the projected area of the rhombus on the plane perpendicular to the light; x^2sina is the area of the rhombus; k is the utilization rate of the photovoltaic panel.

Fig. 11.18 Rhombus unit schematic

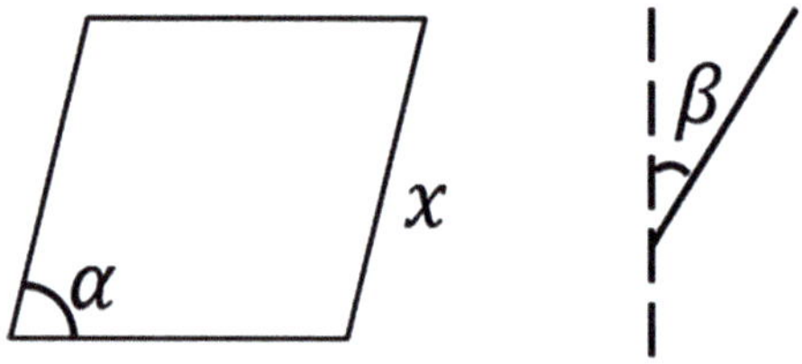

Table 11.2 The relationship between the dynamic envelope system and the season

Solar zenith angle	Angle β	PV panel utilization (%)	Projection length a (cm)	Projection length b (cm)
Winter: 26.5°	19.8°	97.1	37.6	33.9
Summer: 73.5°	39.6°	82.9	30.6	30.3
Spring and autumn: 50°	32.4°	88.6	34.6	33.8

In terms of unit form and scale division, the basic modules are combined into a plane to meet the best effect of the winter solar zenith angle and make the photovoltaic panel reach a larger size. Therefore, the size of each block is set to 4.0 $\times$ 3.2 m. According to the data calculation, the projection size of the rhombus module on the vertical plane is a $\times$ b = 37.6 $\times$ 33.9 cm, and the number of modules can be determined to be 9 $\times$ 12. It is calculated from this that the overall facade is reduced to 3.67 $\times$ 2.72 m at the optimal angle of photovoltaic in summer (Table 11.2).

This dynamic envelope system uses the changing principle of folding to achieve better matching of solar zenith angles in different seasons and improve the use efficiency of photovoltaic panels. In winter, it is completely closed after being unfolded to the best angle, and the vents are closed, and the "greenhouse effect" formed will further play the role of heat preservation and heat insulation. In summer, when it is reduced to the best angle, it will ventilate on the bottom side, forming a "chimney effect" with the upper gap, strengthening natural ventilation, and at the same time play a certain shading effect.

11.5.3 Sliding Atrium Interface Based on the Pursuit of Spatial Adaptability and Ecology

As a climate exchange space inside and outside the building, the atrium space plays an active role in building energy conservation and improving the indoor environment (Li, 2015). However, the climate control effect of the atrium is often a double-edged sword. While introducing light and ventilation, and using the greenhouse effect, there is also a contradiction between high temperature in summer and large amount of heat dissipation in winter (Table 11.3). The existing atrium has designs that incorporate dynamic devices, such as operable shading facilities set on the top of the atrium or inside and outside the curtain wall (Zhao, et al., 2018). But its function is single, and the amount of sunlight in the atrium can only be changed by changing the angle.

Based on the idea of climate control, this research proposes an innovative spatial interface regulation model. In response to climate change, through its envelope interface-the floor slab, climate regulation is carried out through lifting and lowering adjustments, so that the indoors can achieve the purpose of climate resilience.

Table 11.3 Advantages and disadvantages of climate control in traditional atriums

Climate control	Advantage	Disadvantage
Atrium interface	Natural lighting	Excessive temperature in summer
	Greenhouse effect	Excessive heat dissipation in winter
	Chimney effect	Lack of dynamic adjustment

During the winter day, the floor slab is lowered to the bottom of the atrium, and the greenhouse effect is used to fully absorb the solar radiation heat to increase the temperature of the atrium. At night in the winter, the floor slab is raised to be flush with the second floor, and the floor slab with heat preservation ability is converted into a building roof to reduce heat loss. During the summer day, the floor slab is also raised to be flush with the second floor. It has a shading design to block the excess solar heat and prevent the atrium from overheating. At night in the summer, the floor slab is lowered to the bottom of the atrium, and the vertical through space of the atrium forms the "chimney effect", with the help of windows at the top to form buoyancy-driven ventilation to obtain a good passive cooling effect (Fig. 11.19). Therefore, the dynamic atrium interface integrates multiple functions of climate control such as lighting, shading, heat preservation, and cooling.

At the same time, the floor slab integrates the advantages of lighting and thermal engineering, using aerogel glass with low thermal conductivity, long service life, high light transmittance, and a thin and light structure (Fig. 11.20), adopting the construction method of double-layer 8 mm+ toughened glass, 32 mm+ aerogel particles, and 8 mm toughened glass. This can not only meet the efficient lighting needs of the atrium but also ensure a stable thermal environment in the atrium. In addition, the lifting floor slab can also play the role of barrier-free passage and variable space (Fig. 11.21).

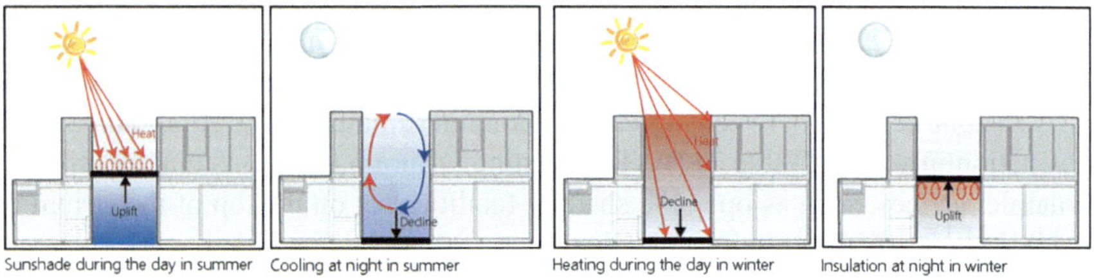

Fig. 11.19 Principles of climate control in the dynamic atrium

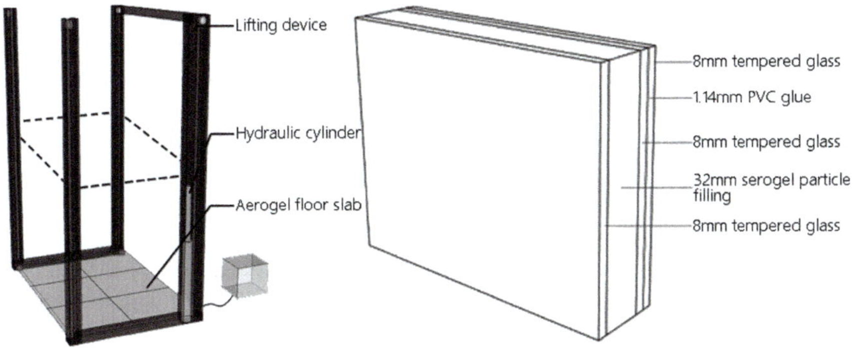

Fig. 11.20 Atrium dynamic system and aerogel floor construction method

Fig. 11.21 The construction process and actual effect of the dynamic atrium interface

11.6 Conclusion

As important focuses of architectural design, the ecological, spatial adaptability and aesthetic value pursuits of dynamic envelope systems gradually emerge in a large number of architectural practices. As an exchange interface inside and outside the building, its ecological value significance has unlimited potential in the context of today's environment. The realization of the dynamic form of the envelope systems, that is, the principle of change and the way of unit organization, will develop more possibilities, so that the construction design will be more integrated into the building facade. Therefore, regional climate adaptability and adjustability of the facade are achieved.

In the future, explore new breakthroughs from the perspective of sustainability between people and the environment, to better combine manual control, passive control, and automatic control, so that the dynamic envelope system has more adjustability. With the improvement of intelligent control technology, new materials, and innovative construction practices, the problems of energy drive and maintenance, and renewal are bound to be solved. The dynamic envelope system of environmental interaction will be widely used in a variety of building types under different climatic conditions.

Notes

1. It was first proposed by Louis Sullivan, a modernist architect of the Chicago School, in "The Tall Office Building Artistically Considered", with the original text "That form ever follows function. This is the law."
2. Theodore Roszak. The Voice of the Earth [M]. New York: Simon & Schuster, 1992.
3. "Biophilia" refers to people's natural and inherent emotional connection with other people and creatures.
4. Optical illusions, refer to judgments and perceptions that are differentiated from objective reality when people observe objects based on psychological factors such as empiricism or false references through geometric arrangements and the laws of visual imaging. The geometric optical illusion phenomenon is the most common.
5. The tessellations refers to the non-overlapping and gap-free merging of one or several polygons in the same plane, which satisfies the sum of the internal angles of all common vertices to be 360°. Tessellations can be divided into periodic and non-periodic types.
6. The chimney effect, or stack effect, refers to the strong convective ventilation of indoor air along the vertical space inside the building, using the temperature and density differences formed at different heights. The intensity of the chimney effect is related to the vertical height and the temperature difference in the space. The upper part of the building is equipped with exhaust vents to exhaust the dirty hot air from the room, while the fresh outdoor cold air is sucked in from the bottom, thus creating a comfortable and healthy indoor environment.

Acknowledgement The project is funded by National Natural Science Foundation of China (Grant Nos. 52078264 and 52078294).

References

Chen, J., & Mo, T. (2008). From architectural surface to surface architecture [J]. *New Architecture, 05*, 50–56.

Cui, Y., & Miao, Z. (2014). An analysis of dynamic skin based on visual consumption culture [J]. *New Architecture,* (05), 115–117.

Deleuze, G. (1993). The fold: Leibniz and the Baroque [M]. U of Minnesota Press.

Elghazi, Y., Hassaan, A., & Mahmoud, A. (2016). Origami explorations a generative parametric technique for kinetic cellular facade to optimize daylight performance [J]. Shape, Form and Geometry.

Eric, B. (2019). The Shed Opens in New York's Hudson Yards [EB/OL]. https://www.archdaily.com/914450/the-shed-opens-in-new-yorks-hudson-yards.2019.

Feng, L. (2004). Historical perspective of surface [J]. *Architect,* (04), 6–15.

Li, J. (2015). Passive Adjustment Performance of Intermediary Space in Buildings [D]. Tsinghua University.

Li, G., Li, B., & Gong, B. (2008). Ecological meaning of building skin [J]. *New Architecture, 02,* 14–19.

López, M., Rubio, R., Martín, S., Croxford, B., Jackson, R. (2015). Active materials for adaptive architectural envelopes based on plant adaptation principles [J]. *Journal of Facade Design and Engineering,* (1).

Miao, Z., & Feng, G. (2016). Juanli Guo. *Design of variable architectural surface in response to external environmental changes [J], 04,* 48–55.

Moloney, J. (2011). *Designing kinetics for architectural facades: State change [M].* Routledge.

Payne, A. O., & Johnson, J. K. (2013). Firefly: Interactive prototypes for architectural design [J]. *Archit Design.*

Roszak, T. (1992). *The voice of the earth [M].* Simon&chuster.

Smart glass [EB/OL]. (2020). https://en.wikipedia.org/wiki/Smart_glass.

UNStudio. (2019). Galleria Centercity department store [EB/OL]. https://www.unstudio.com/zh/page/12104/galleria-Galleria Centercity Department Store.

Venturi, R. (1977). *Complexity and contradiction in architecture [M]* (p. 1). The Museum of Modern Art.

Wang, JL. (2011). Biomimicry· kinetics· sustainability-study on kinetic building envelopes based on biological acclimatization. Tianjin University.

Wikipedia. (2020). Miura fold [EB/OL]. https://en.wikipedia.org/wiki/Miura_fold.

Wikipedia. (2020). Tessellation [EB/OL]. http://www.hdsdjc.com/index.php/jspd/item/660-2018-09-05-15-15-12.

Zhao, J., et al. (2018). A multifunctional adjustable louver roof system [P]. Shaanxi: CN207260399U, 2018-04–20.

Zuk, W., & Clark, R. H. (1970). *Kinetic architecture [M].* Van Nostrand Reinhold.

Chapter 12
Aesthetics and Perception: Dynamic Facade Design with Programmable Materials

Dale Clifford

Abstract Technology is often defined as the application of science to solve pragmatic human problems. Etymologically, the roots of technology are closely related to artful and skillful, and entwined with the idea of beauty. Much of technology teaching and professional architectural practice focus the former definition, with emphasis on energy reduction, measurement and building physics. While technical building metrics are important, they often overshadow the prospect for building technology to appeal to the imagination, engage the senses, and support emotive human connection with the environment. In fact, most building technology is designed to separate us from the environment. Smart materials, those which adjust their properties according to environmental stimulus, have the potential to physically index subtle changes in the environment, generating human awareness of environmental flux. It is proposed that by integrating the properties of smart materials within the façade and interior building surfaces, building technology can become a dynamic link between environmental changes and human sensory experience. Design experiments that expand the thermal and aesthetic properties of dynamic facades, with phase change materials and shape memory polymers, are presented.

Keywords Building performance · Aesthetics · Shape memory materials · Programmable matter · Architecture

12.1 Introduction

It's easy to fixate on technology. Especially if we are working toward a goal with quantifiable metrics, such as computing thermal transfer through a given wall section. Often these metrics become primary drivers in the early stages of the design process and significantly influence, the form and aesthetics of a building, and by extension, all those that come into contact with the building. While there is no single definition of building performance, the quest for performance increasingly impacts the design

---

D. Clifford (✉)
California Polytechnic State University, San Luis Obispo, CA 93407, USA
e-mail: dtcliffo@calpoly.edu

J. Wang et al. (eds.), *Advanced Materials in Smart Building Skins for Sustainability*,
https://doi.org/10.1007/978-3-031-09695-2_12

of the built environment through an ever-widening array of simulation tools targeted to optimize flows of heat, light, sound, air, and information. As such, the building envelope has become a locus for innovation, and a shift from viewing the façade as a wrapper to a medium of exchange between interior and exterior. Architects and engineers are beginning to apply computational tools, emergent materials, and a host of sensing and actuation technology, not just in service of building science, but to advance the role of the sensorial and qualitative aspects of building technology and search for ways that buildings can communicate their operation to building inhabitants.

A position is taken that the built environment can and should more effectively respond to nuances in local environmental conditions and that humans should be aware of how a building adapts in real time. Adaptive technology is becoming more prevalent as the building industry shifts from relatively static materials such as glass and steel, to dynamic materials that exist in multiple equilibrium states or non-equilibrium states and can respond autonomously to their environment. Still in their emergent stages, programmable materials, when applied to the building envelope, have the potential to lower the complexity of dynamic and responsive envelope systems by consolidating the sensing and actuation systems that more mechanical dynamic systems require to operate.

Two projects are presented that bridge material science and architecture through the application of programmable materials. The projects contribute to the current trajectory of work that challenges the notion of fixity in the design and operation of buildings, in favor of relatively simple means of dynamic environmental response. Each project has technological roots in programmable matter, but the architectural applications are consciously low-tech. The projects are intended to lower energy consumption by actively responding to fluctuating environmental conditions, but importantly, they are intended tap into the human need to be connected to the environment. Design examples are given of recent work with an emerging class of organic phase change materials (PCMs) and with shape memory polymers (SMPs) that demonstrate the prospect of dynamic materials to make architecture more thermally and visually responsive to local temperature variation.

12.2 Engaging the Senses

In Thermal Delight in Architecture, Lisa Heschong gives a compelling history of cultural development through human engagement with thermal sources. She speaks to the sensory engagement and human connection that occur through direct contact from thermal sources. Whether a group of humans are immersed in a thermal bath or sitting in a circle around a fire, Heschong proposes that we gather around thermal sources to satisfy our physical need to warm ourselves, our emotive need to be physically engaged, and our cultural need to exchange ideas and tell stories.

Heschong's book is informational and not as immediately apocalyptic as Rachel Carson's truth-based fable Silent Spring, but it carries a foreboding message in terms of blind reliance on technology. Specifically, Heschong warns of the cultural damage brought about by the designed and progressive separation of humans from nuances in the environment. She insinuates that we care less for our surroundings if we are disassociated from them. In thermal terms, if we do not know where the heat in a building is coming from or how it is produced, this lack of connection can beget carelessness. It is arguable that social and sensorial impoverishment can result from environmental detachment, leading to a coarseness with which we relate to our surroundings. As we are sensing creatures, once our physiological response to the environment is stunted, so too is our cognitive awareness. Heschong does not advocate a return to the campfire but is in favor of building technology that enhances social interaction by engaging the user sensorially and emotively. While contemporary building technology is the result of thought, engineering, and skill, in general, it is designed to detach the user from the operation of a building, in favor of semi-autonomous control systems that are often acoustically masked and hidden from view. This following work is part of a resurgence of projects that seek to make environmental technology both visual and visceral.

12.3 Keeping the Good Stuff in and the Bad Stuff Out

Buildings are often designed as a static barrier condition, isolating inhabitants from the environment. This approach approximates that tactic of a latex glove; a strategy designed to keep the good stuff in and the bad stuff out. In the case of buildings, the bad stuff is the environment, even though studies have shown that interior air quality often contains significantly more pollutants than exterior air (U.S. Environmental Protection Agency 1987). This is a concern as the global population is becoming domesticized and spends an increasing amount of time indoors (Wallace, 1987). Intuitively, we know that connection with the environment has been fundamental to our evolution and is central to our physiological and psychological health. Though there is little professional agreement on what connotates good design, researchers have studied the direct and indirect effects of buildings on human health, psyche, and well-being (Evans, 2003). Designers are also beginning to challenge the practice of defaulting to highly insulated air-tight buildings in favor of responsive and dynamic building envelopes that are meet the needs of occupants by adjusting to the environment.

Researchers and practitioners have proven Winston Churchill's observation that "We shape our buildings; thereafter they shape us." R.A. Ulrich, a researcher of evidenced-based health care design, has provided a direct link between design and well-being through study of the acoustic and visual environment on patient stress, anxiety, and recuperation times. In one project, Ulrich studied patient reports from a Pennsylvania hospital, where rooms on a double loaded corridor either had unobstructed views of a stand of trees, or a view of a monotonous brown brick wall. He found that the patients with a view received fewer negative evaluative comments, required less pain medication, and were released earlier, than those subject to the monotony of a brick wall (Ulrich, 1984). Many other studies have shown that increasing occupant awareness of variations in daylight levels within buildings leads to better and longer sleep patterns, positive mood effects, and increased cognitive functioning (Boubekri et al., 2020).

The projects that follow focus on the prospect for architecture to continually attune itself to its surroundings, and visually index thermal environmental change through variations in opacity and surface texture. The objective is to apply the attributes of programmable materials to combine aesthetics and performance, thereby recoding our relationship with the environment through the variability of the building envelope.

12.4 Project 1_ Phase Change Materials

Architectural materials are generally static and designed for stability and durability. Responsive materials, also known as reactive or 'smart' materials, vary their properties in response to external stimuli. These materials are controllable and reversable, and some are programmable. A common property of these materials is that sensing and actuation are embedded withing the material itself, lowering mechanical complexity in comparison to dynamic systems that rely on external sensors and actuation methods. PCMs and SMPs are part of an emerging class of responsive materials with programmable properties that can be tailored to trigger or switch in response to specific environmental stimuli such as magnetism, light, temperature, or humidity.

It makes all the sense in the world to use mass, as opposed to air, to moderate temperature in buildings. Massive materials have long been used to store or release energy to mediate diurnal temperature swings. Historically, these materials have functioned as sensible heat storage (SHS) systems that proportionally absorb or release heat according to increasing or decreasing temperatures. This approach, used in Trombe walls and concrete slabs exposed to the sun, comprise effective and uncomplicated thermal storage banks, though a lot of material is required to stabilize internal temperatures. In SHS systems, thermal storage capacity generally increases with density. Unlike conventional SHS materials, however, when PCMs reach the temperature at which they change phase, referred to as their set point, they absorb or release large amounts of energy and maintain an almost constant temperature. The PCM continues to absorb heat without a significant rise in temperature until all the material is transformed to the liquid phase. When the ambient temperature around the liquid PCM falls, the PCM solidifies, releasing its stored latent heat, as depicted in Fig. 12.1. Water, the PCM we are most familiar with, changes phase (liquid/solid) at 0C. As liquid water crystalizes, it takes on a more ordered structure and absorbs energy from the surrounding environment. As solid water melts, it assumes a less energetic state and releases energy. The reverse is true as water freezes, making water a highly effective thermal storage material as the material changes phase. In both cases, the material maintains a near constant temperature as it proceeds through phase transition. This property is especially useful in moderating temperature swings, whether in buildings or a cold glass of water. Organic PCMs, such as the soybean oil used in this study, operate with the same principle, though the melt/freeze point can be targeted to ambient room temperatures, making the organic oils effective materials for thermal stabilization in buildings.

PCMs in building technology are generally micro-encapsulated or macro-encapsulated. The process of micro-encapsulation entails enclosing phase change particles within a polymer shell. The shell effectively contains the PCM core through its freeze-melt cycles. Micro-encapsulated PCM can be added to building products such as plaster, drywall, concrete, or composite roofing systems to increase their thermal storage capacity. PCMs generally exhibit low thermal conductivity, and micro-encapsulation is a step towards overcoming this limitation. Macro-encapsulation refers to enclosing substantial amounts of PCM within a single container, a process that is becoming more technically and economically feasible with the emergence of organic PCM. Organic materials are generally preferred as they exhibit lower corrosive properties than inorganic salts or metallic compounds, making container composition less exacting and container degradation less of an issue. These properties open the door for more creative and effective packaging solutions. The following projects encapsulate PCM in containers that augment the thermodynamic attributes of the material and to make the phenomena of phase change visible.

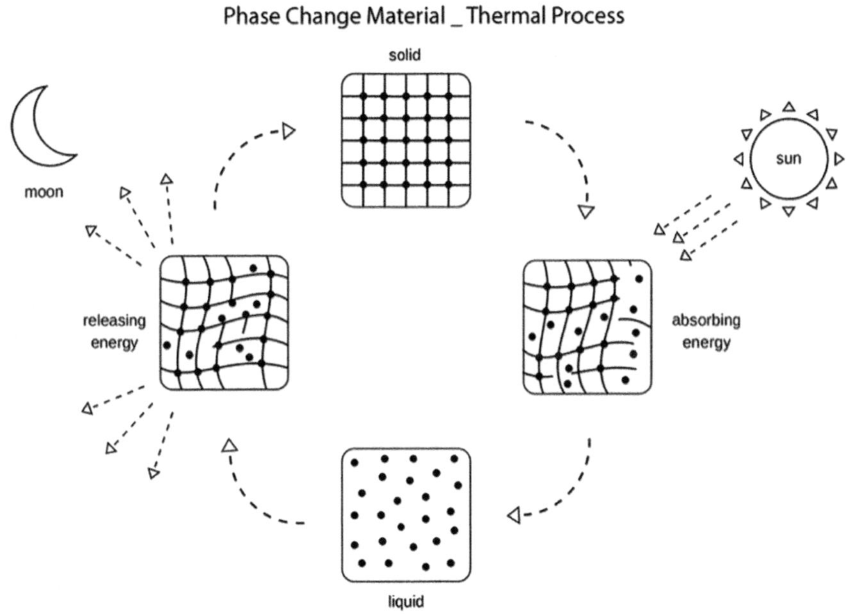

Fig. 12.1 Phase change cycle. This figure shows the cyclical thermal process of a phase change material beginning in a solid state, liquifying as energy is absorbed, then returning to a solid state as energy is released

12.4.1 Rethinking PCM Placement and Operation

Commercially available PCMs are most often packaged in opaque plastic and located above acoustic ceiling tiles or behind drywall. In these locations, the material mitigates internal temperature fluctuations, but it remains hidden from view. The 'Tile' project is an effort to rethink the placement of PCM within a building and the existing methods of packaging to better promote the conduction of thermal energy between the phase change material and the surrounding air and to visually exhibit the freeze/melt transitions to building users. The soybean oil used in this study has relatively low thermal conductivity (0.2 W/m/K) and tile designs with a high ratio of PCM to container surface were made to manage this issue. The first series of tiles (Fig. 12.2) are constructed from thin glass with an interlayer of PCM. The high surface area of the container accelerates the transfer of heat between the environment and PCM. The transparent nature of the tile also exhibits the crystalline structure of the oil as it solidifies and absorbs energy from the surrounding environment. The oil varies optically from opaque when frozen to near clear when melted.

It was observed that the tile acted as a temperature dependent optical switch, registering the thermal condition where the tiles were located. To a building user, the tiles index the relationship between the optical aspects of material change and the thermodynamic conditions of a given space. The tiles are designed to make

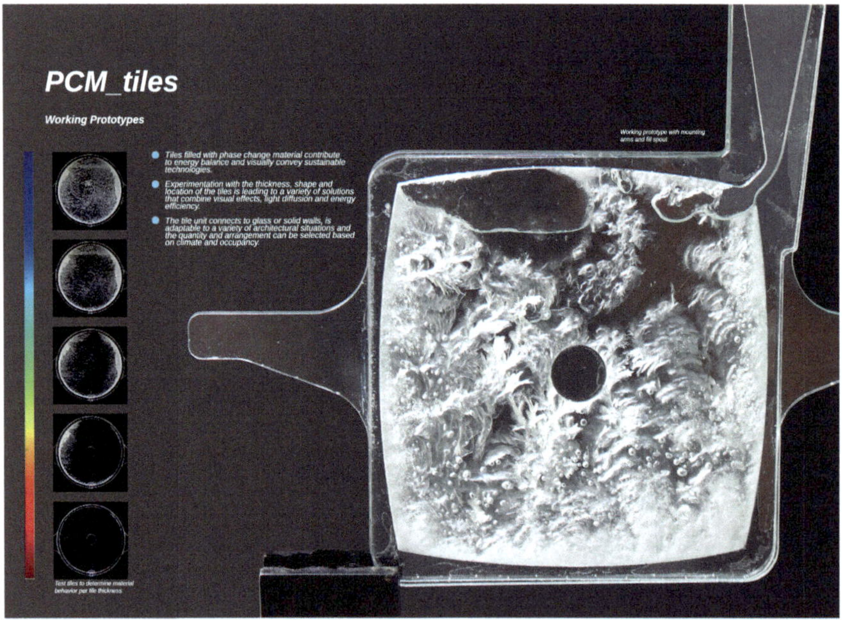

Fig. 12.2 PCM tile. This figure shows the encapsulation of PCM within a glass container. The fill tube and expansion chamber are seen at the upper right of the tile. The images at left demonstrate the spectrum of material opacity, from near clear to opaque, during the freeze/melt cycle

invisible temperature changes visible and alter one's perception of a building as a static entity, in favor of recognizing the energetic exchange between interior and exterior conditions. The intent of the project is to change the way buildings are experienced by communicating through its operations.

12.4.2 Application_ Expanded Wall Section

The tiles were assembled into an array that expanded the 'space' of the façade as they are located directly adjacent to the interior of the glass surface. This zone undergoes significant diurnal temperature swings, and the placement of tiles enables the façade to serve as a thermal storage bank and heat recovery system. Arrayed vertically and horizontally, a thermal bank of 360 tiles was designed to help the Frick Park Environmental Center, in Pittsburgh, PA meet the energy petal of the Living Building Challenge and to explore the communicative potential of sustainable technology (Fig. 12.3). Early energy analyses showed that total building energy consumption would be reduced by 25% with the PCM system proposed.

The following iterations of the PCM tile depart from the planar container and are shaped to increase the volume of PCM per tile and to increase the velocity of airflow across the container surface (Figs. 12.4 and 12.5). The design intent is to increase the effectiveness and the thermal storage capacity of the system by increasing the convective heat transfer to the surrounding air (Clifford et al., 2017). The thermo-formed container is filled with PCM with a variety of set points, to further enhance the reading of the interior facade's response to temperature variation.

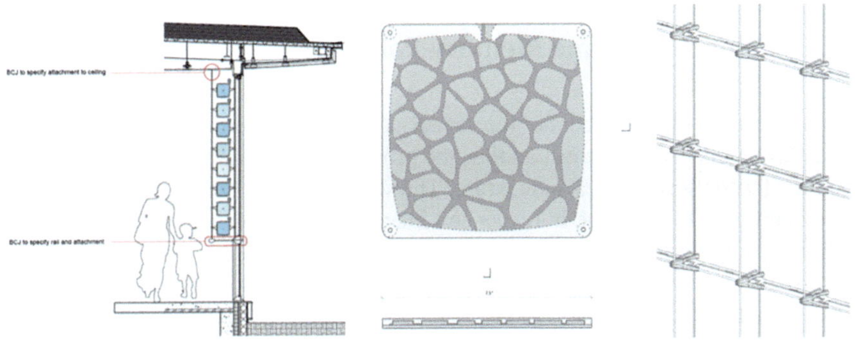

Fig. 12.3 PCM tile application. The image at left shows the placement of the PCM tile array perpendicular to the interior of a glass façade as designed for the Frick Park Environmental Center, Pittsburgh, PA. The image at right is a test for a multi-cellular PCM tile and an installation system that places the tile parallel to the glass façade

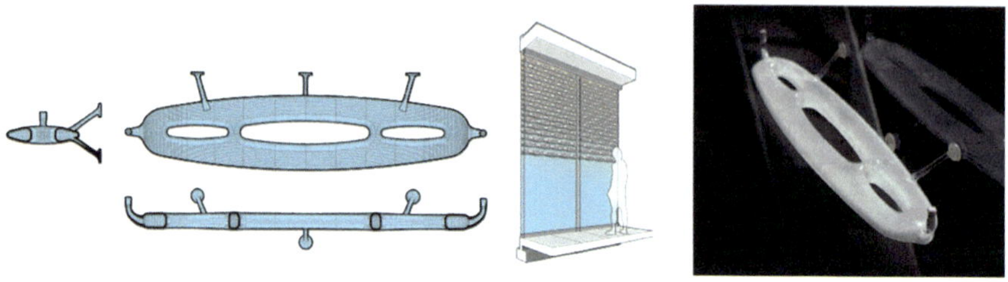

Fig. 12.4 Increasing Airflow. The image at left is a drawing for a PCM filled container designed to increase airflow when arranged in a field (center image) on the interior of a glass façade. The image at right shows a prototype module filled with frozen soy oil based PCM

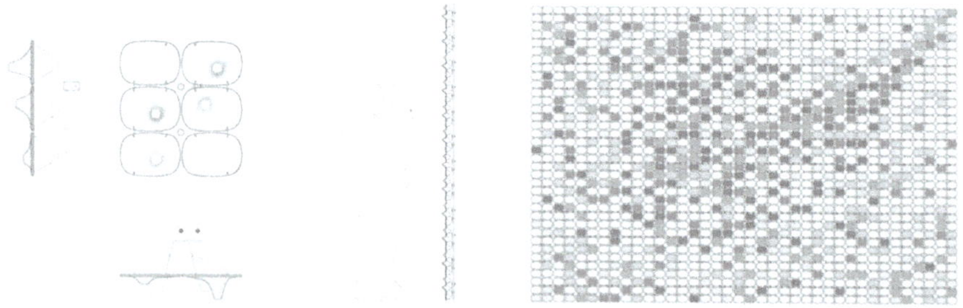

Fig. 12.5 PCM Tile Array. The drawing at left shows a plan and elevation of a PCM containment system that clips onto vertical cables. Three container types were designed with different surface curvature and volumes. The image at right shows the PCM wall system with tiles in different stages of their freeze/melt cycle due to volumetric difference of the tile. The thermal operation of the system is registered in the patterns of translucency

12.5 Project 2_ Shape Memory Polymers

Continuing the intent to enable the façade to adjust to thermal fluctuation in the local environment, this project applies the properties of shape memory polymers to vary the façade surface area and texture. The shape memory polymer tile developed, predictably varies its shape in response to heat and applied air pressure. Operating as a field, the tiles form an articulated surface designed to reduce the transmission of solar radiation through adaptive shading. The tile system is designed to respond to multiple solar insolation states rather than a single optimized state. A secondary goal is to develop a computational design tool capable of determining the local mechanical actuation and thermal activation mechanisms to morph the tile into a shape that maximizes the desired solar-tile interaction with minimal total morphing energy cost. Initial physical models are presented that incorporate knowledge exchange between partners at the University of Pittsburgh that specialize in applied computational mechanics, researchers from the University of Dayton Research Institute with expertise in materials science, and a team from the Cal Poly Department of Architecture that specializes in design process.

12.5.1 Shape Memory Effect

Shape memory refers to a class of materials that can be deformed and then recover to its previous shape when subject to stimuli. Shape memory effect was coined from the study of alloys and was first observed in the 1930's. Shape memory research with synthetic resins dates to 1941 and is attributed to a patent issued to Vernon et al. (1941) who developed a thermoplastic that exhibited elastic memory when heated. Vernon's discovery introduced a body of research into shape memory materials and shape memory effects. Most shape memory research has been with metals, known as shape memory alloys (SMAs) and shape memory polymers, although less studied. A wide variety of organic materials such as wood (Zhang et al., 2021), hair (Wortmann et al., 2021), and collagen also exhibit shape memory properties. Among the shape memory materials under development, SMPs have characteristics that are particularly applicable to building technology and differ from their alloy counterparts as they have a lower density, exert less force upon recovery, and can undergo significantly higher strains. Polymers can be designed to respond to a range of stimuli, including, light, electricity, magnetism, moisture, and ph, but the most studied stimulus is heat. Other desirable properties of SMPs include that they are programmable, exhibit variable stiffness, undergo large recoverable deformation without fatigue damage, and require minimal actuation force to change shape. They also possess good shape fixity and have a relatively low time to recover to their programmed shape (Kang et al., 2018).

12.5.2 Temperature Activated Shape Memory Polymers

The SMP used in this study is a temperature activated polyurethane polymer that can be programmed to attain a single, yet reversible, programmed shape. Above its glass transition temperature (Tg), the polymer softens and can be easily deformed. If cooled, the material stiffens, holding the deformed shape. Upon reheating above the Tg, the material will assume it's programmed shape which is set into the material upon fabrication. The programmed memory shape is achieved either through casting or printing, for example, if a tile with a curved surface is printed, and deformed, the tile will recover to the printed shape upon heating. In our experiments the polymer reached a limit and could undergo the strain of approximately 200% before rupturing, which set a range of maximum surface areas to work within. Temperature activated polymers were used in this study due to their availability and relative ease of fabrication technique. To deform the tiles, heat was applied externally with a heat gun, a significant source of external energy. If the tiles were intended for production, a more ideal SMP would be a light activated polymer which has been shown to use far less energy for activation. Another possibility is to tune the Tg of the polymer more closely to the ambient air temperatures hot environments, thus decreasing the energy cost of activation. In some design iterations, dark mass was applied to the polymer surface to concentrate heat in targeted areas. And further options include

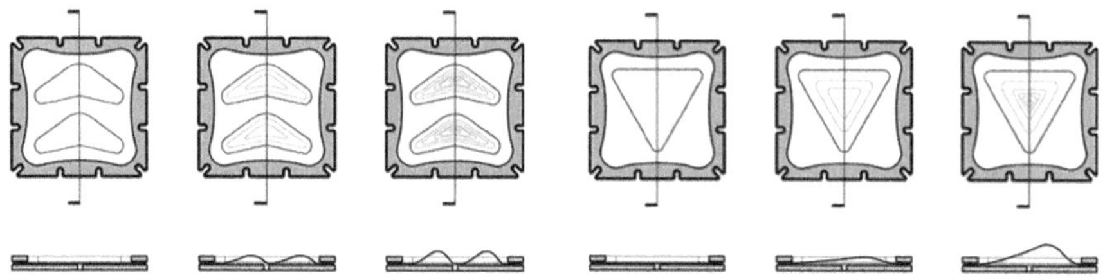

Fig. 12.6 Composite tiles. Variable stiffness composite tiles made from shape memory polymer and ABS plastic showing sequential actuation. Tiles were heated just above the glass transition temperature with constant air pressure. The lower series of drawings indicate the deformed tile shape

the application of multi-stimuli-responsive shape memory polymers programmed to respond to both heat and light.

Two fabrication techniques were used to better understand the dynamic qualities of SMP's, with an end goal of fabrication consistency and gaining predictive control of the deformed state of the tile. The first series of tiles (Fig. 12.6) were cast from a two-part resin with a Tg of 25C with materials purchased from SMP Technologies. They were combined with a stiffer inner material to generate a variable stiffness tile intended to influence overall deformation. The composite tile exhibits variable stiffness as the SMP has a lower elastic modulus than the printed ABS. This effect is pronounced when the tile is heated above the Tg and the polymer enters the plastic state.

The second series of tiles (Fig. 12.7) were made from SMP filament with a Tg of 55C and printed with variable thickness along the cross-section. The tile composed in successive layers oriented 45 degrees from the previous layer forming a cross-laminated tile. Three layers form a 0.06″ base layer and three more layers form a 1.2″ region noted by the shaded area in the drawing and in the more deformed area in the physical model. The thinner regions reach Tg more quickly and thus deformed earlier, exhibiting greater deformation than the thicker region. This experiment gave control over asymmetrical shape generation with even surface heating and constant air pressure. Tiles were arrayed into a larger matrix and air pressure was regulated via a microcontroller to a series of four tiles (Fig. 12.8). Pressure was controlled with a temperature sensor that released air to the tiles when they reach their transition temperature. As noted previously, there is a lot of kits associated with the current experimental setup and the end goal is to rely less on mechanical sensing and actuation devices and more on the inherent responsive properties of the materials.

12.5.3 Applications and Issues

The proposed application is an external shading device activated by air pressure when the Tg is reached. The intended source of heat ideally comes from the sun,

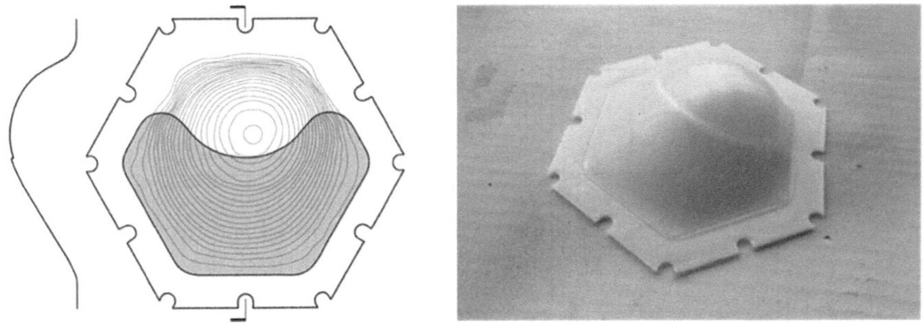

Fig. 12.7 Variable thickness tile. At left is a section elevation drawing of a variable thickness shape memory polymer tile with the shaded region thicker than base thickness. At right is a deformed variable thickness printed tile. Varying the tile thickness gave the team insight into control over tile surface curvature

necessitating a Tg targeted towards specific thermal microclimates. It is anticipated that through selective application of dark mass, the SMP would reach the transition temperature, become plastic and inflate, providing shade to the façade surface. Like shape memory alloys, SMPs require a bias force to reset them to their original preprogrammed shape. This condition requires additional energy input, increasing the cost of operating the system. Energetically, reheating is not viable, and therefore the preferred tile would be composed of light activated polymers.

While two-way shape memory materials are promising in laboratory studies, materials with these properties are not yet commercially available. As described above, the polymer used in this study is a 'one-way' system where a shape is thermally set. Current research teams have produced polymers that can be programmed to attain multiple shape shifting states (Chen et al., 2010). Application of a material with multiple programmed shapes would potentially increase the fidelity of dynamic facades constructed from SMPs and significantly reduce or eliminate the complexity of the actuation system. It is speculated that a façade composed of materials with multiple shape shifting states could operate solely from thermal input from the sun.

12.6 Conclusion

For decades, the directive of environmental control systems in buildings has been directed at human comfort and convenience, and convenience generally meant rendering invisible the systems responsible for internal climate control. An arguable point behind the work presented is that separating humans from the technology that produces comfort has exacerbated our distance from nature, dulled the senses, and by extension, diminished the imagination. This argument supports an expanded definition of building performance to include cognitive, emotive, and communicative aspects of design. This is an old idea, as buildings have long had the ability to communicate to their people through their form, and in some cases, their operation.

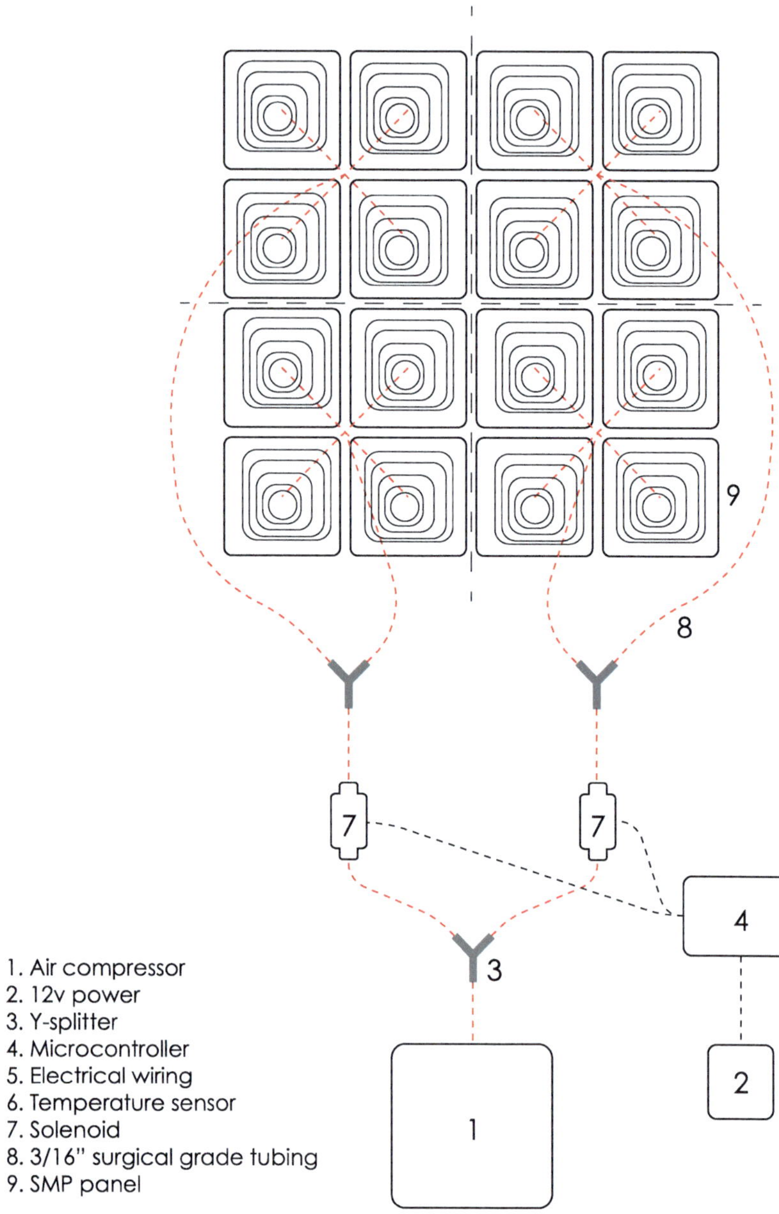

Fig. 12.8 Schematic of pneumatic inflation and control system. Schematic drawing of a 4 × 4 tile layout. Compressed air is distributed to the SMP tiles once their transition temperature is reached, causing the tile to deform. Upon cooling, the tile retains the deformed shape, and the air supply is no longer needed to retain the shape. Upon reheating to the transition temperature, the tile recovers to its' programmed shape

Ornamentation is often viewed as gratuitous, though it can contribute meaning to a building through its symbolism and impart delight, wonder, and sometimes reveal to us that we are part of a larger story. Programmable materials offer a step towards active and functionalized ornamentation that visually and thermally convey their response to local environmental changes. It seems valuable to consider the semiotics of emerging adaptive building technologies as the activated façade can become the literal performance of the building, calling into question the way in which we visually and thermally relate to architecture. The projects presented are a step toward the work of a confluence of artists, engineers, material scientists, and architects that no longer view architecture as a static backdrop which we live our lives against, but a dynamic and communicative condition that continually engages us with environmental flux.

To suggest that architecture alone can change the way one thinks or acts, is overreaching, but the potential for responsive design technology to inspire one to change their associations and relationships with the natural world is possible. Herbert Marcuse argues that "Art cannot change the world, but it can contribute to changing the consciousness ... of the men and women who could change the world." (Marcuse, 1979) Taking a conceptual leap, we could apply Marcuse's thinking towards aesthetics and ecology, and propose that awareness of environmental fluctuations, through the filter of responsive architectural systems, may contribute to a shift in our attitude towards the energy we use and the world we inhabit.

Acknowledgements Thanks to Mark Whittock, Kay Bromley, Craig Kimball, Diana Rodriguez, Nick Taylor, Trevor Larsen, and Professor Jeff Ponitz, Professor John Brigham, Richard Beblo, and Robert Zupan.

Portions of this work were funded by an AIA Upjohn grant and the National Science Foundation.

References

Boubekri, M., Lee, J., MacNaughton, P., Woo, M., Schuyler, L., Tinianov, B., & Satish, U. (2020). The impact of optimized daylight and views on the sleep duration and cognitive performance of office workers. *International Journal of Environmental Research and Public Health, 17*(9), 3219.

Chen, S., Hu, J., & Zhuo, H. (2010). Properties and mechanism of two-way shape memory polyurethane composites. *Composites Science and Technology, 70*(10), 1437–1443.

Clifford, D. T., & Yao, S. C., Carnegie Mellon University. (2017). Devices for modulation of temperature and light based on phase change materials. U.S. Patent 9,797,187.

Evans, G. W. (2003). The built environment and mental health. *Journal of Urban Health, 80*(4), 536–555.

Kang, M., Pyo, Y., young Jang, J., Park, Y., Son, Y.H., Choi, M., Wan Ha, J., Chang, Y.W., & Lee, C.S. (2018). Design of a shape memory composite (SMC) using 4D printing technology. *Sensors and Actuators A: Physical 283*, 187–195.

Marcuse, H. (1979). *Aesthetic dimension.* Macmillan International Higher Education.

Ulrich, R. S. (1984). View through a window may influence recovery from surgery. *Science, 224*(4647), 420–421.

Vernon, L. B., Vernon, H. M., & Vernon Benshoff Co. (1941). Process of manufacturing articles of thermoplastic synthetic resins. U.S. Patent 2,234,993.

Wallace, L. A.(1987). *Total exposure assessment methodology (TEAM) study: Summary and analysis,* Vol. 1. Environmental Protection Agency, Washington, DC (USA). Office of Acid Deposition, Environmental Monitoring, and Quality Assurance.

Wortmann, F. J., Jones, C., Davies, T. J., & Wortmann, G. (2021). Perm-waved human hair: A thermorheologically complex shape memory composite. *Biophysical Journal, 120*(17), 3831–3840.

Zhang, C., Chen, M., Keten, S., Derome, D., & Carmeliet, J. (2021). Towards unraveling the moisture-induced shape memory effect of wood: The role of interface mechanics revealed by upscaling atomistic to composite modeling. *NPG Asia Materials, 13*(1), 1–14.

Chapter 13
Design Research on Climate-Responsive Building Skins from Prototype and Case Study Perspectives

Zhenghao Lin and Yehao Song

Abstract As low-carbon and sustainable development has become a global trend, climate-responsive building envelopes still gain traction in the architecture world. However, relevant design and research remain too disconnected to drive further breakthroughs. Thus, this chapter explores an integrated design-research approach for climate-responsive building skins, especially for the continental climates in China. The study begins with a practical case in China's Cold Zone to demonstrate how architects consider climate responsiveness when designing building skins and reveal the challenges the static structures might face in responding to external dynamic climates. A research prototype of climate-responsive skin was extracted from the Cold Zone case and materialized in a full-scale test platform. A series of comparative experiments are performed to identify the optimal solution for each of the four key design features and their optimal combinations for summer and winter, respectively. The thermal performance impact of each design feature and the energy-saving behaviors of the seasonal optimal configurations are also evaluated quantitatively. In order to integrate the highly differential summer and winter configurations for maximized climate responsiveness, a novel dynamic skin prototype with rotatable triangular blades is further presented. Overall, the study aims to bridge the gap between the research and design of climate-responsive building skins, thereby providing a reference for their application in similar climates.

Keywords Climate responsiveness · Building skin · China · Continental climates

Z. Lin
School of Architecture, South China University of Technology, Guangzhou, China
e-mail: linzhenghao@scut.edu.cn

Y. Song (✉)
School of Architecture, Tsinghua University, Beijing, China
e-mail: ieohsong@tsinghua.edu.cn

J. Wang et al. (eds.), *Advanced Materials in Smart Building Skins for Sustainability*,
https://doi.org/10.1007/978-3-031-09695-2_13

13.1 Climate-Responsive Building Skins

As one of the core topics in architecture, climate responsiveness remains relevant in the context of the global endeavor to cope with climate change and pursue sustainable development (Papadikis et al., 2019). Discussions about the climate responsiveness of buildings involve the impact of climatic elements on indoor environments and comfort levels, requirements for structural safety and material durability, and the possibilities of adjusting user behaviors responding to the environment (Hao, 2016). Among all building components, envelopes play an essential part in allowing buildings to respond to the changing climate. It is because they are the main interface separating the indoor and outdoor environment, a shield or buffer against light, heat, humidity, wind, rain, snow and other climatic factors, and an important carrier for human factors engineering (HFE). In this context, such concepts as "climate responsive skins", "bioclimatic skins", "smart skins" and "adaptive building envelopes" emerge. Currently, research on the climate responsiveness of transparent envelopes represented by double-skin facades has been evolving due to the springing up of high-rise office buildings, while opaque envelopes are relatively less studied (Ibanez-Puy et al., 2017). For most building types, however, the latter tends to cover a better part of the surface area, and their role in climate regulation cannot be ignored. Therefore, this paper selects opaque envelopes as the main object of design and research.

In terms of design practice, architects in different parts of the world have been exploring ways to make building skins more climate-responsive. Considering the tropical climate of Africa, Jean Prouvé equipped his prototype Maisons Tropicale with continuous, adjustable louvered aluminum sunscreens to reflect the intense sunlight, as well as the double roof structure and central roof vent to remove heat quickly. Built in the Chesapeake Bay in the United States, KieranTimberlake's Loblolly House is covered with overlapping cedar boards, which serve as a "filter for rain, wind and solar radiation (Kieran & Timberlake, 2008)". In response to China's hot summer and cold winter, Shanlong Tan used semi-transparent polycarbonate materials to wrap the building at Xiuning primary school, balancing ventilation and insulation through some air vents. Facing the Spanish Mediterranean climate, H Arquitectes and DATAAE adopted polycarbonate shutters as the bioclimatic skins of the ICTA-ICP research center. Controlled by environmental sensors, the shutters automatically shut down during cold and wet weather, and open to increase ventilation when the weather is hot and dry.

In terms of academic studies on the climate responsiveness of opaque building envelopes, researchers focus more on opaque ventilated facades (OVF) and have come a long way. It is observed that most studies about OVF focused on the hot temperate climates in southern Europe and the Mediterranean region (e.g., Italy and Spain), which can be further categorized as a temperate oceanic climate (Cfb type) (Fantucci et al., 2020; Labat et al., 2012) and the hot-summer Mediterranean climate (Csa type) (Aparicio-Fernández et al., 2014; Peci Lopez et al., 2012) according to the Köppen–Geiger classification (Kottek et al., 2006). With the rapid development and global spread of OVF, the research field has gradually expanded to other climate zones like tropical and subtropical areas (Zhang & Yu, 2017; Fernandes Maciel & Carvalho, 2019; Gregorio-Atem et al., 2020) and some particular application scenarios like extremely windy climate (Mammadova et al., 2021). However, studies for continental climates (D type) with hot, humid summers and cold, dry winters, such as in the cold zones in China, are still rare.

Secondly, As a result, most current studies devoted themselves to investigating the thermal behavior of OVF in the dominant hot seasons (Marinosci et al., 2014; Stazi et al., 2014), and it is widely agreed that the shading effect of the external skin and the natural convection of the ventilated air duct can help to remove part of the heat load cross the facade, thus reducing the indoor heat gain in the summer period. Although papers about the winter behavior of OVF are relatively more minor, they have drawn more attention in recent years (De Masi et al., 2021; Peci Lopez & de Adana Santiago, 2015). Some authors proved that installing an OVF brought positive winter energy savings by preventing the outdoor cold winter or recovering the heated air in the cavity for indoor warming. However, winter investigations still acted like supplements for the summer studies due to their less significant energy-saving rates. The winter configurations of OVF often inevitably give way to their summer versions, resulting in nonoptimal performances in the cold seasons. Consequently, the annual benefits of OVF cannot be fully discovered, leading to restrictions on further popularization, especially in regions with significant seasonal variability.

Thirdly, these OVF studies are centered on such typical commercial systems as rainscreen ventilated facades (Marinosci et al., 2014) and open-joint ventilated facades (Sánchez et al., 2017). Less exploration of more skin structures has led to the failure to meet the increasingly diverse needs for design. Furthermore, the gap between practice and research is widened by the fact that architects and researchers, which are by nature two different professions, do not necessarily have shared values, expertise and work methods.

In order to bridge this gap, and with the advantage of being both architects and researchers, the authors aim to explore an integrated design-research approach to climate-responsive building skins, aiming to expand their application border to continental climates with hot, humid summer and cold, dry winter. This paper begins by introducing a practical case in the Cold Zone of China, and elaborating on how the building skin was designed to respond to the dynamic climate, thus revealing the gains and limitations of architects' efforts. On this basis, the authors extracted the opaque ventilated facade with louver cladding as the prototype, materialized it in a Cold Zone-specific test platform, and then conducted comparative experiments on its key design features to obtain the optimal configurations suited to summer and winter conditions respectively. Finally, to integrate the highly differentiated skin structures for summer and winter, the authors propose a new triangular blade prototype with a variable structure, which is expected to outperform traditional static skins and achieve maximum climate responsiveness.

13.2 A Case Study in Continental Climate

13.2.1 Project Description and Climatic Features

Located in Shunyi District, Beijing, CSC Zero-carbon Pavilion (CSC Pavilion), serves as a small community center. It is also a Nearly Zero Energy Building with BREEAM Outstanding and LEED Platinum certifications. The building is composed of three b-shaped building units around the sunken courtyard in the center, which function as the meeting room, exhibition hall and gym respectively, with a total construction area of about 157 square meters (Fig. 13.1).

The CSC Pavilion was developed to respond to the continental climates in China, characterized by hot, humid summers and cold, dry winters. In determining the main building structure, the design team used the prefabricated laminated timber structure and the light-timber framed insulated envelopes, aiming to reduce the embodied carbon in material uses and to restrict the environmental impacts of on-site construction or the potential demolishing in the future. Moreover, to improve the energy efficiency during the operation stage, the CSC pavilion was further designated as an integrated platform for various passive architectural design strategies and active energy-saving building technologies. As one of the promising passive solutions, the concept of climate-responsive building skin was introduced from the very early design stage and successfully implemented in the construction phase, especially for the opaque parts.

Fig. 13.1 The meeting and exhibition units of CSC Pavilion

Beijing is a typical city located in China's Cold Zone, which has a humid continental climate, or Dwa type (Fig. 13.2). In this zone, temperatures vary greatly in different seasons; it is hot and rainy in summer, and cold and dry in winter. The average outdoor temperatures in the two seasons are 27 and −2.9 °C respectively, and the prevailing winds are S and N respectively. This shows that responding to the Cold Zone's climate is more complex and challenging than other climates that were only dominated by hot or cold seasons (Fig. 13.3).

Moreover, Beijing is located in the Normal Zone regarding solar radiation resources in China, with cumulative solar radiation averaging around 1550 and

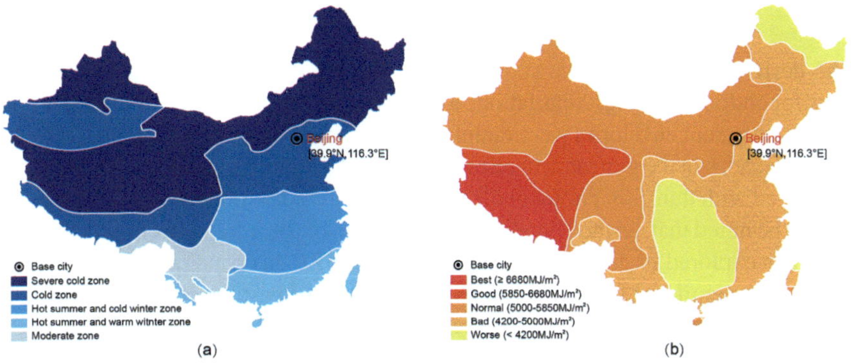

Fig. 13.2 **a** Climate zones for building thermal design and **b** Solar resource zoning of China

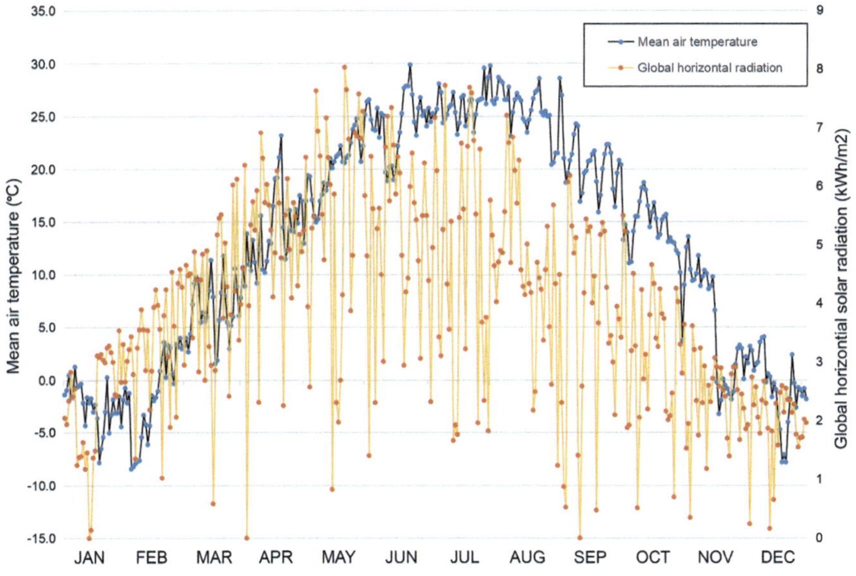

Fig. 13.3 Climatic features of a typical city in the Cold Zone (Beijing)

710 MJ in summer and winter, respectively. However, this relatively intense solar radiation will differently impact the building performance according to the season. During the summer, it usually causes the penetration of undesired solar heat into buildings, bringing indoor overheating risks; during the winter, it turns to serve as a positive heat source for reducing heating demands. As the interface directly contacting with the external climatic factors, the building skin's dynamic regulation capacity for solar radiation becomes a challenging issue.

13.2.2 Design of the Climate-Responsive Skin

The design team defined the south facade of CSC Pavilion as the primary interface for indoor lighting, ventilation, and visual access, meaning that the full- or semi-transparent envelopes were mainly adopted, such as the highly insulated single glazed facades in the gym and meeting units, and the double-glazed ventilated facade integrated with thin-film photovoltaic modules in the exhibition unit. Nevertheless, it should be noted that, as the climatic responsiveness of these transparent-type facades had been explored in the previous studies (Lin, 2018), this paper mainly focuses on the opaque ventilated facade, which was applied to the east, west and north facades of each unit, and a small number of such modules were used on the south side of the gym unit.

In more detail, this OVF system consists of three layers. The external skin is formed by carbonized timber battens and anchored to the internal base wall through a supporting steel frame. The internal light-wood framed wall provides basic thermal insulation, water- and air-tightness. These two layers are separated by a 150 mm-width air cavity, serving as a buffer zone or a ventilation duct.

Considering the inherent difficulties of static skin structures in responding to the changing external climate, the ventilation state of the OVF cavity was changed by adjusting the timber batten's openness for different facade orientations to enable a response to both summer and winter climates in the Cold Zone. For the east, west and south facing facades exposed to direct solar radiation, the utilization of a closed cavity will cause overheating inside and provide cavity temperature even higher than the ambient one, thus bringing extra cooling loads in summer. In this regard, the team learned from the working principle of ventilated facades, opening multiple groups of wooden battens at the bottom and top of the external skin to form air inlets and outlets. By introducing the solar chimney effect, the vertical airflow driven by the thermal buoyancy was successfully established and exhausted the undesired hot air in the cavity. The timber louvers at the vents were opened by rotating at particular angles (40–80°), guaranteeing sufficient areas for air exchange between the cavity and the outdoor environment. As shown in Fig. 13.4, the varieties in angle, number, and position of the rotated timber batten provided more aesthetic possibilities for the whole facade system. It even created icon shapes representing different building functions, like the beating-heart icon beside the entrance of the gym unit (Fig. 13.5).

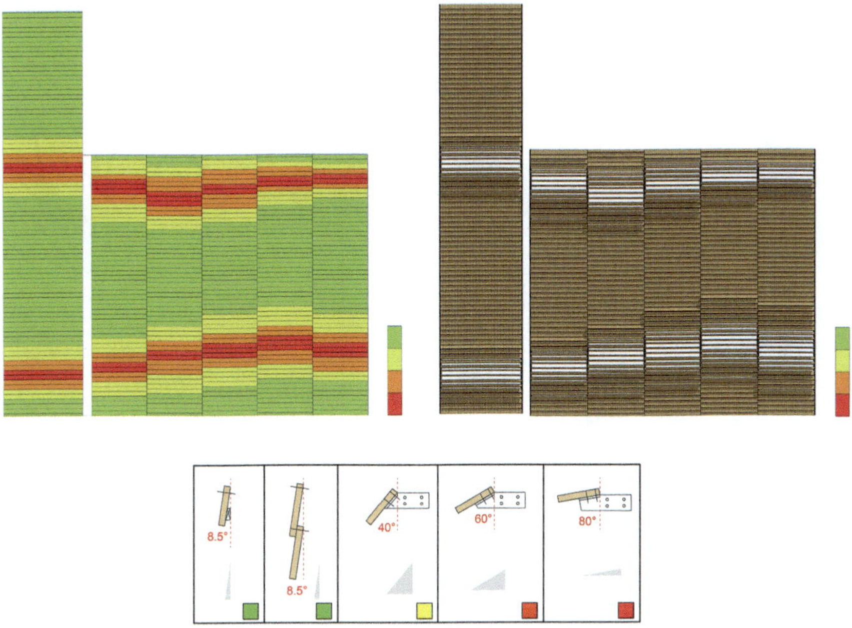

Fig. 13.4 Patterns for CSC Pavilion's timber louvers

However, it should be noted that, further evaluations for the thermal and energy impact of the ventilated facades in the cold winter are also essential.

For the north facing facade, free from direct solar radiation and acts as the primary windward side in winter, a thermal buffer formed by the closed cavity helps prevent heat loss through the internal walls. Thus, in this orientation, the timber louvers are no longer opened but are closely connected to create a continuous interface.

In terms of the physical implementation, the external skin of CSC Pavilion, with a total area of 369 square meters, was divided into 64 prefabricated modules (Fig. 13.6). These modules were supported by a steel frame while the carbonized timber battens were attached to the frame through wood angle adjusters.

For time-saving, the team initially planned to prefabricate the whole skin module in the factory. However, the 1:1 mock-up showed that the regular shape of the modules would be broken when integrated with the timber battens, especially the rotated ones. This not only occupied more transportation resources but was also more prone to damage during the transportation and installation process due to its more complex and fragile structure. Therefore, the architects separated the timber battens, which were carbonized and trimmed only in the factory, from the steel frame, which was prefabricated and integrated along with the angle adjusters. After the prefabricated steel frame was hoisted and fixed on-site, skilled workers would nail the timber battens onto the adjusters in a very short time (Fig. 13.7). This "moderate prefabrication" strategy worked as it capitalized on the complementary advantages of factory

(a) (b)

Fig. 13.5 The western facades of the **a** exhibition and **b** gym units of CSC Pavilion

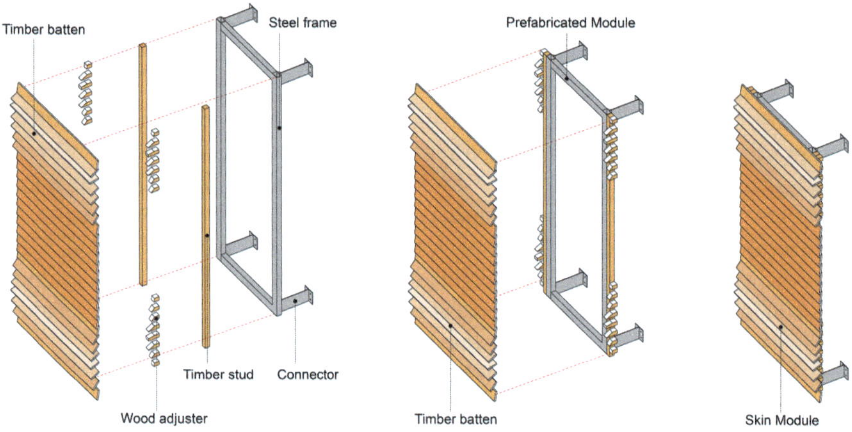

Fig. 13.6 Structure of CSC Pavilion's skin module

prefabrication and on-site assembly. It was a cost-efficient solution that it took only five workers one week to complete the whole installation.

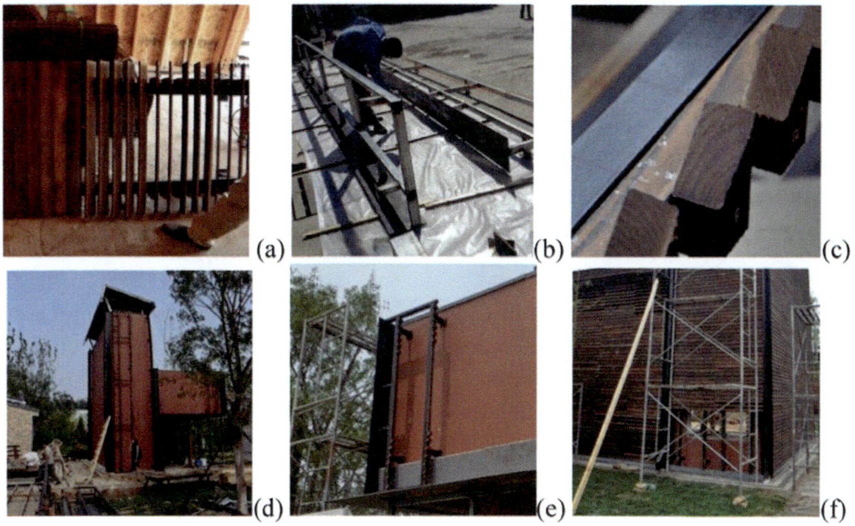

Fig. 13.7 Prefabrication and assembly of the skin modules **a** mock-up module; **b** steel frame; **c** wood adjuster; **d–e** installation of steel frame with adjusters; **f** installation of timber battens

13.2.3 Limitations and Challenges

The practice of the CSC Pavilion shows how Chinese architects make building skins more responsive to the continental climate. Their specific design strategies, construction methods, and performance presentations will provide references for future practice in similar climates. However, exploring climate-responsive skins simply based on practical cases has its limitations.

Firstly, the fast construction pace in China leaves construction teams little time for systematic prototype studies. Their understanding of skin-specific climate responsiveness comes primarily from their knowledge base of thermal engineering and limited meteorological insight.

Secondly, such practical factors as aesthetics, cost, construction time, and the owner's preference often come into play in the design process. The preference for static structures in most projects, for example, results in a skin system that can only respond to the ever-changing external climate in a limited or moderate manner. Even if the testing process is carried out after construction, it will be, more often than not, reduced to the verification of the original design plan due to temporal and spatial constraints: it is difficult to conduct long-term comparative experiments with multiple variables to discover how the building skin responds to the external climate or to reach specific strategies for structure optimization.

Based on this, the authors recognize the necessity of shifting the exploration of climate-responsive skins from sole reliance on case design toward integration with prototype research. Fortunately, the authors' team, which comprises both researcher and architect members, is well-positioned to explore an integrated design-research approach to climate-responsive building skins.

13.3 Prototype Research

13.3.1 Prototype Extraction

Based on the previous practice in China and the analysis of other existing cases, the prototype of opaque climate-responsive skin is not complex and mainly consists of three parts: the internal wall, air cavity and external skin. The innermost wall is responsible for fundamental functions such as thermal and sound insulation, fire prevention, waterproofing and air-tightness, serving as the ultimate barrier between the indoor and outdoor environments. The intermediate cavity acts as a buffer or coupling zone for different environmental factors (heat, moisture, wind, solar radiation, etc.) between the internal wall and the external skin, usually with structural components connecting the two. And the external skin, which may be composed of multiple materials, is directly exposed to the ambient climate, severing for light and solar radiation filtration, wind and rain prevention, thermal and sound insulation, and determines the overall visual image of the building. However, when it comes to the specific configuration for each layer, a multitude of variables could emerge: (1) the material, color, smoothness, opening ratio, inclination, the opening position of the external skin; (2) the geometry of the air cavity, the connection with the indoor and outdoor environments, the internal division, and the application of mechanical ventilation devices; (3) thermal resistance and inertia of the internal wall, as well as its surface material, color and roughness settings. The variation and combination of different parameters offer a wealth of possibilities for the forms and climate responsiveness of the opaque building envelope with a double-layer structure.

On this basis, the authors extracted a simplified skin prototype from CSC Pavilion's timber-batten facade for the following experimental investigation. Its reasons can be attributed to three: First, as one of the general facade techniques, the batten skin shares a worldwide application, making this study more practical and typical; Second, the practice of CSC Pavilion can provide direct guidance and reference on implementing prototype skin physically; Finally, few performance investigations of batten skin are conducted under the typical continental climates, which determines the necessity of prototype exploration.

To simplify the research boundary conditions, the external battens of the prototype skin are no longer rotated but uniformly set parallel to the internal wall. For this prototype, four key design features are further extracted by a comprehensive analysis of previous studies and relevant practices: (1) the color and (2) the joint opening ratio of the external skin; (3) the air vents openness and (4) the cavity depth. From the performance perspective, the first two features are more related to solar radiation control, while the latter two are associated with cavity ventilation regulation. From the design perspective, these four features have the most significant impact on the overall form of the skin, and are thus the focus of architects' attention.

13.3.2 Prototype Experiments

Based on the extracted skin prototype and its key design features, the research team planned to conduct a systematic study of the climate responsiveness of the prototype skin under real climate, so as to reach its optimal design and operation strategies. Supported by Tsinghua University and Tsinghua Redbud Innovation Institute, the team built a full-scale test platform in Tianjin (Fig. 13.8), a city only 90 km away from Beijing with a very similar climate. Consisted of two identical (3000 × 3000 × 3200 mm) and non-interfering (distance >7 m) test rigs, the platform enables comparative experiments and analysis of control variables under the same external climatic conditions. On this basis, the research team materialized the extracted skin prototype as follows:

The eastern and western walls of the test rigs are constructed of a wooden frame with a depth of 140 mm and filled with foam polyurethane insulation. They are covered with oriented strand boards with a thickness of 9 mm, as well as white breathing membranes. With a comprehensive heat transfer coefficient of about 0.22 W/m^2·K, these walls provide the basic thermal insulation.

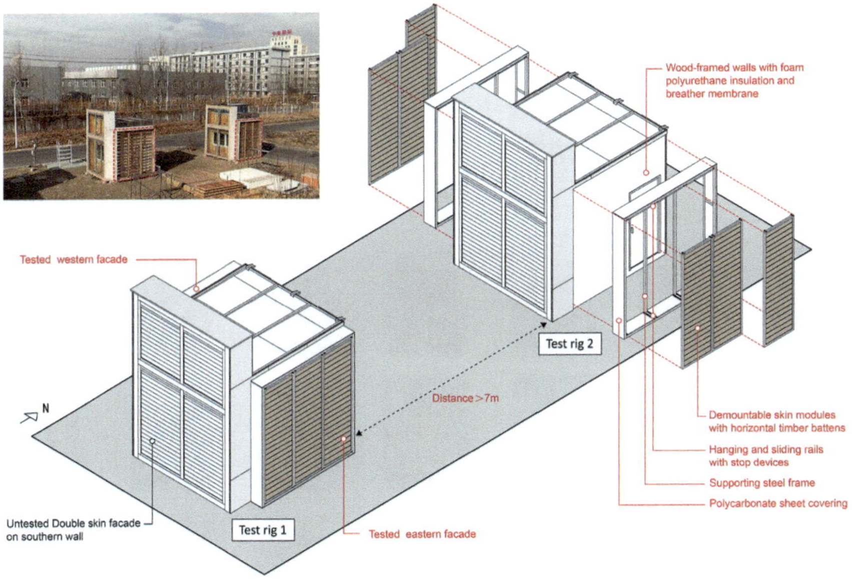

Fig. 13.8 Comparative test platform & structure of the timber batten skin

The external skin consists of three demountable skin modules of the same size (3000 × 920 × 50 mm), which, after being mounted to the rectangular steel frame on the outer side of the internal wall, can be slid and stopped freely perpendicular to the internal wall by means of hanging rails with stop devices, thus creating cavities of different depths ranging from 50 to 350 mm. It is worth noting that the other faces of the steel frame are covered with polycarbonate sheets to avoid unorganized connectivity of the cavity with the ambient air.

Each skin module consists of uniform horizontal wooden battens (820 × 125 × 10 mm) combined with a timber frame. Three design features can be adjusted by changing the patterns and assembly methods of timber battens: The joint opening ratio of the skin can be controlled (0/6.25/12.5/25/50%) by adjusting the distance between adjacent battens; The air vents openness can be changed by retaining or removing battens at the top and bottom; The external cladding color (white or wood) can also be altered by wearing a different coat.

Based the test platform and its variable facades, the research team conducted a series of comparative experiments on the skin prototype under summer and winter conditions (Fig. 13.9).

Firstly, in-field experiments were performed to measure the outer and inner surface temperatures of the internal wall. With the measurement of the internal walls' thermal conductance (C_{wall}), their penetrating heat flux were successfully calculated and transformed into the daily cumulative thermal gain (E_{gain}) and thermal loss (E_{loss}) values for each tested configuration. Through comparing these thermal performance metrics, the optimal solution for respective design features were determined.

Secondly, the impact of different design features on the envelope's thermal performance was compared. As shown in Fig. 13.10, a normalized analysis involving all

Fig. 13.9 Various skin configurations under experiments

skin configurations was conducted using the E_{gain} value of Facade B and the E_{loss} value of Facade A* as the summer and winter thermal benchmarks. In the summer tests, the thermal influence of four design features on resisting heat gain were ranked as follows: joint opening ratio (30.6%) > external cladding color (17%) > air vents openness (7.8%) > cavity depth (4.8%). It is evident that the former two design features are directly linked to the design of external skins, with the joint opening ratio of 6.25% and bright color being the optimal solutions. In other words, it is suggested to maximize the shading effect of the external skin and reduce its solar radiation absorption. The latter two features are tied to the cavity ventilation state, with a cavity depth of 200 mm and additional top and bottom vents promoting vertical airflow as the optimal solutions. Therefore, for the skin to respond to summer climates, the design team should prioritize the resistance to direct solar radiation and optimize the structure to facilitate thermal exchange between the cavity and the external environment. In the winter tests, the thermal influence of four design features on resisting heat loss was ranked as follows: joint opening ratio (19.8%) > cavity depth (12.5%) > cladding color (2.2%). Among them, 0% is the optimal solution for joint opening ratio, aiming to minimize the contact between the external cold air and the internal wall, no additional air inlets and outlets on the skin are needed. In addition, a medium depth for the sealed cavity is best for thermal insulation. Moreover, in winter, the adoption of external skins with dark color and a high absorption rate in solar-rich areas will slightly improve thermal performance. Therefore, under winter conditions, the closed cavity should be used on a priority basis to create an insulation buffer.

Thirdly, the best solutions for four key design features were systematically integrated to obtain optimal configurations responding to summer and winter climates. These optimal configurations with double-layer structures were further compared with the conventional single-layer facade to verify their performance advantage and connect to the actual application scenario. It should be noted that these tests were conducted with the air conditioners activated, and their daily energy consumption data were collected by power meter plugs. Based on the energy use comparison, it is found that the optimal skin configurations, when applied only to the east and west facades of the test rigs, can bring a considerable saving of cooling energy (11.4%) in summer and heating energy (6.7%) in winter.

13.3.3 Prototype Integration

The two optimal combinations discussed above have proven effective in addressing the hot summer and cold winter in China's Cold Zone, improving climate responsiveness from the energy perspective. However, its construction is markedly different in summer and winter. If the skin system is restricted to static structures, it can only be designed to respond to the region's prevailing climate. For example, although the energy-saving ratio of the winter-specific skin prototype is relatively low, its absolute energy saving is 2.3 times higher than that of the summer solution. The optimal combination for winter use should thus be prioritized in optimizing the skin,

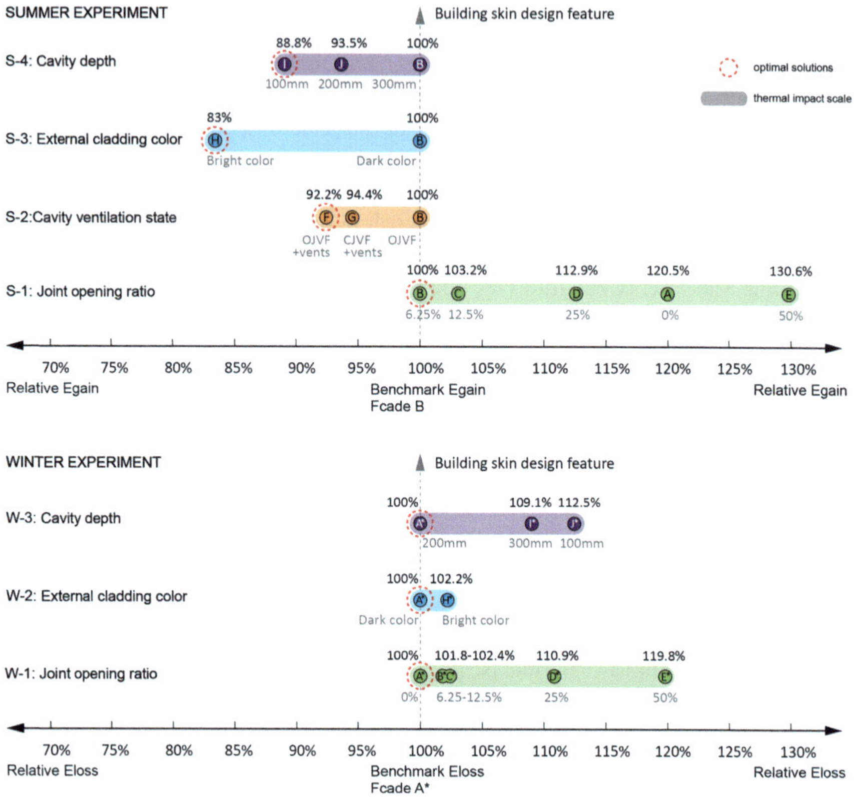

Fig. 13.10 Impact of key design features on thermal performance and respective optimal solutions

which, however, will also lead to weakened responsiveness to the summer climate. Therefore, introducing variable or dynamic structures might be an effective alternative. Nevertheless, new technical difficulties could arise, including the need to (1) integrate all design features on the same skin system; (2) make each feature independently controllable; (3) build a variable structure that is stable and easy to operate; and (4) meet basic performance requirements such as waterproofing, rain proofing, airtightness, and structural durability.

The conventional slatted wooden skins cannot meet the above objectives due to their differentiated structural requirements for summer and winter applications. Inspired by the Roman architect Vitruvius's Periaktoi, a kind of wooden prism device designed to rapidly change theatre scenes in ancient Greek and Rome (Fig. 13.11), the authors propose a variable triangular blade skin system for use in different climate scenes. Unlike Vitruvius's design, these triangular blades are installed horizontally and arranged from top to bottom to form a complete climatic interface. Their inner sides and the internal wall create the cavity space as a climatic buffer. At the same time, each side of the blade can be made of different materials, and the axial rotation

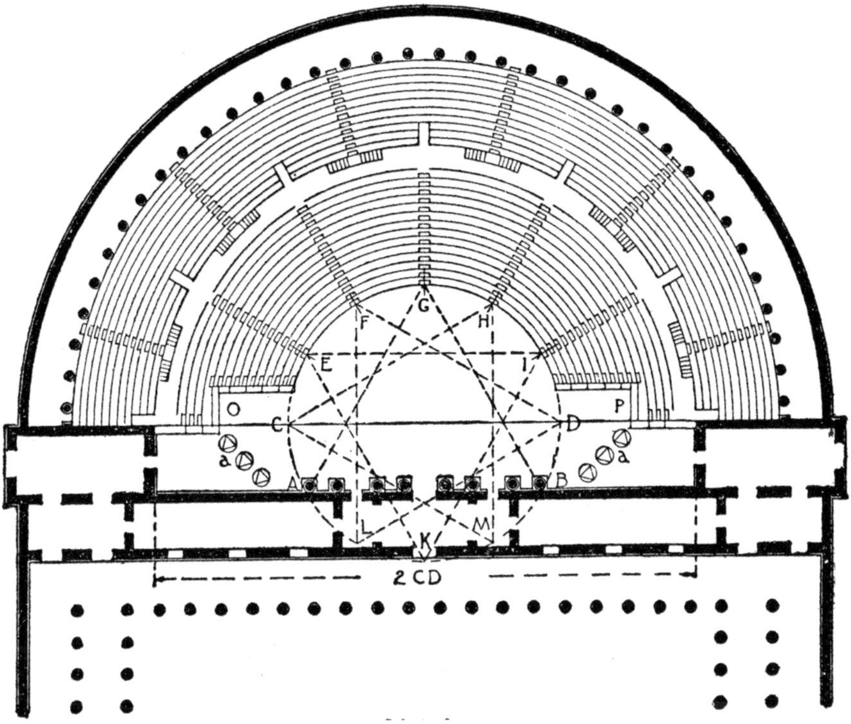

Fig. 13.11 Plan of a roman theater using Periaktoi (Harrison et al., 2013)

of the blade allows the climatic interface to be freely switched, thus making the skin much more climate-responsive and diverse in its appearance. Furthermore, due to the unique cross-section of the triangular blades, when all of them rotate together, changes in their distance will bring different joint opening ratios.

Based on this triangular blade prototype, the research team integrated the summer and winter optimal skin configurations obtained in Sect. 13.3.2. As shown in Fig. 13.12, the three sides (Sa, Sb, Sc) of a triangular blade can be made of dark-colored wood, bright-colored wood and heat reflective materials (e.g., aluminum foil) respectively. Different blades are connected to each other by a hinged drive rod, and synchronously rotated from 0 to 90° around the axis by an electric push rod. And their position is controlled by a baffle fixed at the junction of the Sa and Sc side.

In winter, the Sc faces of the blades are tightly connected to form an interface parallel to the internal wall. In addition to a sealed cavity, the heat-reflective materials on the Sc faces will also enhance the insulation performance of the skin. At the same time, the dark-color material on the Sa side is also rotated to a position that directly faces the sunlight, thus absorbing heat from solar radiation to further warm up the air cavity. In summer, the blades can be rotated 90° inward along the axis. In this case, their light-colored sides turn to face the sunlight and help reduce heat gain by directly reflecting the intense sunlight. The mutual shading effect of the

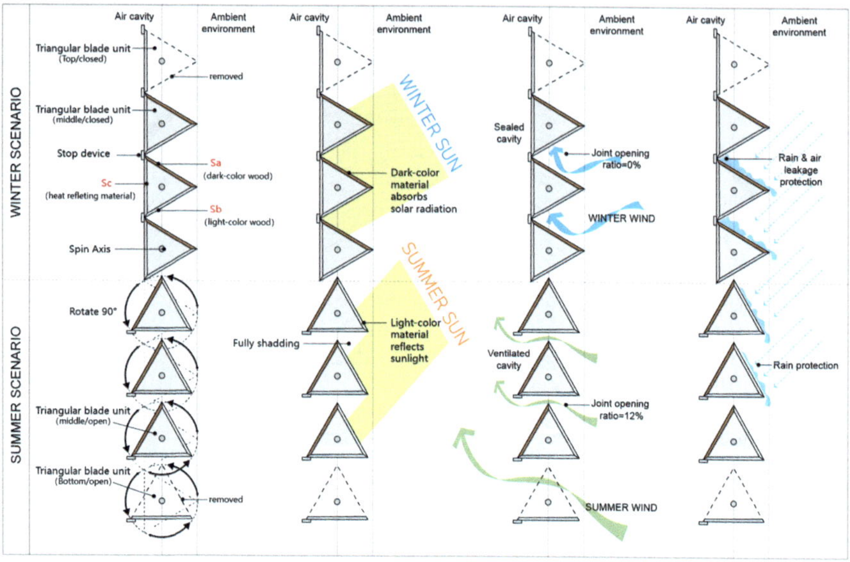

Fig. 13.12 Structure and climate responsiveness of triangular blades

adjacent blades can also prevent solar radiation from entering the cavity. The gaps between the rotated blades also connect the cavity and the ambient air, creating an open-joint ventilated facade (with a joint opening ratio of about 12%). Additionally, to strengthen buoyancy-driven ventilation within the cavity, it is also possible to remove the Sa and Sb sides of the top and bottom blades, leaving the remaining Sc sides to ensure the closure of the cavity in winter and work as vents in summer. Overall, the above triangular blades enable changes in the joint opening ratio, external cladding color and cavity ventilation state under different climatic conditions. Moreover, in winter, the limit plate between the blades equips the skin with an airtight and watertight structure; in summer, its equilateral triangular section forms a rainproof structure, thus ensuring the weather resistance of the prototype. As to the cavity size, it has been proved that the same 200 mm-deep cavity simultaneously brings the best thermal performance in summer and winter, which this triangular blade skin system can directly adopt.

Therefore, the final summer and winter modes for this triangular blade skin were obtained and presented in Fig. 13.13.

13.4 Conclusion

As China pledges to peak carbon emissions by 2030 and achieve carbon neutrality by 2060 (Jia & Lin, 2021), its building sector has contributed to the national decarbonization efforts through active reduction of energy consumption in the operational

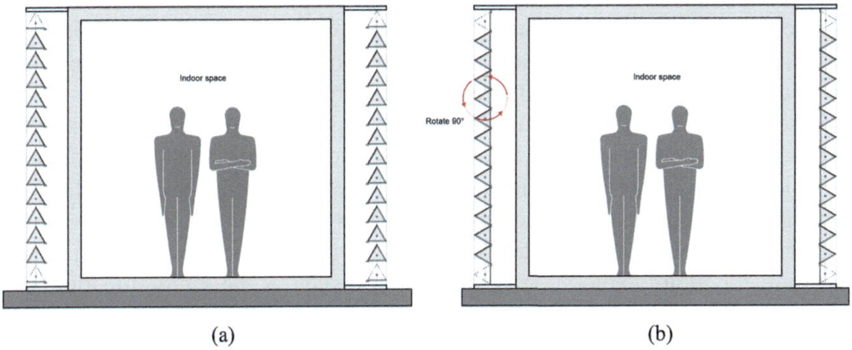

Fig. 13.13 The **a** summer and **b** winter modes of the triangular blade skin

phase. In this process, passive design and technology solutions, including climate-responsive skins, have an essential role to play. However, while both architects and scientists have worked to explore the climate responsiveness of building skins, the differences in their approaches have led to a gap in design and research, making it difficult to realize the full potential of the skins. To narrow this gap and focus on the typical continental climates in China, the authors have tried to explore an integrated design-research approach to climate-responsive building skin through case studies and prototype research.

The case in Cold Zone of China shows how architects have consciously taken climate responsiveness into account in the whole process of constructing the skin system, from planning and design to construction and operation. However, these seasonal climatic differences have become too much for static skin structures to response simultaneously. Architects have no other choice but to resort to such compromise strategies as layered and origination-based controls to make the skin responsive to the prevailing climate, which will inevitably cost the overall optimal performance.

In this context, the authors extracted the prototype of louver skins from the Cold Zone case and materialized it in a full-scale test platform, where comparative experiments were performed to identify the optimal solutions for four key design features (joint opening ratio, cladding color, cavity depth and air vents openness) and their optimal combinations for summer and winter. The team also evaluated the impact of each feature on the thermal performance of the skin quantitatively. The experimental results show that climate-responsive skins help save much more energy than conventional single-layer skins, with energy-saving rates increasing by 11.4 and 6.7% in summer and winter, respectively. In addition, the highly variable structure of the skin, which was designed to respond to climates in different seasons, has confirmed the limitations of traditional static structures and provided guidance on the design of dynamic skins. Driven by the prototype research, architects should try to do more, using their expertise and experience to construct a reliable, operable and affordable skin system that responds dynamically and intelligently to the ever-changing climates. In this regard, the authors proposed a novel skin prototype with triangular blades, and integrated the optimal solutions for summer and winter use by means of the blades' diverse materials and rotatable structure.

Based on a Chinese case, this research attempts to demonstrate the feasibility and necessity of a design-research approach that collects the wisdom of architects and scientists to create more climate-responsive skin solutions and even drive the sustainable development of buildings, with a view to providing some reference for similar work in the future. In the authors' view, the research serves as a point of departure for future exploration of more topics, such as the multi-climatic factor coupling mechanism of the skin system, methods of improving the responsiveness accuracy, the determination of the threshold for dynamic skin operations, and the application of new intelligent materials to dynamic skins.

Acknowledgements This work was funded by National Natural Science Foundation of China (Grant No. 51908358), Postdoctoral Science Foundation of China (2020M672632), and Basic and Applied Basic Research Foundation of Guangzhou Municipal Science and Technology Bureau (202102020985).

References

Aparicio-Fernández, C., Vivancos, J.-L., Ferrer-Gisbert, P., & Royo-Pastor, R. (2014). Energy performance of a ventilated façade by simulation with experimental validation. *Applied Thermal Engineering, 66*(1), 563–570.

De Masi, R. F., Ruggiero, S., & Vanoli, G. P. (2021). Hygro-thermal performance of an opaque ventilated façade with recycled materials during wintertime. *Energy and Buildings, 245*, 110994.

Fantucci, S., Serra, V., & Carbonaro, C. (2020). An experimental sensitivity analysis on the summer thermal performance of an Opaque Ventilated Facade. *Energy and Buildings, 225.*

Fernandes Maciel, A. C., & Carvalho, M. T. (2019). Operational energy of opaque ventilated facades in Brazil. *Journal of Building Engineering, 25.*

Gregorio-Atem, C., Aparicio-Fernandez, C., Coch, H., & Vivancos, J. -L. (2020). Opaque ventilated facade (OVF) thermal performance simulation for office buildings in Brazil. *Sustainability, 12*(18).

Hao, S. (2016). Research on the climate responsive characteristics of vernacular architecture, a case study of southeastern Chongqing. Tsinghua University.

Harrison, G. W. M., Liapēs, V., & Liapis, V. (2013). *Performance in greek and roman theatre*. Brill.

Ibanez-Puy, M., Vidaurre-Arbizu, M., Antonio Sacristan-Fernandez, J., & Martin-Gomez, C. (2017). Opaque ventilated facades: Thermal and energy performance review. *Renewable and Sustainable Energy Reviews, 79*, 180–191.

Jia, Z., & Lin, B. (2021). How to achieve the first step of the carbon-neutrality 2060 target in China: The coal substitution perspective. *Energy, 233.*

Kieran, S., & Timberlake, J. (2008). *Loblolly house: Elements of a new architecture*. Princeton Architectural Press.

Kottek, M., Grieser, J., Beck, C., Rudolf, B., & Rubel, F. (2006). World map of the Koppen-Geiger climate classification updated. *Meteorologische Zeitschrift, 15*(3), 259–263.

Labat, M., Woloszyn, M., Garnier, G., Rusaouen, G., & Roux, J. J. (2012). Impact of direct solar irradiance on heat transfer behind an open-jointed ventilated cladding: Experimental and numerical investigations. *Solar Energy, 86*(9), 2549–2560.

Lin, Z. (2018). Research on design strategy of building envelope for nearly Zero Energy oriented to lightweight prefabricated building. Tsinghua University.

Mammadova, G., Sharifov, A., & Akbarova, S. (2021). Experimental study of air cavity thermal performance of opaque ventilated facades under extreme wind conditions: Case study Baku. *Informes De La Construccion, 73*(561).

Marinosci, C., Semprini, G., & Morini, G. L. (2014). Experimental analysis of the summer thermal performances of a naturally ventilated rainscreen facade building. *Energy and Buildings, 72*, 280–287.

Papadikis, K., Chin, C. S. Galobardes, I., Gong, G., & Guo, F. (2019). Sustainable buildings and structures: Building a sustainable tomorrow. In *Proceedings of the 2nd International Conference in Sutainable Buildings and Structures (ICSBS 2019)*. Suzhou, China, CRC Press.

Peci Lopez, F., & de Adana Santiago, M. R. (2015). Sensitivity study of an opaque ventilated facade in the winter season in different climate zones in Spain. *Renewable Energy, 75*, 524–533.

Peci Lopez, F., Jensen, R. L., Heiselberg, P., & de Adana Santiago, M. R. (2012). Experimental analysis and model validation of an opaque ventilated facade. *Building and Environment, 56*, 265–275.

Sánchez, M. N., Giancola, E., Suárez, M. J., Blanco, E., & Heras, M. R. (2017). Experimental evaluation of the airflow behaviour in horizontal and vertical open joint ventilated facades using stereo-PIV. *Renewable Energy, 109*, 613–623.

Stazi, F., Veglio, A., & Di Perna, C. (2014). Experimental assessment of a zinc-titanium ventilated facade in a Mediterranean climate. *Energy and Buildings, 69*, 525–534.

Zhang, J.-J., & Yu, X.-C. (2017). Preliminary study on open-joint ventilated facade. *Building Energy Efficiency, 45*(10), 48–50.

Index

J. Wang et al. (eds.), *Advanced Materials in Smart Building Skins for Sustainability*,
https://doi.org/10.1007/978-3-031-09695-2

GPSR Compliance

The European Union's (EU) General Product Safety Regulation (GPSR) is a set of rules that requires consumer products to be safe and our obligations to ensure this.

If you have any concerns about our products, you can contact us on ProductSafety@springernature.com

In case Publisher is established outside the EU, the EU authorized representative is:

Springer Nature Customer Service Center GmbH
Europaplatz 3
69115 Heidelberg, Germany

Batch number: 10726135

Printed by Printforce, the Netherlands